Green Energy and Technology

Climate change, environmental impact and the limited natural resources urge scientific research and novel technical solutions. The monograph series Green Energy and Technology serves as a publishing platform for scientific and technological approaches to "green"—i.e. environmentally friendly and sustainable—technologies. While a focus lies on energy and power supply, it also covers "green" solutions in industrial engineering and engineering design. Green Energy and Technology addresses researchers, advanced students, technical consultants as well as decision makers in industries and politics. Hence, the level of presentation spans from instructional to highly technical.

Indexed in Scopus.

Indexed in Ei Compendex.

Michael Nii Addy · Clinton Aigbavboa ·
Titus Ebenezer Kwofie

Decarbonization of Building Energy in Developing Countries

Springer

Michael Nii Addy
Sustainable Human Settlement and Construction Research Centre, Faculty of Engineering and the Built Environment
University of Johannesburg
Johannesburg, South Africa

Clinton Aigbavboa
Faculty of Engineering and the Built Environment
University of Johannesburg
Johannesburg, South Africa

Titus Ebenezer Kwofie
Department of Architecture
Kwame Nkrumah University of Science and Technology
Kumasi, Ghana

ISSN 1865-3529 ISSN 1865-3537 (electronic)
Green Energy and Technology
ISBN 978-3-032-09112-3 ISBN 978-3-032-09113-0 (eBook)
https://doi.org/10.1007/978-3-032-09113-0

This Springer imprint is published by the registered company Springer Nature Switzerland AG
The registered company address is: Gewerbestrasse 11, 6330 Cham, Switzerland

Competing Interests The authors have no competing interests to declare that are relevant to the content of this manuscript.

Contents

Chapter 1
General Introduction

1.1 Introduction

The world is urbanizing quickly, and Africa is no exception [1, 2]. As urbanization increases across the globe, climate change has become a growing concern closely linked to urban growth. Historically, buildings have been created to offer respite from inclement weather and its associated vagaries. This results in the built environment being tailored to suit the environmental conditions of their respective regions. Presently, the building process—spanning design, construction, and operation—plays a pivotal role in contributing to climate change [3, 4]. At the same time, climate change poses a significant threat to the built environment.

The call for climate action is necessitated by the severe impacts of this global change on both environmental health and human life. Interestingly, sub-Saharan Africa is most susceptible to climate change and faces unique challenges in adapting to its impact. The inability of many developing countries, particularly in Africa, to adequately respond to the climate crisis is captured in the Sixth Assessment Report of the Intergovernmental Panel on Climate Change. This report details increase in surface temperature with corresponding increase in drought intensity and exposure to extreme weather conditions that compromise indoor thermal comfort. These elevated temperatures have a corresponding impact on both energy supply and demand in buildings. Studies have indicated a strong positive relationship between changes in temperature and the amount of power used in cooling [5]. This suggests that temperature variations will continue to drive up electricity consumption in future. Moreover, in the tropics and particularly in sub-Saharan Africa, the highly populated informal settlements are expected to be the worldwide battleground for fatal heat [6]. Millions of people worldwide already suffer from the health and well-being effects of extreme heat [7]. Some scholars agree that this is largely underreported in sub-Saharan Africa. Cities experiencing simultaneous rapid urbanization and global warming will be greatly impacted [8]. The interplay between climate change and the built environment is becoming more critical as the pace of urbanization increases.

M. N. Addy et al., *Decarbonization of Building Energy in Developing Countries*, Green Energy and Technology, https://doi.org/10.1007/978-3-032-09113-0_1

There is therefore the urgent need to address these challenges by drawing on multiple perspectives to develop a comprehensive strategy. Decarbonization strategies have become one of the key policy directives in meeting this challenge.

Decarbonization involves reducing greenhouse gas (GHG) emissions, through the use of low-carbon energy sources and technologies, building energy efficiency strategies with the ultimate aim of achieving net-zero emissions [9]. Net-zero energy systems are characterized by significant reductions in GHG emissions, with any remaining emissions offset by an equal amount removed from the atmosphere [10]. This can be achieved through strategies like carbon capture and storage (CCS), reforestation, and other carbon sequestration methods, ensuring that the net impact on the climate is neutral [11]. Due to the various environmental impact, there has been growing focus on decarbonization. However, extant literature shows a larger emphasis on supply-side management of energy with less attention paid to demand-side approaches [12, 13]. Again, while previous studies show that demand-side approaches have been explored largely in the global north, a closer look shows that developing countries are much less well researched [13, 14]. It is against this backdrop that this book seeks to explore demand-side management on decarbonization of building energy within the developing country context. This book explores this subject within the framework of economic and urbanization challenges predominant in most sub-Saharan countries.

1.2 Unpacking the Problem

Demand-side flexibility involves strategies that reduce the need for electricity generation by shifting or reducing energy usage, particularly for non-electric applications, on the demand or end-use side. Although it is theoretically possible to meet the increasing energy demand with additional renewable energy capacity, this requires the expansion of renewable energy generation to outpace demand growth in order to reduce fossil fuel generation and meet decarbonization targets. Given the current limitations in renewable energy supply, the best approach to achieving decarbonization is through enhanced demand-side flexibility rather than blanket electrification. This approach allows for the gradual integration of renewable energy as it grows, while also addressing the rising electricity demand. Proffering demand-side-oriented solutions requires a strategic approach to addressing the complex challenges posed by climate change.

Climate change is an immensely complex challenge to address, not only because of the intricate interdependencies and uncertainties involved but also because the longer it remains unsolved, the more difficult it becomes to tackle. The key to overcoming this challenge lies in continuously striving to understand the problem, which increases the chances of developing effective solutions. However, this requires an objective analysis of the gaps that may hinder progress. Climate change is conceptualized as a super wicked problem. Firstly, the problem of climate change is difficult to define. Secondly, solving climate change issues requires significant structural and economic efforts. Thirdly, time is of the essence, as the opportunity for effective action is running out. Additionally, multiple stakeholders, including businesses, governments, and individuals, are part of the problem, complicating efforts to find a cohesive solution. Most businesses that are in the position to aid in solving it often lack the incentive to do so. A major factor in this is the misalignment between the long-term nature of climate change and the short-term focus of business decision-makers, whose tenure typically does not extend far enough into the future to prioritize climate action. Moreover, institutional interventions addressing the issue are weak or ill-equipped. Lastly, institutional interventions discount the long-term implications of climate change, focusing instead on short-term gains. What makes this further intriguing is the unique problem that faces most developing countries. Developing economies face unique challenges that differ significantly from those encountered by developed countries. Developing economies are faced with weak regulatory systems, increasing urbanization, and struggling economies. As such solving this problem requires a tapestry of research, collaboration, and a political will. Moreover, this unique problem requires that it is not just about introducing regulations. But how should regulation be introduced that makes them effective. What knowledge should be built up and how should the capacity of local experts be enhanced? In this book, building energy decarbonization is explored through a demand-side perspective, addressing these critical questions and outlining pathways to effectively tackle the problem in developing economies.

1.3 The Context

Globally, developing countries are often categorized into distinct growing economic stages. In sub-Saharan Africa, countries can be grouped as low-income, lower-income, middle-income or upper-middle-income countries. Each of these categories has their own unique challenges and opportunities in balancing economic growth with environmental sustainability. These classifications carry significant implications for the various types of sustainable development strategies needed. Moreover, these economic levels reflect differing stages of economic development. Despite this, there are characteristics that are generally common across the economic categorization. Some common attributes that may be used to describe include unemployment, large informal sector, basic infrastructure challenges, poor health, weak institutional and regulatory framework. These challenges reflect different developmental needs that the regions need to grapple with in meeting environmental concerns. For instance, with regard to unemployment, Fox et al. [15] argue that a lack of job opportunities is a distinct feature in most African countries, especially in urban areas. Studies show that there is a high concentration of unemployed people living in Africa and about 300 million people living on less than a US$1 a day. The corollary of this is lack of a general concern towards environmental needs as economic needs take prominence. This is further exacerbated by a large informal sector. The informal sector in Africa accounts for about 70% of the economic activities in developing countries.

The informal construction sector is a sector that is difficult to regulate, and many have argued for different approaches in managing this sector. Aside from these challenges, Africa is urbanizing quickly.

Despite these challenges, Africa is experiencing rapid urbanization—a transformation that is at once a by-product of economic ambition and a potential catalyst for deepening inequality. Urbanization in Africa is not merely a demographic shift; it is a multi-dimensional process encompassing economic restructuring, social change, and environmental transition. Between now and 2040, Africa's population is projected to grow by over 63%, and its economic output is expected to more than double. Cities like Lagos, Nairobi, and Accra are evolving into sprawling megacities, with expanding middle classes, real estate booms, and increased demand for energy-intensive housing, transportation, and infrastructure systems. Yet, this urban expansion is spatially and socially uneven. In many cities, planned formal neighbourhoods coexist with vast informal settlements that lack access to reliable electricity, water, or sanitation. The result is a highly fragmented urban fabric, where the built environment reflects sharp inequalities in resource access, quality of life, and environmental exposure. This dichotomy complicates efforts to introduce universal building energy codes or technologies, as solutions must navigate both high-rise commercial developments and self-built informal dwellings.

As of 2018, Africa's building sector accounted for 61% of final energy use and 32% of energy-related CO_2 emissions—excluding emissions from construction materials like steel, cement, and glass [16]. These figures will only rise unless urgent interventions are made. The twin pressures of urban population growth and economic expansion, if unmanaged, risk locking African cities into a high-carbon future. Hence, any strategy for building energy decarbonization must be grounded in a nuanced understanding of Africa's economic dynamism and social heterogeneity. For instance, the rising demand for air-conditioning in middle-income urban households presents an opportunity to promote energy-efficient cooling systems and passive design. Simultaneously, for informal and low-income households, efforts could focus on affordable insulation techniques, rooftop solar kits, and the integration of vernacular construction methods that leverage local climatic conditions.

In essence, Africa's urbanization presents both a threat and a stimulus for sustainable development. When strategically managed, urbanization can create economies of scale that support the adoption of clean technologies and efficient infrastructure. But if allowed to proceed in an uncoordinated and inequitable manner, it can entrench unsustainable energy use and deepen socio-economic divides. Therefore, achieving building energy decarbonization in Africa requires more than technical fixes; it demands integrated solutions that respond to economic realities, social structures, and spatial dynamics. By contextualizing interventions within Africa's uneven but promising economic trajectory, this book seeks to chart pathways that are both ambitious and attainable—balancing innovation with inclusion and growth with sustainability.

1.4 Objectives and Focus of the Book

This book will seek to unearth contextual realities facing developing countries and provide developmental pathways to overcome the challenges that constantly plague the building energy sector. The study is focused on the demand side. It is embedded within the framework that recognizes Africa's dynamic but unequal economic expansion and increasing urbanization. This volume advocates for affordable, context-appropriate, and innovative sustainable solutions that reflect both the socio-economic and cultural dynamics of the Global South.

Central to the book's approach is the application of socio-technical systems thinking, which acknowledges that building energy performance is not solely a technical issue, but rather the outcome of dynamic interactions between people, technologies, institutions, and the built environment. By integrating multiple viewpoints—including those of policymakers, local communities, engineers, urban planners, energy providers, and informal sector actors—the book moves beyond reductionist models to provide nuanced, inclusive, and systems-oriented pathways for decarbonizing the demand side of the building energy sector in Africa.

Technically, the book explores issues such as building energy simulation, thermal comfort, daylighting, climate response design, and renewable energy technologies, particularly in the context of their relevance and applicability in African urban environments. Yet, beyond technical efficacy, the book critically examines how these technologies can be embedded and fully utilized within specific social contexts, addressing behavioural norms, affordability constraints, acceptance, and institutional capacity. Equally, it addresses how social barriers can be overcome in order to support the deployment and scaling of these energy technologies.

In doing so, the book reveals the intersections between the technical and social domains, forming a unique tapestry of interconnected solutions to an inherently integrated and multi-dimensional problem. Rather than treating energy efficiency and decarbonization as isolated technical fixes, this volume illustrates how technical innovations and social transformation must co-evolve to achieve meaningful, scalable, and sustainable change. The book contributes to a growing body of knowledge that bridges the gap between theory and practice, offering empirically grounded insights and actionable recommendations. Ultimately, it aspires to inform policy, empower communities, and reshape urban energy futures in a manner that is both equitable and sustainable.

1.5 Summary

This chapter has espoused the main idea driving this research book, providing explanations for the focus of the study. A strong case is made for focusing on the demand side of building energy decarbonization within the research context. Building energy decarbonization is a key priority issue due to its global impact. Its centrality to driving the economy of a country and at the same time contributing to global warming continues to make it a topical issue. This chapter unpacks the problems faced in energy decarbonization studies and brings to light the contextual issues. It further builds up a case to proffer demand-side-oriented solutions which is built on a strategic approach to address the complex challenges posed by climate change.

References

1. Adeodun A (2024) African cities lead charge in clean energy revolution. Energy News Afr Plus, 6 May. Available at: https://energynews.africa/2024/05/06/african-cities-lead-charge-in-clean-energy-revolution/. Accessed 5 June 2025
2. Smith T (2024) Accelerating investment into clean energy in cities across Africa. ESI-Africa.Com, 3 May. Available at: https://www.esi-africa.com/industry-sectors/future-energy/accelerating-investment-into-clean-energy-in-cities-across-africa/. Accessed 5 June 2025
3. Chen L et al (2024) Biomaterials technology and policies in the building sector: a review. Environ Chem Lett. https://doi.org/10.1007/s10311-023-01689-w
4. Pomponi F, Moncaster A (2016) Embodied carbon mitigation and reduction in the built environment—what does the evidence say? J Environ Manage 181:687–700
5. Avordeh TK, Gyamfi S, Opoku AA (2021) Quantitative estimation of the impact of climate change on residential electricity demand for the city of Greater Accra, Ghana. Int J Energy Sect Manage 15(6):1066–1086
6. Wilby RL et al (2021) Monitoring and moderating extreme indoor temperatures in low-income urban communities. Environ Res Lett 16(2). Available at: https://doi.org/10.1088/1748-9326/abdbf2
7. Dasgupta P, Dasgupta R, Ebi KL, Sharma G, Chowdhuri M, Shankar S, Pujari S (2025) Harms from heat-health risks: morbidity evidence from India and global learnings for policy action. medRxiv, pp 2025–07
8. Gao S, Chen Y, Chen D, He B, Gong A, Hou P, Li K, Cui Y (2024) Urbanization-induced warming amplifies population exposure to compound heatwaves but narrows exposure inequality between global North and South cities. NPJ Clim Atmos Sci 7(1):154
9. Karakosta C, Papathanasiou J (2025) Decarbonizing the construction sector: strategies and pathways for greenhouse gas emissions reduction. Energies 18(5):1285
10. DeAngelo J, Azevedo I, Bistline J, Clarke L, Luderer G, Byers E, Davis SJ (2021) Energy systems in scenarios at net-zero CO2 emissions. Nat Commun 12(1):6096
11. Raj A, Jhariya MK, Banerjee A, Meena RS, Bargali SS, Kittur BH (2022) CO_2 capture, storage, and environmental sustainability: plan, policy, and challenges. In: Plans and policies for soil organic carbon management in agriculture. Springer Nature Singapore, Singapore, pp 159–189
12. Scott K, Smith CJ, Lowe JA, Garcia-Carreras L (2022) Demand vs supply-side approaches to mitigation: What final energy demand assumptions are made to meet 1.5 and 2° C targets? Glob Environ Change *72*:102448
13. Usman R, Mirzania P, Alnaser SW, Hart P, Long C (2022) Systematic review of demand-side management strategies in power systems of developed and developing countries. Energies 15(21):7858. https://doi.org/10.3390/en15217858
14. Panda S, Mohanty S, Rout KP, Sahu KB, Parida MS (2024) A comprehensive review on demand-side management and market design for renewable energy support and integration. TIDEE: TERI Inf Dig Energy Environ 23(1/2):62–63
15. Fox L, Senbet LW, Simbanegavi W (2016) Youth employment in Sub-Saharan Africa: challenges, constraints and opportunities. J Afr Econ 25(suppl_1):i3–i15
16. International Energy Agency (IEA) (2020) Buildings: a source of enormous untapped efficiency potential. International Energy Agency. Available at: https://www.iea.org/topics/buildings

Chapter 2
Climate-Responsive Building Design

2.1 Introduction

The population of African cities is projected to increase by over 600 million, over the next two decades, a growth rate that outpaces the urban expansion China witnessed during its 20-year period of economic and energy boom [1]. Most African cities are grappling with numerous challenges stemming from over 50% of their populations residing in urban areas and the ongoing surge in urbanization [2]. In the light of the detrimental consequences of climate change on the continent's ecology, it is imperative to understand the relationship between energy use and climate change in order to put sustainable solutions into action and create recommendations for future energy practices [3]. Policymakers must acknowledge the benefits associated with energy-efficient buildings. These benefits extend beyond enhancing building value and resilience, they also improve the quality of indoor air, foster healthier living conditions for inhabitants, and provide opportunities to lower energy costs [4].

Africa is one of the continents, most vulnerable to climate change. This vulnerability and the inability of developing countries to adapt to the challenges posed by climate change were highlighted in the Third Assessment Report of the Intergovernmental Panel on Climate Change (IPCC) [3]. Extreme weather events are predicted to increase in intensity and frequency in this period of climate change, impacting several areas of the socio-economic, ecological, and environmental situations, including human well-being [5, 6]. In the tropics in general and sub-Saharan Africa in particular, the highly populated informal settlements are expected to be the worldwide battleground for fatal heat [7]. Extreme heat affects the supply and demand for energy in buildings, human productivity, heat-related health effects, the difficulties low-income communities experience in surviving, and the quality of the environment in metropolitan areas [8]. It is projected that by 2050, space cooling in buildings will rise to 30%, a significant increase from 18% in 2016 [9]. But, mechanical air-conditioning, which is a readily available solution to the space cooling demand, exacerbates climate change [10]. In a paradoxical feedback loop, efforts to

M. N. Addy et al., *Decarbonization of Building Energy in Developing Countries*, Green Energy and Technology, https://doi.org/10.1007/978-3-032-09113-0_2

mitigate the adverse effects of climate change inadvertently accelerate climate change itself. Therefore, there is a pressing need for climate-mitigating alternatives that do not exacerbate climate change.

Implementing sustainable measures can lead to a reduction of 2.2 °C in global temperatures [11]. One of the sustainable measures for managing the adverse impact of climate change is climate-responsive building design. Climate-responsive building design is an approach that leverages building's existing characteristics to effectively adapt to the local environmental and climatic conditions [12]. Climate-responsive buildings have the added advantage of promoting interactions with the natural environment [13]. Such buildings create a built environment that adequately protects the occupants from the inclement weather. This is quite imperative in increasing global temperature and uncertain climatic changes. The absence of climate-responsive features in building may lead to severe discomfort among occupants of building such as overheating in warmer climates [14]. Climate action has become imperative as a result of growing global population and increasing resource demands [12]. As a result, architects and construction professionals need to be innovative in meeting this challenge and further restrict the impact of the construction process on the environment. In developed countries, there is a concerted effort towards ensuring climatic considerations in contemporary architectural design, this is notably lacking in most developing countries [13]. The building sector has significant energy-saving potential through the application of climate-responsive building design during the design phase. Hence, enhancing building energy efficiency and reducing the corresponding carbon emissions represents a pivotal approach to tackle global challenges, including curbing energy consumption, mitigating climate change, and minimizing the carbon impact of human activities [15].

2.2 Principles of Climate-Responsive Building Design

Climate-responsive buildings are built to enable conditions such as natural ventilation and daylighting [16]. Traditionally, climate-responsive buildings relied heavily on outdoor conditions [17]. However, given the current energy and extreme heat conditions, this approach is not effective in maintaining occupant comfort, which significantly impacts productivity. It is imperative that climate-responsive building designs for contemporary architecture integrate past lessons with present and future needs. As a result, understanding the various principles of climate-responsive building is key in reducing energy consumption and contributing to the fight against climate change [14]. This will further lead to the design and construction of resilient building which is urgently needed considering predicted extreme weather events [5, 6]. A building designed to be responsive to climate conditions must incorporate certain design principles. These principles include passive cooling, natural daylighting, proper orientation of the building, shading devices, and use of local materials [16, 18]. What follows next is an overview of some of the principles.

Passive design strategies, with its formalization from developed countries such as the Germany and US [19], are one of the key principles of climate-responsive design. It is worth pointing out that these principles have long existed with roots deep in ancient architecture. Its formalization has played a pivotal role in enhancing occupants' thermal comfort [20]. The incorporation of passive design strategies at the early design stage is a sustainable alternative towards energy conservation in buildings [21]. These design strategies can reduce annual energy demand from 112 to 68 kWh/m^2 in new construction and from 112 to 85 kWh/m^2 in refurbished buildings [19].

In warm-humid climate, passive design strategies heavily influence the energy consumption of buildings due to high demand for thermal comfort of occupants. [22] showed an overall reduction in energy usage of about 37% after employing strategies such as shading devices and efficient glazing on buildings in Ghana. Similarly, Zoure and Genovese [23] reinforce the significance of passive design strategies after an optimized building model including natural ventilation, shading, and insulation achieved a 43% reduction in annual cooling load. Also, the findings from the optimization of passive design strategies such as WWR, orientation, and façade material portray the impact of the strategies on reducing energy consumption and improving thermal and visual comfort of occupants in warm-humid climate regions [24]. Passive design strategies comprise natural ventilation, daylighting, thermal mass, building orientation and layout, shading systems and building envelope design. Climate-responsive design optimizes these design strategies in the provision of energy-efficient and thermally comfortable built environment.

2.2.1 Natural Ventilation

Natural ventilation is the process where outdoor air is introduced into a building [23]. This process is facilitated through two main mechanisms: wind-induced airflow and temperature gradient effort (buoyancy-driven ventilation) [25]. The ventilation of buildings is achieved through two main strategies: single-side ventilation and cross-ventilation. Single-sided ventilation involves situating all openings on a single façade, irrespective of their number or position [26]. On the other hand, cross-ventilation involves situation all openings on two different façades of a building [27]. This process allows airflow to enter from upstream opening, circulate within the indoor space, and exit through the downstream opening. Natural ventilation system is more energy-efficient alternative to conventional air-conditioning systems [25]. However, its cooling efficacy diminishes as outdoor temperature rise over time [14]. Therefore, careful consideration to natural ventilation systems through climate-responsive design maintains its potential as a sustainable cooling design strategy in warm-humid climates.

2.2.2 *Daylighting*

The International Energy Agency reports that about 20% of worldwide electricity is attributed to lighting [28]. Investigating daylight system to a building is therefore crucial opportunity towards energy decarbonization in the built environment. Passive daylighting systems are eco-friendly techniques of increasing the infiltration of natural light into a space which significantly reduces the dependence on artificial lightings [29]. Passive daylighting system lowers heat gain and demand for air-conditioning [30]. Also, this system is important for human functioning, regulating circadian rhythms, and stimulating the production of hormones and vitamins.

2.2.3 *Thermal Mass*

Thermal mass is another alternative to explore passive design potentials in a building design, significantly reducing building energy load during peak load time [31]. In a study by Kuczyński et al. [32], it was demonstrated that an increase in thermal mass from light to very heavy resulted in a reduction of the average peak temperature by 3.7 K in a day and 1.2 K by night. This is very effective in reducing maximum daytime temperatures. Kuczyński and Staszczuk [33] concluded that the use of cellular concrete walls instead of lightweight timber frame walls reduced the demand for cooling energy by 67–75% depending on assumed temperature threshold. However, the material used for thermal mass must be carefully checked to reduce the increase of carbon footprint.

2.2.4 *Building Orientation*

Building orientation and layout involve strategically positioning structures on-site to harness regional climatic features, such as prevailing southwest winds and sun paths, thereby enhancing natural cooling within the building [34]. Architects prioritize building orientation during the early stages of design, as it directly impacts the amount of solar radiation a building receives [35]. In hot and humid climates, building orientation significantly impacts thermal performance, with south and west orientations posing particular challenges due to their exposure to intense solar radiation, even at higher altitudes [36]. Also, in the northern hemisphere, orientations due south, southeast, and southwest are strategically chosen to ensure maximum exposure to direct sunlight, especially during the winter months when the sun's path is lower in the sky [37]. The orientation of a building significantly influences the wind ventilation parameter, which, in turn, enhances residents' comfort and satisfaction by optimizing the air change rate within the range of comfortable velocities [38]. The building envelope comprises external walls, doors, windows, roof, ground, and internal walls that demarcate conditioned and non-conditioned spaces [39]. Based on how building envelope materials interact with the climate, earth's human settlement environments can be classified into heat preservation priority and heat insulation priority climate zones [40]. In heat preservation priority climates, building materials prevent external heat gain (hot zones) or internal heat loss (cold zones) due to significant temperature differences between indoor and outdoor spaces. Conversely, heat insulation priority climate zones include tropical and subtropical regions where high humidity, temperatures, and solar radiation pose major challenges. Hence, integrating high-density and high-heat capacity materials like bricks, concrete, and tiles into buildings can substantially reduce cooling or heating expenses, in contrast to lightweight materials such as timber, which have lower thermal mass [34].

2.2.5 Shading and Insulation Systems

Shading and insulation systems are integrated into building designs to mitigate excessive daylight, which can lead to elevated internal temperatures and unwanted illumination [41]. Achieving optimal effectiveness in the application of this concept necessitates a thorough comprehension of both the macro- and microclimates specific to the area [42]. Shading devices are categorized based on their installation location to interior shading devices and exterior shading devices [43]. The environmental performance of shading systems encompasses two key objectives [44]: (i) optimizing visual performance by preventing glare and maintaining appropriate indoor illuminance [45], and (ii) ensuring acceptable thermal conditions through effective control of solar heat gains from both direct and diffuse sunlight [46]. Shading devices significantly contribute to maintaining thermal comfort in buildings by either absorbing or reflecting solar radiation [47].

2.2.6 Local Building Materials

The use of local building materials in construction activities requires less energy to produce and transport which is a significant step towards more sustainable practices [48]. This approach significantly reduces greenhouse gas emissions. The adoption and adaption of local materials are vital initiative towards sustainable architecture [49]. The adoption of local materials reinforces community identity and resilience. Choosing local, indigenous materials is an important step to becoming more sustainable in the construction industry [50]. As far back as 1980s, Ofori et al. [51] has been advocating for the adoption of local materials for the Ghanaian construction industry. This advocacy is furthered by Agyekum et al. [102] on vernacular architectural principles as sustainable development alternatives in the industry. However, the demand is still low [50].

2.2.7 *Roof Design*

Another option in passive design strategies is the consideration of roof design. This is because the roof is receiving solar radiation directly [52], making them the main source of heat stress in a building [53]. With a growing demand for indoor quality, there is a need to enhance the design of roofs to adapt to the increasingly severe climate conditions [54]. This will improve the use of roofs as just protection for physical assets to self-cooling building features. Reference [21] argued that the impact of green roof in energy conservation and thermal management of buildings has been widely established. Reference [55] concluded that the adoption of green roof on Egyptian school buildings resulted in 40% energy savings and 4 °C reduction in room temperature. In another study by [56], it was reported that the inclusion of a green roof led to an average reduction of 9.88% in the energy consumption. Adjustments to a conventional flat roof led to a significant reduction in thermal discomfort hours by 45.29% [53].

2.3 Benefits of Climate-Responsive Building Design

Climate-responsive buildings have the potential to generate localized microclimates under specific climatic conditions [107]. This is an urgent act to implement measures to reduce greenhouse gas emissions aimed at safeguarding human and environmental well-being [57]. Climate-responsive building design addresses the adverse effects of climate change, improving thermal comfort within spaces and enhancing energy efficiency. In response to energy scarcity and initiatives towards clean energy, the principles of climate-responsive building design serve as an opportunity to reduce cooling energy demand while ensuring a comfortable indoor thermal environment [106]. The principles of climate-responsive building design embody thoughtful considerations of building orientation, shape, position, site microclimate conditions, and the selection of building envelope materials that impact heat gains and losses [34].

Zoure and Genovese [23] are confident that sub-Saharan African countries can benefit from climate-responsive building design concepts to ensure passive solutions in future design projects. Zhang et al. [105] concluded that natural ventilation has gained recognition in temperate and tropical climates as alternative to reducing building energy consumption and CO_2 emissions while ensuring desired indoor thermal comfort. Climate-responsive designs harness the potential of shading to minimize cooling energy demand while ensuring indoor thermal comfort, by reducing solar radiation penetration and enhancing indoor natural ventilation [42]. When properly designed to suit a given climate, shading devices facilitate passive cooling and heating that creates healthier and more comfortable conditions for occupants [58]. In dry and hot climates, shading devices have been demonstrated to be effective [43], extending the number of hours of comfort by 27% and simultaneously reducing overall cooling demand by 31% [59]. Climate-responsive designs also take into consideration the orientation and envelope materials of the building. Reference [35] asserted that optimal building orientation has the potential to reduce cooling demand in warm climates. Also, climate appropriate selection of building materials has significant impact on energy consumption [60] and is effective in achieving standard indoor thermal comfort [40]. The adoption of climate-responsive building design leads to enhanced health of occupants due to improved indoor air quality, comfort, and satisfaction [61].

2.4 Challenges to Climate-Responsive Building Design

Despite its widely established benefits, climate-responsive building design faces significant barriers in Africa. With the evidence of a combination of rapid population, increase in demand for new construction, and its consequent impact on furthering climate change [62], there is an obligation of the research to better understand challenges to climate-responsive building design. This is critical in developing countries where the concept is often overlooked [50]. This section of the chapter discusses the challenges to the uptake of climate-responsive building design. The discussions will cover lack of localized knowledge and expertise, weak regulatory frameworks, cultural and social barriers, climatic data, limited financial resources, technological barriers, and availability of local materials.

2.4.1 Knowledge and Expertise

The absence of adequate knowledge and professionals on the climate-responsive design is a major barrier to its uptake. A study in Ghana revealed that inadequate training and education, unfamiliarity with green technologies, and higher initial costs of green construction practices and materials are the key challenges that hinder project management teams' implementation of sustainable building processes [63]. Also in Tanzania, lack of training and professional development programmes serves as a significant barrier to climate-responsive designs [64]. This is evident of the lack of understanding of the sustainability of buildings within the professionals [10]. This fact is affirmed by majority of respondents of a study conducted in Ghana [50]. On the barrier imposed by professionals, [61] further the discussions in terms of lack of communication and collaborations across all stakeholders, weak organizational structure to support passive design strategies, lack of awareness of expertise, and lack of education. The challenges posed by the lack of knowledge and expertise hinder the successful implementation of sustainable principles in the built environment.

2.4.2 Regulatory Framework

Weak regulatory framework is another challenge towards the uptake of climate-responsive building design. It is reported that there are limited policies and regulations towards sustainable building in Ghana [50]. This situation hinders the uptake of climate-responsive building design principles, such as the use of local materials [65]. Reference [61] explored weak regulatory framework in terms of the absence of guiding policy to improve technical capability of consultants, lack of enforcement from government, lack of relevant laws and regulations, and absence of an official sustainable design body. A weak regulatory framework towards climate-responsive building design creates a vacuum in the efforts to achieve a sustainable built environment.

2.4.3 Culture and Social

Cultural and social barriers are considered as major challenges to the implication of climate-responsive building design. These barriers refer to the perception of stakeholders towards climate-responsive design options [66]. The general lack of awareness and interest in environmental issues is forefront of obstacles in implementing climate-responsive designs in developing countries [67]. A study by Johansson and Winge [50] describes it as "people resistance to change from conventional way of building. The resistance comes from both the people working in the construction industry and the public". According to the study, majority of the buildings in big cities like Accra are built with concrete, a sign of economic success. Hence, the adoption of mud houses considered poorly built. Majority of clients, unaware of the benefits of climate-responsive building design, are unwilling to incorporate passive design strategies into their building designs [61].

2.4.4 Economic and Financing

Sustainable buildings are expensive. This is an assertion is well established in literature discussing environmentally friendly alternatives in the construction industry [68]. The case is no different for climate-responsive building design. This challenge is more eminent in dire economic situations such as recent times, with more interest on profits and financial gain rather than on environmental concern. According to Johansson and Winge [50], there is lack of funding and support when working with sustainable building solutions. The study further showed that it is labour-intensive to work with sustainable materials, which will increase the cost. Africa is undergoing a rapid population and urbanization which further increases the need for housing. But limited financial resources have resulted in self-built structures with minimal climate-responsive features. In Nigeria and Cameroon, about 80 and 97% respectively of the housing stock are self-built and often from low-income households [69]. Thus, addressing the barriers imposed by the cost of climate-responsive building designs is an imperative to advancing sustainable architecture in Africa.

2.4.5 *Technology*

Technological barriers to climate-responsive building design consider lack of standards and limited technological advancements. Research reports that the built environment of sub-Saharan countries is prone to applying inappropriate Western world design standards which has led to increased dependence on air-conditioning [70]. This is mainly due to the absence of standards to facilitate the adoption of climate-responsive building design options [71]. Also, there are limited demonstration projects to facilitate knowledge sharing about sustainable building and benefits [50]. Climate-responsive building design is a response to the changing climate of a locality. To effectively achieve this, there should be access to reliable climatic data to advice on design options. However, the situation is different in Africa [72]. This is a major barrier because lack of detailed and localized climatic records leads to generalized assumptions by designers [13]. Generalized assumptions and unreliable climatic data can result in failure of buildings designed to leverage natural ventilation and other passive systems [73]. And with contemporary buildings predicted to be subjected to increasingly warmer and more volatile climates in future [74], there is the need to prepare buildings with more accurate and reliable climatic data.

In order to reach full sustainability in the construction industry, the industry has to focus on country-specific context. This situation is different in Africa where most of the countries such as Ghana, Nigeria, and Kenya largely depend on imported building materials [75, 76]. Johansson and Winge [50] reported that this is caused by the unavailability of local materials. The reason is largely being the inability to upscale production of local building materials to meet the increasing needs of the construction industry. The use of local materials is well reported to contribute to environmentally friendly and climate-suited built environment. There is therefore the importance of ensuring that barriers to obtaining local materials are well managed.

2.5 Pathways to Climate-Responsive Building Design

The term "climate-responsive" design denotes the goal of transforming climate challenges into opportunities while addressing both climate adaptation and mitigation [77]. In the twenty-first century, sustainable buildings have emerged as a critical concern for both developing and developed nations [78]. It is therefore crucial to adopt construction methods more suited to the conditions of developing countries

[70]. This section of the chapter discusses strategies for promoting climate-responsive building design in the African context.

In developing countries, cost is the topmost driver for the adoption of climate-responsive designs [65]. Ideally, plans should be straightforward, reasonably priced, and a component of long-term strategies and an integrated plan in order to maximize the likelihood of effective execution [79]. There should be promotional schemes, tax rebates, and low-cost financing to entice developers and homeowners towards demand of sustainability features [80]. Dosumu and Aigbavboa [65] suggested that market incentives for adopters and financial approval for sustainable projects are significant drivers for sustainable construction in Africa. This will create an attractive economic system for developers and clients to consider climate-responsive design initiatives. This strategy is not stand-alone. Its implementation will be realized through policy decisions, public education, and advancement of technology.

Government policies and regulations are significant drivers for climate-responsive design in Africa [65]. All stakeholders must be included and involved in order to achieve climate-responsiveness in the built environment [81]. To promote a culture of sustainability, there is a need for local policies which incentivize energy-efficient practices [82]. Also, to ensure the smooth implementation of government policies and interventions, there is the need to raise awareness and build the capacity of construction industry professionals towards the integration of sustainable principles within procurement processes [83]. These policies should provide sufficient flexibility, support in terms of incentives, and guidance to help designers develop strategic climate-responsive design solutions that fit with local contexts [79]. Adequate government incentives such as tax rebates and expedited permitting will encourage the adopting of climate-responsive designs [84]. Unegbu et al. [104] advocate for public mechanisms to support the adoption of sustainable and climate-responsive designs.

Further, policies should encourage interdisciplinary collaborations and global best practice knowledge transfer to accelerate localized sustainability solution development [80]. Policies and regulations are the medium through other strategies for climate-responsive building design can take shape.

Research and innovation are essential in advancing climate-responsive designs in Africa. They help in the development of effective policies, whereas development of sustainable materials and practices [85]. Policymakers should utilize the wide array of urban planning and design tools at their disposal to more aggressively include climate-responsive urban design methods [79]. Technological advancements, coupled with relevant research, ensure that benefits of climate-responsive building design have long-term effects [82]. Hence, principles of climate-responsive building designs should be integrated into architecture and engineering curricula as a vital tool to enhance climate literacy and development of skilled workforce in addressing climate challenges. Currently, the curricula of some African universities are modelled to suit the western systems which usually overlook local construction practices and climate conditions [86]. Mba et al. [80] recommends updating curricula, mandating the integration with climate-responsive principles and strengthened educational programmes in Nigeria as a strategic pathway towards climate-responsive building designs. In East Africa, principles of climate-responsive building designs have been integrated in architectural education to equip student on the need for climate-responsive designs [86]. Additionally, due to the interdisciplinary nature of urban climate resilience, [87] emphasized the need for co-creation of knowledge coupled with interdisciplinary approach to achieve a climate-responsive design in their study across Ghana, South Africa, and Zimbabwe. This is an effective strategy to identify socio-cultural barriers to climate-responsive designs. Research and innovation into local climate data which significantly enhances the effectiveness of climate-responsive policy instruments [79].

Diverse solutions for climate-responsive designs remain either undocumented or not easily accessible. Hence, open-access repositories should be developed to manage all tested strategies that enhances the climate-responsiveness of a building. These will further improve and accelerate the transfer of knowledge between academia and industry. The repositories can be institutionalized through partnerships such as universities and industry. In Senegal, knowledge sharing and development of local repositories are useful to aid decision-making and enhance accelerated climate adaptation [88].

The adoption of climate-responsive designs requires practical evidence of feasibility of implementation and success. Hence, there is a need for pilot projects and development of living labs and centres of excellence to promote innovation and local relevance through the demonstration of design projects. Demonstration of project and living labs plays a crucial role in bridging the gap between the theoretical climate-responsive designs and the practical implementation where technologies and diverse strategies can be tested. Living labs in Tanzania integrate local knowledge with foreign expertise to serve as a platform for co-creation of local innovations [89]. Although living labs are few in Africa, few countries, such as Libya, South Africa, Kenya, and Ghana through their lab, have successfully piloted localized technologies to support climate resilience. Reference [90] suggest that living labs in Kenya should be expanded to accommodate more people and enough support and supervision for innovation should be provided. The concept of living labs is growing steadily in Africa and, when harnessed, could serve as innovative hub for development of climate-responsive designs [91].

Similarly, Regional Centres of Excellence serve as a backbone for capacity building, interdisciplinary collaboration, and an agent of advocacy for climate-responsive practices. These include the African Centre for Cities (ACC) and African Cities Lab (ACL). They integrate climate-responsive designs into city system and pivotal in the exchange and dissemination of knowledge through conferences, open-access publications, and postgraduate workshops.

To ensure these interventions are data-driven, there is a need for support for the development of climate specific simulations and modelling in African cities. Tools such as energy modelling tools, daylight modelling, and tools for optimization of passive design strategies are essential to accurately predict the energy consumption, daylight performance, thermal comfort, and visual comfort in African climates. Gradually, some studies in Africa have used these tools to predict the comfort levels of occupants. With the use of local meteorological data, a simulation study analysing the heat stress of urban schemes in Nigeria was conducted [92]. The findings of the study underscore the importance of climate specific simulations to enhance accuracy [92]. With the use of climate-based daylight metrics, Akubah et al. [103] assessed the impact of passive design strategies on a daylight and thermal performance of a school building in Ghana. The study further reiterates the importance of climate specific simulations to achieve energy efficiency and increased occupancy comfort. Hence, investment into African specific datasets such as weather files will greatly impact the accuracy of simulation results.

Another significant pathway to climate-responsive building design in Africa is the adoption and adaptation of vernacular housing styles and concepts. Research has argued that 200-year-old meet modern bioclimatic strategies and comfort requirements of current occupants [93]. This claim is confirmed by global examples of vernacular architecture demonstrating thermal comfort and energy efficiency potentials [94]. Traditional houses in Iran were originally designed with climate-responsive solutions suitable for its climate [95]. Traditional Egyptian architecture is effective in managing heat gain [96]. It was demonstrated that earth-sheltered structures are a viable and cost-effective alternative in managing adverse impact of climate change [97]. The Eastgate Building in Zimbabwe another evidence of the ingenuity and appropriateness of vernacular architecture [98]. Using the concept of biomimicry, a key concept of vernacular architecture [99], the Eastgate Building has managed to reduce energy cost by 35% while eliminating the use of heating, ventilation, and air-conditioning systems [100]. For developing countries, vernacular architectural knowledge is advantageous in developing climate resilient built environments with adaptability and sustainability [101].

The pathways to climate-responsive design in Africa are deeply rooted in the continent's diverse climatic conditions, local building practices, and resource availability. By combining traditional knowledge with modern sustainable technologies, Africa has the opportunity to create innovative, affordable, and resilient built environments. Fostering policy support, community involvement, and technological innovation will be crucial for making climate-responsive design more accessible and scalable across the continent. This approach can significantly contribute to reducing the impact of climate change, improving living conditions, and fostering sustainable development.

2.6 Summary

Climate-responsive design is a crucial step in achieving more sustainable buildings [68]. It is more beneficial to developing countries like Ghana with an ever-growing population and increasing demand for housing infrastructure. This chapter mainly discusses the challenges to realizing climate-responsive building design in Africa, coupled with pathways to mitigating these challenges. Climate-responsive building design is proven to be effective in finding localized sustainable alternatives for the built environment. Despite its plethora of benefits, it is plagued with challenges with implementation in Africa. Predominant of these are lack of localized knowledge and expertise, weak regulatory frameworks, cultural and social barriers, climatic data, limited financial resources, technological barriers, and availability of local materials. This bouquet of challenges hinders the effective and efficient implementation of principles of climate-responsive building design in Africa.

This chapter however highlights pathways to overcome these challenges. The discussions focus on creating a viable and healthy economic system to facilitate the adoption of the design principles. The development of government policies and regulations to serve as a medium to promote the other pathways. Using research and innovation, coupled with advancement in technology, to inform policy formulation and implementation of regulations. Advocating the need to ensure the availability of local and sustainable materials to facilitate easy and reliable uptake. There was also the need for educating and training professionals responsible for the development and implementation of climate-responsive building design. Finally, learning and understanding from vernacular architecture to enhance the advocacy for climate-responsive building. These pathways present an opportunity for African to design and build with our local climate in mind.

References

1. Marriette MS, Mertens O (2021) Developing a pathway to sustainable residential cooling in west African cities: the case study of Tema, Ghana
2. Mohammadpourkarbasi H et al (2022) Evaluation of thermal comfort in library buildings in the tropical climate of Kumasi, Ghana. Energy Build 268:112210. https://doi.org/10.1016/j.enbuild.2022.112210.
3. Kelly AM, Radler RDNN (2024) Does energy consumption matter for climate change in Africa? new insights from panel data analysis. Innov Green Dev 3:100132
4. Silvero F et al (2019) Impact assessment of climate change on buildings in Paraguay—overheating risk under different future climate scenarios. Build Simul 12(6). https://doi.org/10.1007/s12273-019-0532-6
5. Ebi KL et al (2021) Hot weather and heat extremes: health risks. The Lancet. https://doi.org/10.1016/S0140-6736(21)01208-3
6. Mengistu AG et al (2023) Assessing the impact of climate change on future extreme temperature events in major South African cities. Theor Appl Climatol [Preprint]. https://doi.org/10.1007/s00704-023-04712-w
7. Wilby RL et al (2021) Monitoring and moderating extreme indoor temperatures in low-income urban communities. Environ Res Lett 16(2). https://doi.org/10.1088/1748-9326/abdbf2

8. Feng J et al (2023) Annual review of environment and resources overheating of cities: magnitude, characteristics, impact, mitigation and adaptation, and future challenges. https://doi.org/10.1146/annurev-environ-112321
9. Kuhn S, Kutzner F, Thøgersen J (2022) How to make energy efficiency labels more effective: Insights from discrete choice experiments in Ghana and the Philippines. Energy Res Soc Sci 84. https://doi.org/10.1016/j.erss.2021.102320
10. Ohene E, Hsu SC, Chan APC (2022) Feasibility and retrofit guidelines towards net-zero energy buildings in tropical climates: a case of Ghana. Energy Build 269. https://doi.org/10.1016/j.enbuild.2022.112252
11. Shukla PR et al (2022) Climate change 2022 mitigation of climate change working group III contribution to the sixth assessment report of the intergovernmental panel on climate change summary for policymakers edited by. Available at: www.ipcc.ch.
12. Saifudeen A, Mani M (2024) Adaptation of buildings to climate change: an overview. Front Built Environ. https://doi.org/10.3389/fbuil.2024.1327747
13. Amos-Abanyie S, Gyimah KA, Adjei EA (2021) Towards climate responsive building design: bio-climatic design features of residential building typologies in the warm-humid climate of Ghana. J Build Constr Plan Res 09(02). https://doi.org/10.4236/jbcpr.2021.92012
14. Liu S et al (2020) Effectiveness of passive design strategies in responding to future climate change for residential buildings in hot and humid Hong Kong. Energy Build 228:110469. https://doi.org/10.1016/J.ENBUILD.2020.110469
15. Feng W et al (2019) A review of net zero energy buildings in hot and humid climates: experience learned from 34 case study buildings. Renew Sustain Energy Rev 114. https://doi.org/10.1016/j.rser.2019.109303
16. Koranteng C, Simons B, Essel C (2019) Climate responsive buildings: a comfort assessment of buildings on KNUST campus, Kumasi. J Eng Des Technol 17(5). https://doi.org/10.1108/JEDT-03-2019-0054
17. Manu S et al (2019) Performance evaluation of climate responsive buildings in India—case studies from cooling dominated climate zones. Build Environ 148. https://doi.org/10.1016/j.buildenv.2018.10.063
18. Lamsal P, Bajracharya SB, Rijal HB (2021) Article guidelines for climate responsive building design in three regions of Nepal. ACCESS Online J JBEE J-J Build Environ Eng 2(1)
19. Chang H et al (2022) Feasibility study and passive design of nearly zero energy building on rural houses in Xi'an, China. Buildings 12(3). https://doi.org/10.3390/buildings12030341
20. He D, Isa MHM (2024) Investigation of indoor thermal comfort of heritage buildings in hot summer and cold winter zone of China: a case study. Case Stud Therm Eng 53. https://doi.org/10.1016/j.csite.2023.103820
21. Abdul Mujeebu M, Bano F (2022) Integration of passive energy conservation measures in a detached residential building design in warm humid climate. Energy 255
22. Koranteng C, Simons B, Gyimah KA (2021) Potential measures towards the reduction of cooling loads of office buildings in Ghana. Front Eng Built Environ 1:161–172
23. Zoure AN, Genovese PV (2023) Implementing natural ventilation and daylighting strategies for thermal comfort and energy efficiency in office buildings in Burkina Faso. Energy Rep 9. Available at: https://doi.org/10.1016/j.egyr.2023.02.017
24. Addy MN, Amos-Abanyie S, Tetteh BK, Kwofie TE, Aigbavboa C (2024) Design optimisation of fenestration and orientation for daylighting and energy performance in ghana using genetic algorithm. In: Adinyira E, Amoako C, Kwofie TE, Aigbavboa C, Agyekum K, Addy M (eds) Sustainable and resilient infrastructure development in Africa's changing climate. ICIDA 2023. Springer, Cham. https://doi.org/10.1007/978-3-031-69606-0_22
25. Zhang H et al (2021) A critical review of combined natural ventilation techniques in sustainable buildings. Renew Sustain Energy Rev. https://doi.org/10.1016/j.rser.2021.110795
26. Pan W et al (2019) A model for calculating single-sided natural ventilation rate in an urban residential apartment. Build Environ 147. https://doi.org/10.1016/j.buildenv.2018.08.047

27. Seifert J et al (2006) Calculation of wind-driven cross ventilation in buildings with large openings. J Wind Eng Ind Aerodyn 94(12). https://doi.org/10.1016/j.jweia.2006.04.002
28. Onubogu NO, Chong KK, Tan MH (2021) Review of active and passive daylighting technologies for sustainable building. Int J Photoenergy. Hindawi Limited. https://doi.org/10.1155/2021/8802691
29. Teo YH et al (2023) A simulation-aided approach in examining the viability of passive daylighting techniques on inclined windows. Energy Build 282. https://doi.org/10.1016/j.enbuild.2022.112739
30. Brzezicki M (2024) Enhancing daylight comfort with climate-responsive kinetic shading: a simulation and experimental study of a horizontal fin system. Sustainability (Switzerland) 16(18). https://doi.org/10.3390/su16188156
31. Chen Y et al (2020) Experimental investigation of demand response potential of buildings: combined passive thermal mass and active storage. App Energy 280. https://doi.org/10.1016/j.apenergy.2020.115956
32. Kuczyński T et al (2021) Effect of thermal mass, night ventilation and window shading on summer thermal comfort of buildings in a temperate climate. Build Environ 204. https://doi.org/10.1016/j.buildenv.2021.108126
33. Kuczyński T, Staszczuk A (2020) Experimental study of the influence of thermal mass on thermal comfort and cooling energy demand in residential buildings. Energy 195. https://doi.org/10.1016/j.energy.2020.116984
34. Jegede OE, Taki A (2022) Optimization of building envelopes using indigenous materials to achieve thermal comfort and affordable housing in Abuja, Nigeria. Int J Build Pathol Adapt 40(2). https://doi.org/10.1108/IJBPA-01-2021-0009
35. Kohansal ME et al (2022) Investigating the simultaneous effects of building orientation and thermal insulation on heating and cooling loads in different climate zones. Archit Eng Des Manage 18(4). https://doi.org/10.1080/17452007.2021.1901220
36. Alwetaishi M et al (2020) User thermal comfort in historic buildings: evaluation of the potential of thermal mass, orientation, evaporative cooling and ventilation. Sustainability (Switzerland) 12(22). https://doi.org/10.3390/su12229672
37. Ghamari M, Sundaram S (2024) Solar wall technology and its impact on building performance. Energies. https://doi.org/10.3390/en17051075
38. Elshafei G et al (2021) Towards an adaptation of efficient passive design for thermal comfort buildings. Sustainability (Switzerland) 13(17). https://doi.org/10.3390/su13179570
39. Far C, Far H (2019) Improving energy efficiency of existing residential buildings using effective thermal retrofit of building envelope. Indoor Built Environ 28(6). https://doi.org/10.1177/1420326X18794010
40. Jannat N et al (2020) A comparative simulation study of the thermal performances of the building envelope wall materials in the tropics. Sustainability (Switzerland) 12(12). https://doi.org/10.3390/SU12124892
41. Mohammed A et al (2022) Reducing the cooling loads of buildings using shading devices: a case study in Darwin. Sustainability (Switzerland) 14(7). https://doi.org/10.3390/su14073775
42. Yusoff WFM et al (2023) Review of openings with shading devices at naturally ventilated buildings. Archit Eng Des Manag. https://doi.org/10.1080/17452007.2022.2095553
43. Heidari A, Taghipour M, Yarmahmoodi Z (2021) The effect of fixed external shading devices on daylighting and thermal comfort in residential building. J Daylight 8(2). https://doi.org/10.15627/JD.2021.15.
44. Al-Masrani SM, Al-Obaidi KM (2019) Dynamic shading systems: a review of design parameters, platforms and evaluation strategies Autom Constr. https://doi.org/10.1016/j.autcon.2019.01.014
45. Szokolay SV (2008) Introduction to architectural science: the basis of sustainable design. J Am Coll Radiol JACR
46. Baker N, Steemers K (2014) Daylight design of buildings: a handbook for architects and engineers, daylight design of buildings: a handbook for architects and engineers. https://doi.org/10.4324/9781315073750

47. Lim T, Yim WS, Kim DD (2020) Evaluation of daylight and cooling performance of shading devices in residential buildings in South Korea. Energies 13(18). https://doi.org/10.3390/en13184749
48. Sonebi M et al (2022) Trends and opportunities of using local sustainable building materials in the Middle East and North Africa region. RILEM Tech Lett 7. https://doi.org/10.21809/rilemtechlett.2022.169.
49. Firoozi AA et al (2024) Emerging trends in sustainable building materials: Technological innovations, enhanced performance, and future directions. Results Eng. Elsevier B.V. https://doi.org/10.1016/j.rineng.2024.103521.
50. Johansson E, Winge E (2023) Barriers and motivational drivers for sustainable building in a developing country: the case of Ghana. Sweden
51. Ofori-Boateng K, Adams S, Ohemeng W (2020) Coping strategies of the urban poor: a case study from Ghana. Poverty Public Policy 12(3). https://doi.org/10.1002/pop4.286
52. La Roche P, Yeom DJ, Ponce A (2020) Passive cooling with a hybrid green roof for extreme climates. Energy Build 224. https://doi.org/10.1016/j.enbuild.2020.110243
53. Athmani W et al (2023) The potential of using passive cooling roof techniques to improve thermal performance and energy efficiency of residential buildings in hot arid regions. Buildings 13(1). https://doi.org/10.3390/buildings13010021
54. Ling HM et al (2024) Analyzing recent active and passive cool roofing technology in buildings, including challenges and optimization approaches. J Build Eng. Elsevier Ltd. https://doi.org/10.1016/j.jobe.2024.109326.
55. Ragab A, Abdelrady A (2020) Impact of green roofs on energy demand for cooling in Egyptian buildings. Sustainability (Switzerland), 12(14). https://doi.org/10.3390/su12145729
56. Lin Y et al (2021) Design optimization of a passive building with green roof through machine learning and group intelligent algorithm. Buildings 11(5). https://doi.org/10.3390/buildings11050192
57. Osman MM, Sevinc H (2019) Adaptation of climate-responsive building design strategies and resilience to climate change in the hot/arid region of Khartoum, Sudan. Sustain Cities Soc 47. https://doi.org/10.1016/j.scs.2019.101429
58. Homod RZ et al (2021) Effect of different building envelope materials on thermal comfort and air-conditioning energy savings: a case study in Basra city, Iraq. J Energy Storage 34. https://doi.org/10.1016/j.est.2020.101975
59. Diz-Mellado E et al (2023) Energy-saving and thermal comfort potential of vernacular urban block porosity shading. Sustain Cities Soc 89. https://doi.org/10.1016/j.scs.2022.104325
60. William MA et al (2021) Building envelopes toward energy-efficient buildings: a balanced multi-approach decision making. Int J Energy Res 45(15). https://doi.org/10.1002/er.7166
61. Juffle NAH, Rahman MM (2023) An overview of motivators and challenges of passive design strategies. In: IOP conference series: earth and environmental science institute of physics. https://doi.org/10.1088/1755-1315/1195/1/012039
62. Kalu Arachchi KACS et al (2025) A framework for sustainable tropical design: converging circular economy into passive design strategies. Constr Innov. https://doi.org/10.1108/CI-06-2024-0166
63. Ayarkwa J et al (2022) Sustainable building processes challenges and strategies: the relative important index approach. Clean Eng Technol 7. https://doi.org/10.1016/j.clet.2022.100455
64. Mahame C, Kikwasi GJ, Baruti MM (2024) Barriers to the effective selection of sustainable materials for residential building projects: a qualitative study. Sustainability 16:9526
65. Dosumu O, Aigbavboa C (2021) Drivers and effects of sustainable construction in the South African construction industry. Acta Structilia 28(2):78–107. https://doi.org/10.18820/24150487/as28i2.4.
66. Eze EC, Sofolahan O, Omoboye OG (2023) Assessment of barriers to the adoption of sustainable building materials (SBM) in the construction industry of a developing country. Front Eng Built Environ 3(3):153–166. https://doi.org/10.1108/febe-07-2022-0029
67. Addy M, Adinyira E, Danku JC, Dadzoe F (2020) Impediments to the development of the green building market in sub-Saharan Africa: the case of Ghana. Smart Sustain Built Environ 10:193–207

68. Alkali M et al (2020) Hindrances to the utilization of climate responsive architecture principles for residential design in northeast Nigeria. Int J Sci Eng Res 11(9). http://www.ijser.org.
69. Asafo DM (2022) Fragile and compromised housing: implications of land conflicts on housing development in peri-urban Accra, Ghana. Hous Stud 39:1340–1363
70. Genovese PV, Zoure AN (2023) Architecture trends and challenges in sub-Saharan Africa's construction industry: a theoretical guideline of a bioclimatic architecture evolution based on the multi-scale approach and circular economy. Renew Sustain Energy Rev. Elsevier Ltd. https://doi.org/10.1016/j.rser.2023.113593
71. Adabre MA, Chan APC, Edwards DJ, Mensah S (2022) Evaluation of symmetries and asymmetries on barriers to sustainable housing in developing countries. J Build Eng 50:104174
72. Dinku T et al (2022) The climate data tool: enhancing climate services across Africa. Front Clim 3. https://doi.org/10.3389/fclim.2021.787519
73. Rastogi P et al (2022) Projected climate data for building design: barriers to use. Build Cities 3(1). https://doi.org/10.5334/bc.145
74. Yau YH, Hasbi S (2013) A review of climate change impacts on commercial buildings and their technical services in the tropics. Renew Sustain Energy Rev. https://doi.org/10.1016/j.rser.2012.10.035
75. Akadiri PO (2018) Investigating factors influencing building materials selection in Nigerian construction industry. Am J Civ Eng Architect 6:154–157
76. Chan APC et al (2018) Critical barriers to green building technologies adoption in developing countries: the case of Ghana. J Clean Prod 172. https://doi.org/10.1016/j.jclepro.2017.10.235
77. Daniel R et al (2023) Performance of urban climate-responsive design interventions in combining climate adaptation and mitigation. Build Environ 236. https://doi.org/10.1016/j.buildenv.2023.110227
78. Hafez FS et al (2023) Energy efficiency in sustainable buildings: a systematic review with taxonomy, challenges, motivations, methodological aspects, recommendations, and pathways for future research. Energy Strategy Rev. https://doi.org/10.1016/j.esr.2022.101013
79. Brandsma S et al (2024) Implementation of urban climate-responsive design strategies: an international overview. J Urban Des [Preprint]. https://doi.org/10.1080/13574809.2024.2314760
80. Mba EJ, Okeke FO, Igwe AE, Ozigbo CA, Oforji PI, Ozigbo IW (2024) Evolving trends and challenges in sustainable architectural design; a practice perspective. Heliyon 10:e39400
81. Samadkulov M (2024) The urbanization economy: difficulties and approaches for sustainable growth. http://www.academicpublishers.org
82. Kaghembega WSH et al (2024) Paper: modeling and scenario analysis of residential building energy conservation in cities of different weather. Energy Rep 11:2670–2684. https://doi.org/10.1016/j.egyr.2024.02.025
83. Ogunsanya OA et al (2022) Barriers to sustainable procurement in the Nigerian construction industry: an exploratory factor analysis. Int J Constr Manag 22(5). https://doi.org/10.1080/15623599.2019.1658697
84. Simpeh EK, Smallwood JJ (2023) Incentive mechanism for promoting the uptake of green building in South Africa. Open House Int 49:340–357
85. Obianyo II et al (2021) Overcoming the obstacles to sustainable housing and urban development in Nigeria: The role of research and innovation. Clean Eng Technol. https://doi.org/10.1016/j.clet.2021.100226
86. Olweny M (2018) Introducing sustainability into an architectural curriculum in East Africa. Int J Sustain High Educ 19:1131–1152
87. Thondhlana G, Mubaya CP, McClure A, Amaka-Otchere ABK, Ruwanza S (2021) Facilitating urban sustainability through transdisciplinary (TD) research: lessons from Ghana, South Africa, and Zimbabwe. Sustainability 13:6205

88. Biconne R (2014) Knowledge sharing on climate change as a resource for adaptation processes: the case of Malika, Senegal. In: Macchi S, Tiepolo M.(eds) Climate change vulnerability in southern African cities. Springer Climate. Springer, Cham. https://doi.org/10.1007/978-3-319-00672-7_8
89. Hooli L, Jauhiainen JS, Lähde K (2016) Living labs and knowledge creation in developing countries: living labs as a tool for socio-economic resilience in Tanzania. Afr J Sci Technol Innov Dev 8:61–70
90. Ondiek MA, Moturi C (2019) An assessment of the sustainability of Living Labs in Kenya. Innov Manag Rev 16:391–403
91. Mbatha SP, Musango JK (2022) A systematic review on the application of the living Lab concept and role of stakeholders in the energy sector. Sustainability 14:14009
92. Obe OB, Morakinyo TE, Mills G (2024) An assessment of WRF-urban schemes in simulating local meteorology for heat stress analysis in a tropical sub-Saharan African city, Lagos, Nigeria. Int J Biometeorol 68:811–828
93. Tamaskani Esfehankalateh A et al (2022) Bioclimatic passive design strategies of traditional houses in cold climate regions. Environ Dev Sustain 24(8):10027–10068. https://doi.org/10.1007/s10668-021-01855-6
94. Pardo JMF (2023) Challenges and current research trends for vernacular architecture in a global world: a literature review. Buildings 13(1). https://doi.org/10.3390/buildings13010162
95. Shadmand A, Arslan Selçuk S (2022) Lessons from traditional architecture in energy efficient building design: the case of traditional Tabriz houses. Int J Environ Stud 79(2):245–264. https://doi.org/10.1080/00207233.2022.2033492
96. Alhmoud SH (2024) The impact of environmental climatic conditions in the mediterranean (a comparative between Egypt and Spain). Int J Sustain Dev Plan 19(3):843–851. https://doi.org/10.18280/ijsdp.190304
97. Mirabi E, Akrami Abarghuie F (2023) Investigating the climate-adaptive design strategies of residential earth-sheltered buildings in Iran. Int J Build Pathol Adapt 41(5):1029–1048. https://doi.org/10.1108/IJBPA-05-2021-0076
98. R. Fattah T, Hama Radha C, Baranyai B (2025) Biomimicry-based approach to enhance energy efficiency in an office building. In: Proceedings of the 3rd international conference on engineering and innovative technology. Salahaddin University-Erbil. https://doi.org/10.31972/iceit2024.041
99. Sipahioğlu Kara S, Gökmen Erdoğan B (2023) Searching biomimicry in the vernacular architecture. In: Sustainability, conservation and ecology in spatial planning and design: new approaches, solutions, applications
100. Chigwenya A, Zhakata TI (2020) Int J Real Estate Stud INTREST Adopt Green Build Technol Sustain Dev Insights Harare, Zimbabwe. www.utm.my/intrest
101. Ahmed N (2023) Re-exploring vernacular architecture from the lens of regenerative thinking: a case study Gharb Sohail Village in Egypt. J Sustain Archit Civil Eng 32(1):58–76
102. Agyekum K, Kissi E, Danku JC (2020) Professionals' views of vernacular building materials and techniques for green building delivery in Ghana. Sci Afr 8:e00424
103. Akubah JT, Amos-Abanyie S, Simmons B (2022) Simulation-based exploration of daylighting strategies for a public basic school in a hot-dry region of Ghana. In: Applied Research Conference in Africa. Springer, Cham, pp 215–233
104. Unegbu HCO, Yawas DS, Dan-Asabe B, Alabi AA (2025) Measures for overcoming sustainable construction barriers in the Nigerian construction industry. Discov Civil Eng 2:26
105. Zhang M, Han W, He Y, Xiong J, Zhang Y (2024) Natural ventilation for cooling energy saving: typical case of public building design optimization in Guangzhou. China. Appl Sci 14(2):610
106. Li Z, Zou Y, Tian M, Ying Y (2022) Research on optimization of climate responsive indoor space design in residential buildings. Buildings 12(1):59

107. Zhu J, Feng J, Lu J, Chen Y, Li W, Lian P, Zhao X (2023) A review of the influence of courtyard geometry and orientation on microclimate. Build Environ 236:110269

Chapter 3
Building Energy Simulation and Energy Decarbonization

3.1 Introduction

The growing concern for energy-efficient and sustainable buildings is driven by major environmental concerns such as global warming. The development of sustainable buildings is prioritized as a key in the drive towards achieving a near-zero built environment [33]. In achieving this, there is the need to completely overhaul the way buildings are designed and constructed. Innovative technologies play a pivotal role in achieving this while allowing for testing of new ecological building material and building construction process. It is accepted that significant gains can be achieved in the preliminary phase of a building design with little cost implications. As such, current knowledge recommends the increase focused on developing sustainable design early in the building projects phase.

Building energy simulation (BES), one of the most significant and modern methods for carbon mitigation in the building sector, has grown in popularity as a useful and encouraging approach for energy-efficient building designs, operations, and retrofits with the goal of enhancing energy performance and lowering carbon emissions [31]. BES is aimed at evaluating alternative designs, comparing and selecting systems and subsystems; allocating annual energy budgets; achieving compliance with energy standards and economic optimization during the building design process.

BES tools are very resourceful, especially in driving the charge against building energy decarbonization [35]. The main aim of these tools is quantifying the performance of the building which are relevant to the design, operation, and control of the building. As a result, the output of these tools provides the resources needed to make an informed decision in climate mitigation strategies and environmental policy formulation. The expected outcome is also useful in building energy decarbonization providing key information and details for energy transition planning and energy impact assessment [11]. The use of these tools can be further declassified into different subdomains such as thermal simulation, lighting simulation, and airflow

M. N. Addy et al., *Decarbonization of Building Energy in Developing Countries*, Green Energy and Technology, https://doi.org/10.1007/978-3-032-09113-0_3

modelling. This is achieved using various bespoke simulation software such as EnergyPlus, Environmental Systems Performance–Research (ESP-r), Integrated Environmental Solutions Virtual Environment (IES VE), and OpenStudio. These bespoke tools are key in the building energy decarbonization as they are useful in detailing energy transition plans and energy impact assessment.

By simulating a building's energy use, project teams can pinpoint inefficiencies and implement strategies that yield significant cost reductions. Adjustments such as upgrading insulation, fine-tuning Heating Ventilation Air-conditioning (HVAC) systems, optimizing lighting controls, or incorporating renewable energy technologies can drastically cut energy expenses over the building's life cycle. Beyond these immediate savings, BES empowers teams with data to make informed decisions about materials and construction techniques, fostering long-term economic sustainability. BES extends its influence beyond energy savings and compliance, directly benefiting the people who live and work in the spaces it helps design. Energy efficiency is not just about cutting costs or minimizing environmental impact; it's about fostering environments that prioritize comfort and well-being. By leveraging advanced simulations, designers can maximize natural lighting, regulate indoor temperatures, and optimize ventilation systems, creating spaces that are as pleasant as they are efficient. High indoor air quality, combined with a balanced thermal environment, enhances occupant health and comfort. This contributes to productivity and satisfaction, making the building a better place to inhabit. The harmony between occupant well-being and environmental responsibility underscores the win–win nature of energy-efficient design—where innovation meets the needs of both people and the planet.

Navigating the labyrinth of building codes and regulations poses a significant challenge for any construction project. In this scenario, BES proves indispensable, aligning designs with current energy efficiency mandates and environmental standards. Integrating BES early in the design process enables professionals to spot and address compliance issues proactively, saving time, reducing costs, and avoiding legal entanglements. Even when adopted later, BES delivers noteworthy benefits to enhance project outcomes.

BES excels in analysing existing buildings to detect inefficiencies. Energy simulation is able to detect HVAC systems, energy-draining components, or insulation gaps that can be upgraded for better performance. BES provides strategies to cut energy consumption and operational costs for buildings in operation. It can simulate optimized scenarios like improved lighting schedules, more efficient HVAC settings, and the incorporation of energy-saving technologies. BES aids in ensuring continued compliance with evolving standards. This is achieved through the assessment of existing structures against the latest requirements, identifying necessary updates, and safeguarding projects from legal repercussions. Buildings often face evolving demands. This includes fluctuating occupancy, new space functionalities, or advanced technology integrations. BES assists in the optimization of energy efficiency to accommodate these changes without compromising performance. BES enables stakeholders to measure return on investment with data-driven projections. This helps quantify costs, long-term savings, and environmental benefits, enabling informed decisions about potential retrofits or system optimizations.

Sustainability is a major force that drives the value of businesses. Environmentally friendly buildings are attractive to occupants who are nature-conscious and who can also support the increase in asset value. BES is the tool that clears a good space for the project team to market a building's sustainability levels, thus profiting from a unique competitive position. Projects with high energy performance not only reduce operational costs but also enhance market competitiveness by appealing to environmentally conscious occupants and attracting investors who support sustainable practices. It is obvious to builders and financiers that green buildings are beneficial to the environment. Energy-efficient models raise attraction to the market, bring in more profit in resale, and allow the users to retain their best financial state while at the same time being environmentally friendly. These sustainable construction benefits that is identified by BES users through well-shaped numerical data, can be sustainability oriented and, at the same time, go far beyond energy saving only. BES has extended its influence from saving of energy and compliance with the Energy Efficiency Obligations (EEO) to the final beneficiary of the space. Efficiency in the use of energy is not only about saving costs or reducing the negative impact on the environment. It also addresses the first and foremost needs of an individual's comfort and well-being. With the help of sophisticated simulations, designers can optimize natural lighting, indoor temperature control, and even tailor ventilation systems to make spaces not only aesthetically appealing, but also functional. Balanced thermal conditions have a positive effect on the health and comfort of the occupants, especially when combined with high indoor air quality. This increases their productivity and satisfaction, which improves their quality of living within the building. The balance between the well-being of the inhabitants and the caring for the environment highlights the win–win situation regarding energy-saving designs; the essence of innovation coming together to fulfil people's needs and the needs of the world.

3.2 Overview of Building Energy Simulation

Over the past few decades, BES has increased in popularity among building construction stakeholders but usage within the sub-Saharan Africa is uneven and relatively nascent. These tools handle various complexities associated with the construction design process while offering useful design support. The rise in popularity is also occasioned by the corresponding increase in technological capacity such as higher efficient computering systems. Building energy simulation is entirely computed-based which provides a virtual space for modelling a building design [15]. Buildings are built within a particular region, and as such it is important that the climate is also modelled within the software. These softwares allow for modelling the weather data with some taking this information from various meteorological centres. The way the building is used, i.e. the usage pattern and building materials are all part of the input for building simulation. Consequently, a building or a group of building can be modelled and simulated. However, more details within the building model imply that more multiple inputs will be modelled increasing the complexity and the computering load. Expected outputs from these computer models include building energy consumption, thermal comfort, and various building performance outputs. The resulting outputs are useful in assessing an array of design inputs and corresponding building performance. In addition to the above, these tools are useful in building research and standards developments, leading to improved standards and industry best practice.

Energy simulation is a valuable tool in building design. Its main benefit is allowing architects and engineers to compare different design ideas. They can evaluate, through simulations, how different architectural designs can maintain a comfortable indoor environment for people, through heating or cooling, while reducing energy use. These simulations allow for data-driven decisions. Even if some initial assumptions are not perfect, it does not affect the outcome much because all designs undergo similar testing, making fair comparisons possible. This helps professionals make informed decisions, ensuring the building is energy-efficient through its operational phase.

BES is used for a plethora of purposes. While standards and best practices exist in building design orientation, BES can be a good resource when exploring design constraints in orienting the building. It is well-established that the amount of sunlight received in a building depends heavily on the building orientation. This consequently affects the amount of energy used in either cooling or heating the buildings. By leveraging BES, designers can maximize a building orientation to ensure less artificial lighting and reduced cooling loads. Another design constraint is the shape and aspect ratio of the building. This invariably heavily influences the building energy consumption. By evaluating these design variables against a set constraint, building designers can create a design that maximizes functional and aesthetics goals while minimizing energy usage. Studies show the critical role building envelope play in thermally controlling the building and its corollary effect on operational energy used. The BES tools allow architects and engineers to explore the building envelope performance, allowing for evaluating different materials and components. By utilizing these tools, an in-depth analysis is conducted by exploring how construction materials have an effect on lighting, thermal, and airflow. This leads to a better appreciation of these materials and influences the choice or otherwise of these materials. The end result is a better efficient design of the building envelope such as proper insulation, efficient window-to-wall aspect ratio, and appropriate use of shading devices. Building energy systems such as HVAC and lighting are key components that critically contribute to energy consumption in a building. Here again, BES is an invaluable tool in assessing the performance and efficiency of such systems. The existence of various tools in BES enables the efficient modelling and optimizing of these components. This provides an opportunity to select a system that minimizes energy consumption while maximizing building performance. With focus shifting to environmental concern and carbon-related emissions, life cycle assessment has become very topical. BES provides the needed tools to evaluate the life cycle energy and lead to a more environmental design. Life cycle assessments also provide valuable insights into the long-term cost-effectiveness of different materials and systems. BES requires a methodical approach, underpinned by extensive data collection. To ensure a smooth modelling process, it is advisable to gather all necessary data in beforehand. The basic required information is shown in Table 3.1.

Simulation techniques enable the identification of the optimal combination of zone layouts and construction strategies to complement advanced automation systems or achieve highly effective climate-responsive designs [8]. Multiple simulations are performed to develop zoning strategies that not only meet functional requirements but also support sophisticated multi-zone control systems and the redistribution of excess energy. Some simulations may focus on selecting construction materials and determining their optimal placement within layered assemblies to enhance load balancing and temperature regulation. Additionally, alternative strategies for facade fenestration, air circulation, and shading control can be evaluated to optimize both comfort and cost efficiency. Once a robust and well-tested design is established, validated under a range of anticipated operating conditions, various alternative control strategies can be simulated. For instance, basic control studies can inform decisions about optimal start/stop controls, suitable setback temperatures, the effectiveness of weather-responsive adjustments, sensor placement, and the interaction between thermal and visual comfort parameters. Other simulations could also explore how smart control systems can be used to leverage a better building performance based on occupancy usage or daylighting metrics such as luminance intensity.

3.3 Building Energy Simulation Tools and Methods

BES tools provide key outputs such as total energy consumed in a building and thermal performance of a building. Developing a highly energy-efficient building will lead to various trade-offs such cost implications or lower thermal comfort. BES tools provide a good platform to ensure a highly optimized building. Generally, the process of model parametrization is susceptible to errors, as the preparation of building input files may involve inaccurate assumptions [42], or inadvertent user mistakes, leading to unrepresentative results. As common in computers, the accuracy of the final results will rely heavily on the quality of the data used. Energy simulations are based on thermodynamic equations and principles, but because thermal processes in buildings are complex and not fully understood, energy simulation programs approximate their predictions with qualified equations and methods. Because of this, if certain assumptions are not met in the simulation or in real life, the results may be arbitrarily incorrect. As shown in Fig. 3.1, the key inputs include the building's geometry, internal loads, HVAC systems and components, weather conditions, operational strategies and schedules, as well as parameters specific to the simulation process.

Table 3.1 Basic input options

Inputs	Description
1. Zone configuration	(a) One zone confined to a single side of the building
	(b) Four-sided building with well-mixed air, OR
	(c) Four-sided building with air unmixed between zones
2. Building	Location: select the nearest city for climate data Building Dimensions: NS and EW lengths
3. Room	Primary Façade orientation: N, S, E, or W
	Room dimensions: depth, width, and height
4. Window	Type: single-, double-, triple-glazed, or double-skin façade Coating: clear, low-emissivity, etc.
	Glazed Area: as percentage of wall area
5. Wall	Specify: low, medium, or high insulation
6. Thermal mass	Specify: low, medium, or high thermal mass
7. Occupancy	Occupant density: # people per floor area
	Equipment load: watts per floor area Min. lighting Req.: Lux
	Occupancy schedule: hours of occupancy
8. Ventilation	(a) Mechanical system
	(b) Natural ventilation only, OR
	(c) Hybrid mechanical and natural ventilation
Advanced input options	
9. Thermostat	Set upper and lower bounds on room temperature. Set nighttime setback temperatures
10. Ventilation rate	Specify the volumetric fresh air rate: L/s/person
11. Lighting controls	(a) Lights always on
	(b) Lights dim together to supplement daylight
	(c) Lights dim independently to supplement daylight
12. Blinds	Dimensions: width of blinds and spacing select colour: shiny, painted, etc.
	Angle of blinds when closed
	Daytime and nighttime schedules: always opened or closed, responding actively to temperature or solar intensity

(continued)

Table 3.1 (continued)

Inputs	Description
13. Double skinned façades and airflow windows	Depth of airflow cavity
	Airflow rates through the façade
	Location of supply and exhaust vents (interior or exterior)

Source Bryan (2004)

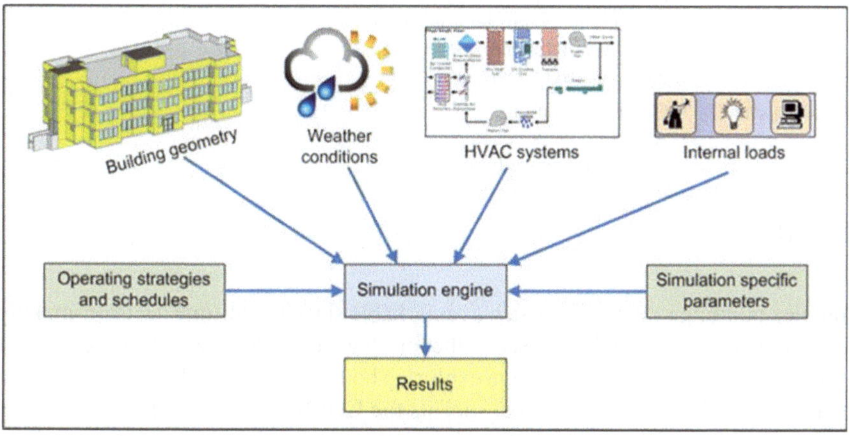

Fig. 3.1 General data flow of simulation engines. *Source* [26]

3.3.1 *Evolution of Building Energy Simulation Tools*

BES tools are being used more frequently in both research and are increasingly applied in building design, construction, commissioning, and operation. They help accelerate and enhance the design and planning process, optimize building performance, develop building controls, test new products, and assess the market potential of innovative concepts. New versions and updated tools are constantly being released. The evolution of building energy simulation tools (as summarized in Table 3.2) reflects an increasing sophistication in addressing energy consumption and efficiency concerns. This is seen from the early 1960s that relied heavily on the degree-day approach to the early 1970s that saw the emergence of the automated energy simulations. The second generation of simulation tools emerged during the global energy crisis in 1973–74 [43]. This era saw the use of dynamic thermal performance to evaluate buildings on an hourly basis. This led to an increase in the energy demand conservation and paved the way for increased functionality and higher accuracy in energy simulation models. The third generation of BES tools saw the improvement of energy flow pathways, which was achieved using numerical techniques. This saw more efficient results coming from simulating various energy behaviours and

Table 3.2 Evolution of building energy models

1st generation	• Piecemeal • Simplified • Handbook oriented	• Indicative • Application limited • Difficult to use • Predictive • Generalized • Easy to use
2nd generation	• Dynamics important • Less simplified • Still piecemeal	
3rd generation	• Field problem approach • Numerical methods • Integrated view of energy subsystems • Heat and mass transfer considered • Better user interface • Partial CABD integration	
4 and beyond	• CABD integration • Advanced numerical methods • Intelligent knowledge-based • Advanced software engineering	

Source [10]

complex building systems. This saw a remarkable shift in real-time decision-making with simulations. The tools released in the fourth generation were tools that enabled a more comprehensive building design exploring both generative and predictive tools in energy usage. European initiatives, such as the COMBINE project [44], exemplify the trend of integrating computational design with sustainable building practices. Modern tools like EnergyPlus and DesignBuilder continue this legacy, allowing for detailed evaluation and optimization of energy use and supporting global efforts to align with sustainability goals like the United Nations' Sustainable Development Goals (SDG 11).

3.3.2 Operation of BES Tools

BES tools enable the creation of a simulation program which allows for the simultaneous interaction of the building's geometric model with external conditions, occupancy, and system usage to predict various loads in the building on an hourly basis. Calculations are based on fundamental laws of physics and energy balance equations. Moreover, the energy consumption required for system operations, including heating and other loads, is calculated within the same time frame. Most BES tools can model energy flows through various building components on an hourly basis, including the transient effects of the building envelope and systems. The simulation is able to account for the effects of thermophysical properties of materials and the performance of different systems under changing external and internal environmental conditions in energy calculations. Most BES tools do not require specialized computing power and can be run on standard desktop computers or laptops [15].

3.3.3 Model Complexity

The tools can be categorized based on two key factors: the complexity of the mathematical models (which depends on the specific goals and focus of the tool) and the ability to access the BES source code (a critical aspect in the research domain). When the source code is available for modification, users can customize the equations to fit the specific case study, enhancing the level of detail. However, this also results in longer simulation times, increased modelling effort, and a higher requirement for technical expertise. Dynamic simulation tools like TRNSYS, Modelica/Dymola, IDA ICE, EnergyPlus, and Simulink focus on analysing transient system behaviours and allow for in-depth studies. However, their use demands advanced user skills and a thorough understanding of numerous input parameters. Moreover, the modelling and simulation process can be computationally expensive and laborious. The appropriateness of a certain model complexity or tool can only be determined based on the specific goals and constraints of the simulation study.

As mentioned earlier, simulation programs are inherently prone to errors. To prevent decisions from being influenced by inaccurate simulation data, it is essential to validate the model to ensure it accurately represents the real-world system. Validation involves rigorous testing of a program, including its theoretical foundation, software implementation, and user interface, under various conditions representative of its intended use. This is a critical step in the simulation process, ensuring that the model's results are reliable and meaningful. There are seven sources of error according to Judkoff et al. [22]. This includes differences between the actual weather surrounding the building and the statistical weather input used with BES. The results are affected by various physical approaches, each with different levels of detail and numerical methods. For instance, the thermal zone can be represented using a 2-star or star-node approach, while the wall model can be simulated using either the finite difference method or a transfer function. It is important to validate the models to build confidence in the building simulation. According to Judkoff et al. [21], the results of BES tools can be validated in different ways. This includes analytical verifications (analytical solutions), empirical validation (by comparing them to real-world measurements), or code-to-code comparisons (cross-checking them against results from other codes). Empirical validation tackles measurement uncertainties and is often constrained to a limited dataset over a defined period. On the other hand, cross-validation helps minimize input uncertainties and facilitates the analysis of various configurations with a high level of detail. However, a key challenge of this approach is the difficulty in precisely defining a set of reference results. Through the use of energy simulation tools, project teams are able to achieve identified cost savings through defined strategies. In addition to these, BES gives teams the data necessary to make informed decisions regarding materials and construction methods, which supports economic sustainability in the long run.

3.4 Challenges and Opportunities

Despite the embedded utility within BES, there are some factors that mitigate against its performance. It should be noted that BES are based on various computational tools, and as such their accuracy depends on the variables and assumptions used. This and other factors account for the performance gap that may exist between actual building performance and simulated performance. This section details some of the challenges in the use of these tools within the African milieu. Despite the impressive applications of BES, it's essential to view it as a guide rather than a guarantee. These models are rooted in predictive analysis and depend on variables and assumptions that can shift over time which are the main reason of the performance gap registered between simulated and operational performance. While not infallible, energy simulation provides a robust framework for optimizing designs, anticipating challenges, and aligning projects with sustainability goals. The tool's strength lies in its ability to inform smarter decisions and foster innovative solutions while accommodating real-world uncertainties.

3.4.1 Climate and Environmental Data

Yu et al. [45] contend that one of the highest risks in BES lies with the uncertainty associated with the weather file. The weather data available for the prediction are generally historical data for a site other than that of the building being analysed [22]. The data do not perfectly reflect the microclimate at the building site, adversely affecting the prediction of building energy consumption. The information contained in weather datasets is completely out of the control of the person in charge of the simulation that may only assume them as a boundary condition. Unfortunately, not all the weather databases show the same level of data accuracy, moreover, they may refer to a climate that has substantially changed in the last decades.

The use of climate data in simulations is also underexploited in terms of airflow modelling. Airflow modelling involves understanding how wind moves around a building's orientation, structure, and surroundings. However, airflow for ventilation is limited in its climatic data for modelling such that, it is difficult to account for the construction methodology of fenestrations and only assumes opening for air inflow into buildings. This makes it difficult to account for the inflow of mosquitoes to prevent the spread of vector-borne diseases (dengue and malaria) predominantly seen in warm-humid climates such as Ghana and other African countries. This is because most of these BES are designed based on other climatic conditions, failing to tackle public health issues of the African region [20]. Having detailed climate datasets about how fast and in what direction the wind blows is very important. These records are necessary to properly consider and include health factors in planning or decision-making processes [41]. Understanding these wind patterns helps to better prepare for and respond to health-related concerns tied to the weather and climate.

Reaching this goal is, however, tough since it relies on using advanced modelling methods along with detailed environmental data that are mostly not readily available for the African context.

3.4.2 Cost and Access to Software

A significant barrier to the usage of BES tools in Africa is the cost and difficulty in gaining access [46]. The high cost of subscription and installation of these modern tools often deters many professionals and institutions from developing countries from buying [17], leading to the use of cracked version or inferior software that may not give accurate results. It is important to note that the financial burden of using these BES tools is not a onetime payment fee, it can extend to regular update fees, cost of buying a high package for more features and functionality and even hardware updates to accommodate the computing power needed. For example, software like EnergyPlus or IES VE has excellent features, but the licenses and computer power needed are expensive for smaller companies or independent workers.

This is a huge challenge in Africa as many institutions and organizations are reluctant to offer support to students, researchers, and building professionals to get access to these BES tools [32]. The repercussion of this challenge is seen in the high rate of software piracy of BES tools. In many developing countries where a high percentage of software piracy is recorded [3], the challenge of data security and privacy is an issue. The usage of cracked versions of BES tools exposes organizations to cyber attracts and hacks which can cause unexpected disruptions and financial burdens. These cracked versions do not support full functionality and easy integration with other software making them unreliable for rigorous analysis. Many professionals may also prefer to use cracked version of this software since they are free to download even though it may be somewhat affordable for them. The cost and access are indeed a primary challenge, but in addition, the attitude of professionals towards the purchase of BES tools needs to be improved.

3.4.3 Model Validation and Calibration

Making BES models accurate is very important. Validation and calibration processes are undertaken to check and adjust for accuracy [27]. Model validation is done by comparing the simulation results with real-world data. This shows us if the model is an accurate representation of reality [14, 37]. Model calibration simply involves changing the model's settings to make simulations match observed performance. As important as model validation and calibration are, the acquisition of data needed for the validation process can be very difficult [24]. This is due to the lack of storing building performance data. Moreover, it is difficult to acquire data even when it is available. Many owners of homes, offices, and other building types are reluctant to give yearly electricity usage records for validation of models [5]. Where there is the possibility of receiving the needed data, a very long waiting time can be spent due to bureaucracy. If validation is not possible, simulations and further analysis are hindered, thus making the usage of BES tools difficult. In other instances, occupants of buildings are unwilling to allow researchers and professionals to carry out a data collection exercise if the records have not been kept. Many occupants are uncomfortable with the placement of data loggers to record indoor environmental temperatures for the validation process. Building energy simulations must represent real-world behaviour accurately to be reliable. If they are not accurate, they cannot analyse the building performance accurately. Thus, validating and calibrating models are vital to ensure predictions match the actual building performance [37].

3.4.4 Lack of BES Awareness, Knowledge, and Expertise

Lack of awareness, according to the study by Mahmoud et al. [46] on opportunities and limitations of BES tools, is the most significant barrier to BES adoption. His study was based in the United Kingdom, a developed country, where there is advancement in technology as compared to developing countries in Africa. Evidently, the lack of awareness in developing countries is much higher. This is revealed in the study by Addy et al. [1], where they state the minimal awareness of architects of energy efficiency and building energy simulation within the subregion. Knowledge, awareness, and technical expertise is a challenge that is seen in the adoption of IT and modern tools in Africa in general [4]. BES tools require technical understanding about the how buildings behave in different climates for different construction scenarios. However, many design and construction professionals in Ghana and Africa at large have limited or zero formal training in the use of building simulation tools [1]. This can be attributed to the absence of building performance simulation courses in academic curriculums of institutions.

A study on the use of BES tools for green building design in South Africa by [32] emphasized the need for training, education, and creation of awareness. They further noted that without the needed education, practitioners find it difficult to access energy-efficient or green building's designs of real-world projects. The lack of awareness and knowledge of these BES tools and their benefits worsen the challenge faced. Many stakeholders, professionals, and policymakers are not aware of the benefits of BES tools to identify cost savings and energy-efficient strategies, improve indoor thermal conditions, and ensure adherence to environmental standards at the design stage [7]. This affects the domain of policymaking, where energy efficiency is not strongly enforced or incentivized for compliers [34]. It is important that design and construction professionals are regularly educated and made aware of the tools available to aid in building efficiency designs towards decolonizing building energy.

3.4.5 Absence of Policy and Regulation

Policy instrument is the key to fostering the implementation of BES tools in buildings. The world energy consumption is estimated to reach 820 quadrillion Btu by 2040 U.S. EIA [47], which represents a 56% increase in consumption by 2040 [48]. This report further suggests that developing countries will contribute up to 85% of the surge as a result of economic growth pursuit in response to population growth. Most governments globally, in response to the challenge, have established energy-efficient targets to alleviate the issue. Policies and regulations, however, to enforce energy-efficient practices to keep in step with these targets, especially in developing countries, are lacking. Among developing countries in Africa, policies for energy-efficient practices establishment are either poorly documented or not documented at all [49]. Gaum and Laubscher [58] undertook a study to review the status of building energy code implementation in Africa. In their study, it was found that out of 26 African countries selected, only 7 countries had building energy codes, as shown in Table 3.3.

There are no standardized building energy codes that demands energy-efficient designs by using BES tools at the design stages. Although a country like Ghana has set up some assessment frameworks by adapting to South African Green Star system, its enforcement is loose and seems voluntary. However, [23] unless climate-specific simulation models are used, the recommendations of the Ghana Green Building Council (GHGBC) from the green star system alone cannot bring about optimum energy performances since it is generally not suited to Ghana's peculiar climate and urban dynamics. Further, Iwaro and Mwasha [49] undertook survey research to review building energy regulation and policy in developing countries, which included 20 African countries. Gaum and Laubscher [58] considered 13 similar African countries in their study out of the 20, out of which 11 had no building code in 2010. The status of these 11 countries in 2010 and 2022, as revealed by Iwaro and Mwasha [49], and Gaum and Laubscher [58], are compared in Table 3.4.

The above table shows that, over 12 years, only 5 of the 11 countries under study took steps to establish or strengthen building energy codes, whereas 6 countries do not have building energy codes that are legally enforced. Figure 3.2 shows a visual presentation of the table.

In spite of the official report on observed and projected energy consumption, emission and sustainable development goals established to mitigate climate change, many developing countries in Africa have no regulations to enforce energy-efficient practices. More so, as population continue to increase, estimated at 75 million per year [50], and consequently economic growth, developing countries will apparently contribute more to carbon emissions, and thus, it urgently demands that policies be put in place to control such developmental issues. Some developing countries have building codes, but they are voluntary [2, 16]. This means it is not legally binding and cannot be enforced by law. Such policies are not helpful since they do not mandate people to comply. Even for countries in Africa where building codes exist, some have no information on the implementation of BES tools. Ghana Building Code, for

Table 3.3 List of African countries and their building energy code status as of 2022

No	African country	Status of building energy code (BEC)
1	Algeria	Mandatory energy efficiency building code
2	Angola	No building energy code
3	Burkina Faso	No building energy code
4	Cameroon	No building energy code
5	Congo (DRC)	No information
6	Cote d'Ivoire	No building energy code
7	Egypt	Mandatory energy efficiency building code
8	Ethiopia	No building energy code
9	Ghana	Mandatory energy efficiency building code
10	Kenya	No building energy code
11	Madagascar	No building energy code
12	Mali	No building energy code
13	Morocco	Mandatory energy efficiency building code
14	Mozambique	No building energy code
15	Niger	No information
16	Nigeria	Mandatory energy efficiency building code
17	Senegal	No building energy code
18	Sierra Leone	No building energy code
19	Somalia	No information
20	South Africa	Mandatory energy efficiency building code
21	Sudan	Mandatory energy efficiency building code
22	Tanzania	No building energy code
23	Tunisia	Mandatory energy efficiency building code
24	Uganda	No building energy code
25	Zambia	No information
26	Zimbabwe	No building energy code

Source Gaum and Laubscher (2022)

instance, has information on energy efficiency regarding HVAC systems and lighting and electrical appliances but lacks regulations to establish the use of BES tools. There is a need for countries in Africa that lack policies and regulations pertaining to energy efficiency and, more specifically, the use of BES tools to establish such policies.

Table 3.4 Comparison of building code status of 11 African countries from Iwaro and Mwasha [49] and Gaum and Laubscher [58]

No.	Country	2010 status (Iwaro and Mwasha 2010)	2022 status (Gaum and Laubscher 2022)	Change
1	Algeria	Proposed	Technical regulations document (DTR C 3-2, mandatory)	Progress
2	Angola	No code	Non-existent	No change
3	Cameroon	No code	Non-existent	No change
4	Ethiopia	No code	Non-existent	No change
5	Ghana	No code	Ghana building code (GS1207:2018, mandatory)	Progress
6	Kenya	No code	Non-existent	No change
7	Morocco	Proposed	General construction Regs 2014 (mandatory)	Progress
8	Nigeria	Mixed/voluntary	National building energy efficiency code 2017 (mandatory)	Progress
9	South Africa	Mixed/voluntary	South African National Standard (SANS 10,400:2011 & SANS 204:2011, mandatory)	Progress
10	Tanzania	No code	Non-existent	No change
11	Zambia	No code	No information	Likely no change

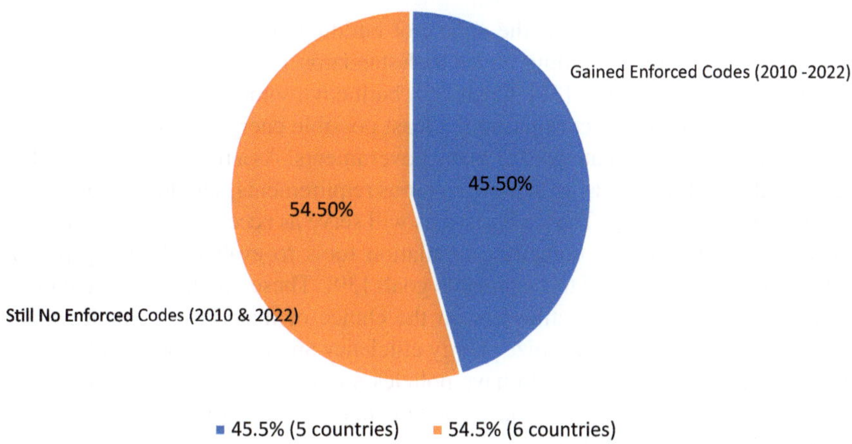

Fig. 3.2 Chart showing progress made by 11 African countries between 2010 and 2022. *Source* Gaum and Laubscher [58]

3.5 Uptake of Energy Simulation in Building Design

BES originated in the 1960s and steadily gained traction over the next decade. However, its rising popularity faced significant hurdles during the 1973 oil crisis [18]. Later after this incident, BES regained traction when the US Department of Energy allocated more than $1 billion to Research and Development projects on energy conservation and renewable energy. Several initiatives have since been underway to encourage the practical use of BES. Consequently, organizations have been established to address challenges encompassing the adoption of Building Performance Simulation (BPS) in practice. For instance, the International Building Performance Simulation Association (IBPSA) was created to advance BPS by offering platforms for discussing issues related to the development, testing, and implementation of simulation tools, through national and international conferences and various publication activities [9]. The IBPSA publishes a wide range of papers and activities. These publications are sufficient proof that the field of building performance simulation is developing and growing [25]. Simultaneously, a number of global research initiatives launched by the International Energy Agency (IEA) under the Energy Conservation in Buildings and Community Systems (ECBCS) program have progressed the use of BPS to assist in building and community design [19].

3.6 Pathways to the Uptake of Building Energy Simulation

3.6.1 Role of Policy and Regulations

An important pathway towards the uptake of building energy simulation in Africa is the establishment of building codes and standards for the usage of BES tools for energy-efficient designs [28]. Creating a built environment that meets the living needs of occupants while consuming the least possible energy and operation costs, and resources is the ultimate goal of many governments. As a result, building performance codes and standards, with more rigorous requirements, should be enacted and continually updated [39]. These regulations will serve as accelerators for promoting and enforcing the use of computer simulation tools to evaluate building performance towards compliance to sustainable goals [39]. These policies and regulations are framework to guide practitioners on the choice of materials, technology, and mythologies employed to optimize energy efficiency and improve occupant comfort [12]. Countries in Africa that do have policies and regulations and codes but still make compliance voluntary may have good intent but poor enforcement strategy and hence limit its impact [38]. The transition from voluntary codes to legally binding codes while including regulations to establish the adoption of BES tools is important. To help with this transition, these countries could establish implementation supports such as municipal training and incentives to encourage compliance [13]. Based on the analysis in Table 3.5, it is realized that policy implementation is slow in developing countries and Africa in spite of growing risk and urbanization. Therefore, provision of policies and regulations on BES implementation should be pushed using continental platforms and campaigns.

3.6.2 Need for Data and Context-Specific Software Development

In many African nations, the tools used for energy modelling often lack the dataset typical to the African context. This is because almost all BES tools were constructed in developed countries with considerations given to the climate, design configurations, energy mix and systems, and economic assumptions specific to those countries [51, 52]. According to Debnath and Mourshed [51], many energy modelling tools do not take the local challenges of developing countries into account. When these tools are applied in settings they were not designed for, they often produce unreliable results, such as underestimating or overstating energy consumption. Their review of 34 models showed that most overlook the everyday realities of low- and middle-income countries. Though BES tools have been adopted by some developing countries, they lack significant considerations of issues pertaining to their context, for instance, traditional energy use, urban–rural divide, low electrification, unreliable power supply, climate data, and skill needs. For instance, 2015 IEA research conducted in Nigeria found that households' power use was greatly understated since backup diesel and gasoline generators were not taken into consideration [52]. One major issue is that many of these models were developed with stable electricity grids and well-established systems in mind, which is a condition common in wealthier countries but not in much of Africa. These models often ignore or struggle to handle key challenges such as frequent power outages, irregular grid access, or the complex nature of informal energy networks [53, 54]. Dioha and Kumar [54] have noted that the limited use of these models in policymaking reflects their limited relevance. Ineffective models for capturing and analysing local settings frequently fail to persuade decision-makers and cast doubt on the analysis's potential policy consequences [51, 53]. Therefore, there's a clear need for BES tools that are built with Africa in mind from the start or modified that suit the regional contextual characteristics. As Avila et al. [59] suggest, new tools should be designed to work within African energy systems. These tools must be able to account for factors like informal settlements, climate impacts, and the growing role of decentralized renewables. However, in recent years, there have been increased attempts to solve this issue, with programs such as the Open-Source Energy Modelling System (OSeMOSYS), which provides statistics specialized for African nations [53]. According to Gebremeskel et al. [53], OSeMOSYS is an easily accessible open-source model with modifiable code. Due to its cost-effectiveness, it is often used in African countries like Ghana, Tunisia, and Egypt. Long-range Energy Alternatives Planning System (LEAP) is also suggested by Gebremeskel et al. [53] as suitable for developing country contexts because it captures important features relevant to developing countries. However, the features captured by these models are not enough for the required accuracy for simulation within the African context. Unless these appropriate tools are developed and adopted, energy simulation across the developing countries in Africa will be abysmal, leading to inaccurate analysis, proposal of wrong policies, and reduced privilege to expand access and build a more sustainable energy future [53].

3.6.3 Training and Capacity Building

The evolution of current technologies and the advent of new technologies for the complete decarbonization of the buildings sector by 2050 demands and calls for upskilling and reskilling of current professionals and the creation of new topics from higher education modules to be able to address the needs of these challenges [55, 56]. The limited adoption of BES tools in developing countries, especially in Africa, is significantly due to skill shortage, lack of awareness, and inadequate technical infrastructure [29]. Investing in students and construction professionals in order to build a skilled workforce has the potential to promote energy efficiency in Africa. Koukou et al. [55] conducted a study to come up with a strategy to train professionals in the construction sector towards decarbonization goals. They aimed at developing a project called BUILD2050 with a transnational educational approach which responds to the need to develop new skills and equip construction professionals with skills necessary to promote energy efficiency and sustainability, which included topics on how energy tools can be used to promote energy efficiency. Roxana-Valentina et al. [56] presented an education and training program called "Solutions and Actions for Companies' Decarbonization" to equip participants with required knowledge and skills for energy efficiency implementation. The program covered six units, with one unit focusing on the use of technology for practical application of the program. Many programs like these are needed to improve the knowledge base of professionals in Africa and increase their technical capacity.

3.6.4 Academia and Industry Collaborations

Collaborative efforts between industry and academia are key for developing cutting-edge technologies and training the next generation of decarbonization professionals. Academic and industry collaboration plays an important role in the uptake of BES tools in Africa by filling the gap between theoretical research and industry applications [36]. Research done on building cases for energy efficiency to bring out optimum design configurations can be taken by industry to implement and test in real-world project. This will help validate results received from building performance simulations to encourage its uptake. So far, in Africa, there is a weak cooperative relationship between education and industry [30]. Either new academic discoveries become the headline for a period and die out without actual implementation by organizational stakeholders, or they end with applause after researchers have been acknowledged for their discovery. Universities have engaged in research over the years in energy efficiency and have had results which are worthy of implementation. Unfortunately, because of the disconnect between industry and education, practice in the industry, more often than not, does not benefit from this research. Universities are important governance actors that provide their expertise and specific know-how and contribute to regional development. They have the capacity to impart certain knowledge that can improve upon existing practices and processes [57].

This collaboration can also foster effective funding opportunities to take care of the high costs of buying these tools [6]. Industry and government can collaborate to make available these tools for academic institutions and organizations to aid in accurate building energy performance simulations [40]. This effort will reduce the reliance on cracked software versions and limit the likelihood to cyberattacks and hacks of organizations. When university-industry relationship is strengthened, knowledge transfer into the industry and possible funding opportunities will be effective, speeding up the implementation of new ideas that are able to spearhead the adoption of BES tools and, consequently, energy efficiency goals.

3.7 Summary

This chapter examines building energy simulation (BES) as an essential tool for promoting energy efficiency and decarbonization in the built environment. The chapter starts by contextualizing the building sector's massive impact on worldwide carbon emissions and emphasizes the importance of novel solutions in building design and operations. BES is a significant method that allows the simulation of complex systems to forecast energy performance, optimize design parameters, and minimize environmental effects. Softwares such as EnergyPlus, IES VE, and OpenStudio provide functionalities such as thermal, lighting, and airflow simulation, all of which facilitate data-driven decision-making throughout a building's life cycle. The chapter describes how BES increases economic, environmental, and occupant performance. By simulating building behaviour, project teams are able to identify inefficiencies, maximize thermal comfort, and improve air quality while complying with energy standards and regulations. BES software also facilitates operational optimization, compliance with evolving energy codes, and cost-effective planning for retrofits and system upgrades. They also measure return on investment, making them an essential tool for convincing stakeholders of the long-term value of buildings. In spite of its advantages, there are barriers to BES application in Africa. They encompass the high cost of the money, the absence of training, insufficient climate data, and limited support from policies. Limited technical expertise and weather dataset issues, and insufficient application of airflow modelling also affect the accuracy of simulation. Model calibration and validation remain a challenge because actual-world data continue to be inconsistent, and there is a lack of technical skill.

Open-source platforms like OpenStudio can provide solutions. However, there is lack of awareness of its existence but is unavailable and lack awareness. Because of these barriers, there is minimal use of BES tools in professional and educational work on the African continent. Policy and regulatory establishment for training programs aimed at BES, as well as workforce upskilling, is necessary for wider use. Moreover, the development of BES tools that are specific to the African context is necessary for successful deployment. BES possesses transformational potential to provide energy-efficient buildings. The potential of BES in Africa requires the development of simulation tools that have datasets which are in the context of Africa. Also, training and capacity building, establishing policies and regulations, and collaboration between universities and industries can spearhead the adoption of BES tools.

References

1. Addy M, Adinyira E, Danku JC, Dadzoe F (2021) Impediments to the development of the green building market in sub-Saharan Africa: the case of Ghana. Smart Sustain Built Environ 10(2):193–207
2. Ahmed I, Kabir H (2021) Building codes and regulations for disaster resilience in Bangladesh: the case of Dhaka. Int J Disaster Resil Built Environ
3. Aleassa H, Pearson JM, McClurg S (2011) Investigating software piracy in Jordan: an extension of the theory of reasoned action. J Bus Ethics 98(4):663–676
4. Amankwah-Amoah J (2019) Technological revolution, sustainability, and development in Africa: OVERVIEllw, emerging issues, and challenges. Sustain Dev 27(5):910–922
5. Bass B, Ezell E, New J, Copeland B (2022) Using measured building energy data to infer building type for urban building energy modeling. Oak Ridge National Laboratory (ORNL), Oak Ridge, TN (United States)
6. Bikard M, Vakili K, Teodoridis F (2019) When collaboration bridges institutions: the impact of university–industry collaboration on academic productivity. Organ Sci 30(2):426–445
7. Bottino-Leone D, Balest J, Cittati VM, Pezzutto S, Fraboni R, Beltrami F (2024) Review of existing tools for the assessment of European building stock energy demand for space heating and cooling. Sustainability 16(6):2462
8. Clarke J (2007) Energy simulation in building design. Routledge
9. Clarke J (2015) A vision for building performance simulation: a position paper prepared on behalf of the IBPSA Board. J Build Perform Simul 8(2):39–43
10. Clarke JA, Maver TW (1991) Advanced design tools for energy conscious building design: development and dissemination. Build Environ 26(1):25–34
11. Coakley D, Raftery P, Keane M (2014) A review of methods to match building energy simulation models to measured data. Renew Sustain Energy Rev 37:123–141
12. Degelman LO, Soebarto VI (1996) Whole building energy performance-simulation and prediction for retrofits. Department of Architecture, Texas A&M University
13. Emere C, Aigbavboa C, Oguntona O (2025) Critical regulatory characteristics for sustainable building construction in South Africa. Sustainability
14. Fumo N (2014) A review on the basics of building energy estimation. Renew Sustain Energy Rev 31:53–60
15. Garg V, Mathur J, Bhatia A (2020) Building energy simulation: a workbook using design-builder™. CRC Press
16. Groesser SN (2014) Co-evolution of legal and voluntary standards: Development of energy efficiency in Swiss residential building codes. Technol Forecast Soc Chang 87:1–16

17. Hajare A, Elwakil E (2020) Integration of life cycle cost analysis and energy simulation for building energy-efficient strategies assessment. Sustain Cities Soc 61:102293
18. Hong T, Chou SK, Bong TY (2000) Building simulation: an overview of developments and information sources. Build Environ 35(4):347–361
19. Hong T, Langevin J, Sun K (2018) Building simulation: ten challenges. In: Building simulation, vol 11, pp 871–898. Springer Berlin Heidelberg. https://www.anl.gov/article/modeling-a-net zero-future-energy-experts-harness-simulation-for-global-decarbonization
20. Jindal A, Rao S (2017) Agent-based modeling and simulation of mosquito-borne disease transmission. In: Proceedings of the 16th conference on autonomous agents and multiagent systems, pp 426–435
21. Judkoff R, Wortman D, Burch J (1983) Measured versus predicted performance of the SERI test house: a validation study (No. SERI/TP-254-1953; CONF-830702-16). Solar Energy Research Inst., Golden, CO (USA)
22. Judkoff R, Wortman D, O'doherty B, Burch J (2008) Methodology for validating building energy analysis simulations (No. NREL/TP-550-42059). National Renewable Energy Lab. (NREL), Golden, CO (United States)
23. Koranteng C, Simons B, Gyimah K, Nkrumah J (2021) Ghana's green building assessment journey: an appraisal of the thermal performance of an office building in Accra. J Eng, Des Technol
24. Koranteng C, Simons B, Gyimah K, Amos-Abanyie S (2022) The idealized and the realistic validation studies of building simulation models in Ghana. J Green Build
25. Lamberts R, Hensen JL (eds) (2011) Building performance simulation for design and operation. Spoon Press
26. Maile T, Fischer M, Bazjanac V (2007) Building energy performance simulation tools-a life-cycle and interoperable perspective. Center for Integrated Facility Engineering (CIFE) working paper, vol 107, pp 1–49
27. Magni M, Ochs F, de Vries S, Maccarini A, Sigg F (2021) Detailed cross comparison of building energy simulation tools results using a reference office building as a case study. Energy Build 250:111260
28. Mohammed BU, Wiysahnyuy YS, Ashraf N, Mempouo B, Mengata GM (2023) Pathways for efficient transition into net zero energy buildings (nZEB) in Sub-Sahara Africa. Case study: Cameroon, Senegal, and Côte d'Ivoire. Energy Build 296:113422
29. Oke AE, Arowoiya VA (2022) Critical barriers to augmented reality technology adoption in developing countries: a case study of Nigeria. J Eng Des Technol 20(5):1320–1333
30. Owuondo J (2023) Bridging the gap between curriculum theory and industrial innovation practice: The necessity of augmenting industrial skills training in the Kenyan education system. *Available at SSRN 4548528*
31. Pan Y, Zhu M, Lv Y, Yang Y, Liang Y, Yin R, Yang Y, Jia X, Wang X, Zeng F, Huang S (2023) Building energy simulation and its application for building performance optimization: a review of methods, tools, and case studies. Adv Appl Energy 10:100135
32. Pillay TL, Saha AK (2024) Framework for a simulation learning tool to optimize green star buildings in south Africa. Sustainability 16(4):1565
33. Sajjad M, Hu A, Alshehri AM, Waqar A, Khan AM, Bageis AS, Elaraki YG, Shohan AAA, Benjeddou O (2024) BIM-driven energy simulation and optimization for net-zero tall buildings: sustainable construction management. Front Built Environ 10:1296817
34. Sanusi Y, Ohadugha C, Martins V, Olaide S (2022) Analysis of energy governance in Southern Africa. Polityka Energetyczna Energy Policy J
35. Scamman D, Solano-Rodríguez B, Pye S, Chiu LF, Smith AZ, Gallo Cassarino T, Barrett M, Lowe R (2020) Heat decarbonisation modelling approaches in the UK: an energy system architecture perspective. Energies 13(8):1869
36. Stephen S, Aigbavboa C (2025) Enhancing academia–industry partnerships for sustainable building: a change framework for research and innovation integration in Sub-Saharan Africa. Sustainability (2071–1050)*, 17(9)

37. Tomrukcu, G., Kizildag, H., Avgan, G., Dal, O., Ganic Saglam, N., Ozdemir, E. and Ashrafian, T., 2024. A systematic approach to manual calibration and validation of building energy simulation. Smart and Sustain Built Environ
38. Twum-Darko M, Mazibuko P (2015) Compliance and enforcement challenges a case of the national building regulations' processes in South Africa. J Gov Regul 4:679–684
39. Wang H, Zhai ZJ (2016) Advances in building simulation and computational techniques: a review between 1987 and 2014. Energy Build 128:319–335
40. Yeatts DE, Auden D, Cooksey C, Chen CF (2017) A systematic review of strategies for overcoming the barriers to energy-efficient technologies in buildings. Energy Res Soc Sci 32:76–85
41. Young H, Cornforth R, Petty C (2019) Climate and health in Africa: research and policy needs
42. Hopfe CJ, Hensen JL (2011) Uncertainty analysis in building performance simulation for design support. Energy Build 43(10):2798–2805
43. Gajdzik B, Jaciow M, Hoffmann-Burdzińska K, Wolny R, Wolniak R, Grebski WW (2024) Impact of economic awareness on sustainable energy consumption: results of research in a segment of Polish households. Energies 17(11):2483
44. Augenbroe G (1992) Integrated building performance evaluation in the early design stages. Build Environ 27(2):149–161
45. Yu J, Chang WS, Dong Y (2022) Building energy prediction models and related uncertainties: a review. Buildings 12(8):1284
46. Mahmoud R, Kamara JM, Burford N (2020) Opportunities and limitations of building energy performance simulation tools in the early stages of building design in the UK. Sustainability 12(22):9702
47. U.S. Briefing (2013) International energy outlook 2013. U.S. Energy Information Administration, pp 506–507
48. Shen L, Sun Y (2016) Review on carbon emissions, energy consumption and low-carbon economy in China from a perspective of global climate change. J Geogr Sci 26(7):855–870
49. Iwaro J, Mwasha A (2010) A review of building energy regulation and policy for energy conservation in developing countries. Energy Policy 38(12):7744–7755
50. Hussain M, Bhat N (2018) World population statistics: some key findings. Res J Soc Sci 9:16–25
51. Debnath KB, Mourshed M (2018) Forecasting methods in energy planning models. Renew Sustain Energy Rev 88:297–325
52. Dioha MO, Mutiso RM (2023) Generating meaningful energy systems models for Africa. Issues Sci Technol 39(3):54–57
53. Gebremeskel D, Bekele G, Ahlgren EO (2020) Energy system modeling tools: review and comparison in the context of developing countries. 2020 IEEE PES/IAS PowerAfrica, pp 1–5
54. Dioha MO, Kumar A (2020) Exploring sustainable energy transitions in sub-Saharan Africa residential sector: the case of Nigeria. Renew Sustain Energy Rev 117:109510
55. Koukou MK, Lucas S, Justino J, Rafael S, Livieratos AD, Carriço N, Konstantaras J, Vrachopoulos MG, Benedetti AC, Mazzoli C, Ferrante A (2024) Training for sustainable and healthy building for 2050: new methodologies for an integrated and transnational education approach targeting skills development for the transition toward ZEB and PEB buildings. Buildings 15(1):67
56. Roxana-Valentina B, Denisa Ş, Levente C, Ştefan C, Dacian J, Alexandru M (2024) Empowering energy professionals: a novel educational and training program on energy transition audits and decarbonization. In: 2024 IEEE International Conference and Exposition on Electric and Power Engineering (EPEi). IEEE, pp 400–405
57. Sedlacek S (2013) The role of universities in fostering sustainable development at the regional level. J Clean Prod 48:74–84
58. Gaum T, Laubscher J (2022) Building energy codes: reviewing the status of implementation strategies in the Global South. Int J Built Environ Sustain 9:39–53
59. Avila N, Carvallo JP, Shaw B, Kammen DM (2017) The energy challenge in sub-Saharan Africa: a guide for advocates and policy makers. Generating Energy for Sustainable and Equitable Development 1:1–79

Chapter 4
Thermal Comfort

4.1 Introduction

The attempt to meet the demands for shelter of a growing populace has advanced various construction practices in the Architectural Engineering and Construction (AEC) industry. Some of these practices have been found to be detrimental to the soundness of the environment despite contributions to socio-economic development of numerous economies [5, 30]. Characterized by activities ranging from material extraction, transportation to construction, then to demolition and waste generation, the industry contributes about 40% of the global greenhouse gas emissions and consumes about 55% of global electrical energy [4, 6, 28, 61]. The industry is therefore saddled with the responsibility to promote practical techniques consistent with the construction of adaptive and energy-efficient buildings without compromising further the environment and other natural resources [28, 40].

It is essential to recognize the role indoor environmental quality plays in achieving this feat. Indoor environmental quality (IEQ) refers to a spectrum of conditions that exists within a building, which have the ability to potentially influence occupants in a holistic manner [41, 93]. These conditions include indoor air quality, temperature, relative humidity, visual/lighting comfort, acoustic/noise comfort, thermal comfort [46, 60]. Research reveals a positive correlation between the factors of IEQ and the physical health, comfort, productivity, and satisfaction of building users [7, 91, 92]. Nonetheless, thermal comfort has been ranked the most important and remains the most studied factor of IEQ [25, 39, 60].

Thermal comfort is defined as "the heat exchanges that occur between a person and the thermal environment and the physiological conditions that are needed for human comfort" [45]. The American Society of Heating, Refrigeration, and Air-Conditioning Engineers (ASHRAE) defined thermal comfort as a mental condition a person exhibit to show satisfaction with his thermal environment [15, 16]. Figure 4.1 illustrates the factors that affect thermal comfort. This includes relative humidity, air flow, mean radiant temperature, humidity, rate of metabolism, and clothing [34, 76].

M. N. Addy et al., *Decarbonization of Building Energy in Developing Countries*, Green Energy and Technology, https://doi.org/10.1007/978-3-032-09113-0_4

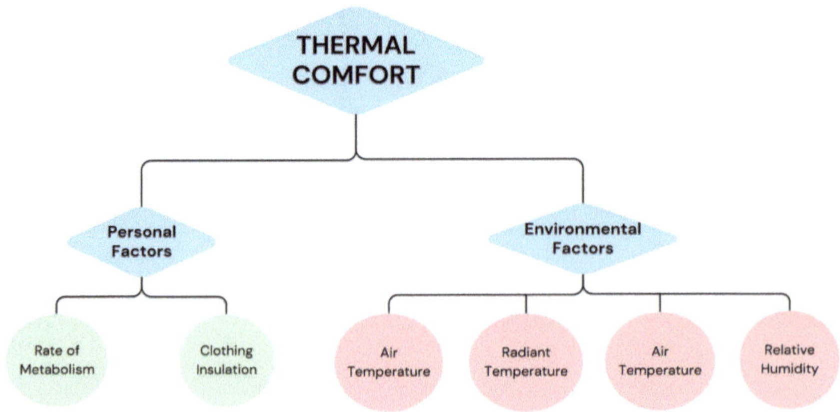

Fig. 4.1 Factors of thermal comfort

At present, building occupants have been recognized as active functionaries within their indoor environment because of the presence of a thermoregulatory system in their bodies [25]. The function of this system is to bring the human to experience a thermal equilibrium between his internal body and his external environment. This tends to influence the conscious and unconscious behaviour of building users [10]. Example include opening or closing shades or windows, turning on fans, or adjusting temperature of air-conditions, among others.

A major portion of the total energy the built environment is used to create thermally conducive indoor environments [2]. As a result, building users depend mostly on mechanical or active thermal systems (e.g., heating, ventilation, and air-conditioning (HVAC)). There are two main methods adopted for indoor thermal comfort assessment: (1) Static model or Heat balance and (2) Adaptive approach. Additionally, the American Society of Heating, Refrigeration, and Air-Conditioning Engineers (ASHRAE) standard 55, the European Standard (EN) 15251 and International Organization for Standardization (ISO) 7730 are some prominent standards for assessing thermal comfort. For air-conditioned indoor spaces, the static model is most appropriate, whereas for naturally ventilated buildings, the adaptive model is most effective [44].

Due to the geographical positioning of countries in the world, each country tends to experience a certain climatic condition, which can be further classified into microclimatic conditions. It is worth noting that the construction of sustainable buildings and ensuring indoor environmental quality can be advanced by reconciling geographical atmospheric data and building designs [77].

4.1.1 Thermal Comfort in Warm Climates

A thermal environment refers to an area or space which is influenced by various meteorological factors like temperature, humidity, air flow, and radiation [101]. Warm climates or warm-humid climates are marked by extended periods of warm weather, in which semi-outdoor and outdoor spaces are considered more comfortable than indoor environments, particularly when there is no air-conditioning [18]. Geographically, these climates can be found in tropical and subtropical regions, and they are characterized by high humidity and temperature with distinct wet and dry seasons throughout the year (see Fig. 4.2). Many developing economies in Africa, Southeast Asia, Central and South American, and Pacific Islands are concentrated in this particular climatic zone [101]. Accordingly, these are tagged as the countries in the "Global South".

The warm-humid climatic zones, due to their location with respect to the equator, are exposed to direct sunlight and radiation. These zones are characterized by high rainfall, relative humidity levels (above 90%) and an average temperature greater than 18 °C [87]. Given the warmth nature of this climate, people's behaviour and culture are impacted. Moreover, Odim [71] maintain that the prevailing climatic situation presents several challenges. For instance, the high temperatures can create uncomfortable indoor spaces, lead to health issue like heatstroke, excessive heat and humidity can affect various socio-economic activities [55]. In view of this, adapting to this climate requires that people develop several personal, architectural and technological strategies.

Human beings desire to dwell in comfortable environments, as a result, they tend to exhibit certain behaviours intended to aid them attain and maintain their thermal comfort. As a result, building occupants adapt their activities (e.g. running or sitting), vary clothing styles, take in hot/cold foods, adjust fan speed, check the nature of equipment being used in the space, change rooms, take shower or wet the face [20, 97, 100, 103].

Buildings within the warm climates need to embrace the concept of bioclimatic architecture. It refers to the design of buildings while adapting the building to harmonize to the local climatic conditions, techniques and materials available to achieve sustainable indoor environment [13, 63]. This concept seeks to use eco-friendly materials and leverage on the nature of the ecological environmental setting to promote sustainable building condition [102]. To do this, buildings designs should have (1) low heat storing capacity (2) efficient and continuous ventilation (3) appropriate design features that dispel radiant heat in indoor environments (4) sufficient protection from weather conditions like rain [43, 72]. Various studies have highlighted various forms that portrays the vernacular concept of buildings in the warm climate. For instance, [8] revealed a number of vernacular architectural styles in Bangladesh including courtyard (for aeration and daylighting), veranda, *jali* (screen for shading, ventilation), *dochala* roof or slit roof (ventilation, rain protection, shading). Within the Brazilian context, buildings have large roof overhangs, verandas, vegetation, asbestos-cement, or aluminium roofing sheets [63]. Others also rely on bamboo, wood, grass, and

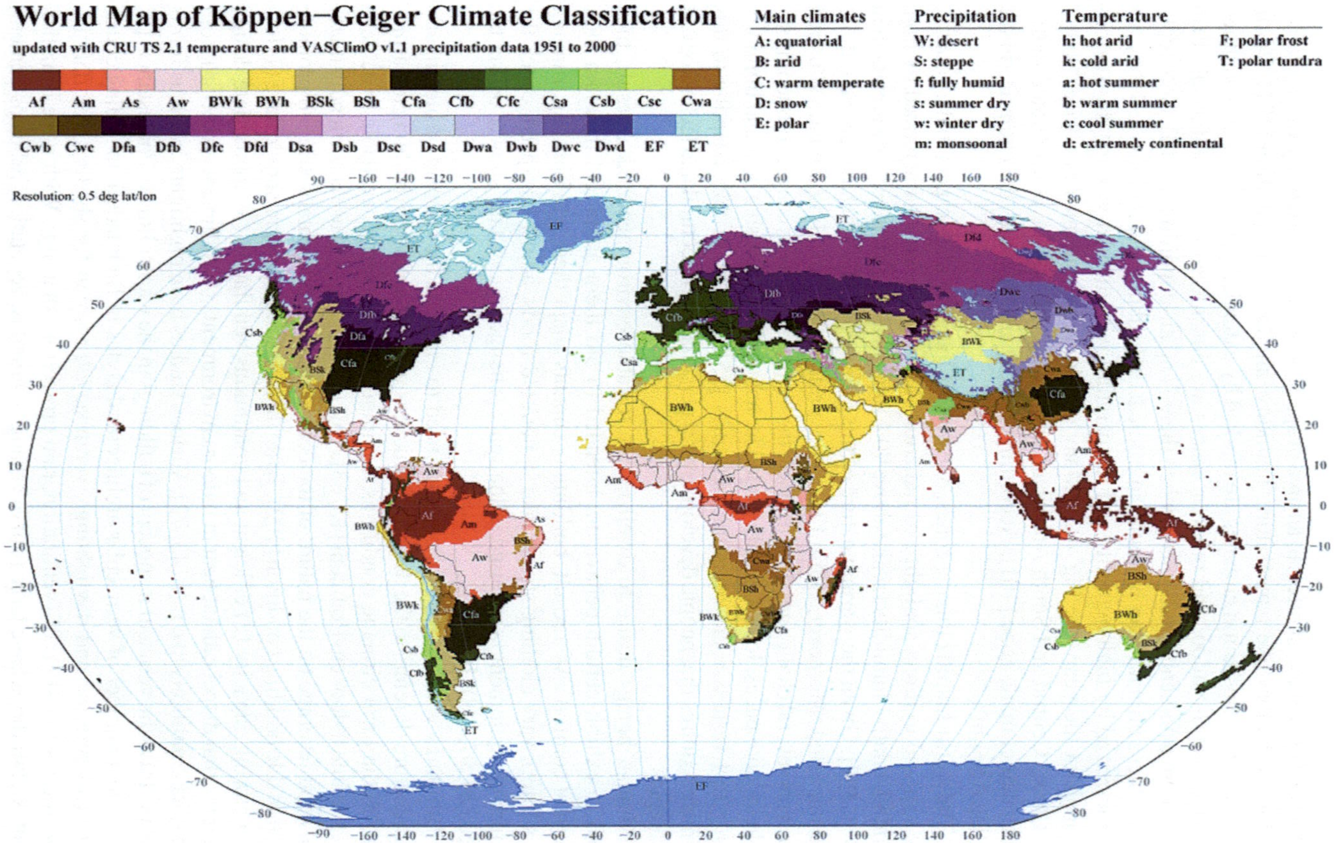

Fig. 4.2 World map of Köppen-Geiger climate classification with precipitation and temperature data from 1951 to 2000 [19, 54]

rammed earth bricks [44, 97]. Conversely, the infiltration of modern architecture seems to be a rising problem within this climatic context [52, 53]. However, issues concerning energy demand and inconsistent power supply, greenhouse gas emission and climate change requires that the built environment embraces the concept of bioclimatic architecture [4, 6, 34, 52, 53].

Technological strategies in this regard, refers to the developing and application of innovative materials, tools, systems, and techniques to design or improve building efficiency. Cody [29] revealed that these strategies can be categorized into two, that is, low and high level technologies. These differ from each other based on their level of sophistication. The low technological strategies encompass the use of traditional and simple materials and passive techniques to provide a thermal comfortable environment [48]. Examples include the size of windows and the materials they are made up of and shading. The high technological strategies involves the use of advanced construction methods, materials, or intelligent tools to enhance buildings' efficiency [29, 48]. This include using polyurethane foam to insulate walls, smart buildings (installing Cyber physical systems and Internet of things (IoT) in buildings), and water walls. Others also consider technological strategies as passive and active strategies [11, 23, 70].

In recent times, several studies have been conducted to understand occupants' perception and responses in the different climatic zones [11, 14, 49, 55, 71, 83, 95]. This notwithstanding, a few countries have developed national standards for promoting thermal comfort in buildings. For instance, India has stipulations in their National Building Code, Eco-Niwas Samhita 2018 and ISHRAE 10001:2019 [90]; Egypt has the Green Pyramid Rating System (GPRS) [1]; Jordan uses the Jordanian Green Building Guide (JGBG) [84]; Saudi Arabia uses the Mostadam [17], etc. Trebilcock et al. [94] maintain that though these standards were founded based on international standards (ISO Standard 7730, ASHRAE Standard 55, or the EN 15251), it is not obligatory to follow their stipulations.

4.2 Static and Adaptive Models

Thermal comfort, also known as hygrothermal comfort, according to the ISO 7730 is a mental condition which expresses satisfaction with a thermal environment [74, 80]. Studies have shown controlling thermal comfort in individuals is a complex activity because of its subjectivity [35, 79]. Nonetheless, Rînjea et al. [80] contend that thermal comfort can be controlled by adjusting environmental parameters like air velocity and humidity, engaging in a certain activity; controlling clothing styles; so on. The development of certain indices and models by scholars like Fanger [37], de Dear and Brager [31], and Humphreys and Nicol [45] among many others have advanced various scientific and systematic modalities for predicting and assessing thermal comfort in various environmental settings. These are broadly classified into two, namely the static and adaptive models.

The static model or the heat balance model or the steady-state thermal comfort model emphasizes on a set of physiological factors—environmental and personal factors—related to heat exchange between humans and their environment, with the assumption that the human is a passive participant in the thermal environment [9, 82]. The heat balance model, is based on regulated climatic conditions in laboratory setting, has brought about the quantitative description of qualitative thermal preferences [22, 104]. The model developed by Fanger is a prominent model in thermal comfort research. Consequently, provisions have been made in standards like the ASHRAE Standard 55 and ISO 7730 based on this model [80]. This model contains two indices that interpret thermal comfort based on Predicated Mean Vote (PMV) and Predicated Percentage of Dissatisfaction (PPD) [27].

The PPD is a quantitative estimation of the percentage of thermally dissatisfied people at each PMV scale [35]. The PMV index, on the other hand, is an analytical estimation that is able to predict the mean thermal sensation vote of a group of people based on the factors that influence thermal comfort [47]. It uses a seven-point scale to predict thermal sensation based on the environmental and personal factors as presented in Fig. 4.3. PMV-PPD can be computed using this equation:

$$\begin{aligned}\mathrm{PMV} = {} & [1.30\exp(-0.306M) + 0.028]\big\{(M - W) - 3.05 \times 10^{-3} \\ & \times [5733 - 6.99(M - W) - P_a] - 0.42[(M - W) - 58.15] \\ & - 1.7 \times 10^{-5} \times M(5867 - P_a) - 0.0014M(34 - T_a) - 3.96 \times 10^{-8} \\ & \times f_{cl}\big[(f_{cl} + 273)4 - (T_r + 273)4\big] - f_{cl}h_c(T_{cl} - T_a)\big\}\end{aligned} \tag{4.1}$$

$$\mathrm{PPD} = 100 - 95 \cdot \mathrm{e}^{(-0.03353 \cdot \mathrm{PMV}^4 - 0.2179 \cdot \mathrm{PMV}^2)} \tag{4.2}$$

where

M = Rate of metabolism (W/m^2);
W = Effective mechanical power (W/m^2);
I_{cl} = Insolation thermique (m^{2}K/W);
h_c = Convection heat transfer co-efficient (W/m^{2}K);
f_{cl} = Clothing insulation; T_a = Indoor air temperature (°C);
T_{cl} = Clothing surface temperature (°C);
T_r = Mean radiant air temperature (°C);
RH = Relative humidity (%);
P_a = Water vapour pressure (Pa);
v_a = Air flow (m/s).

Other static models include the Physiological Equivalent Temperature (PET), Universal Thermal Climate Index, Effective Temperature and Standard Effective Temperature, and so on [35]. Despite being ideal for predicting efficient air-conditioned spaces, it is widely documented that the Fanger's model is not without certain deficiencies since it ignores certain adaptive characteristics of people and their environmental settings [22, 47, 104]. This led to the discovery of adaptive thermal model.

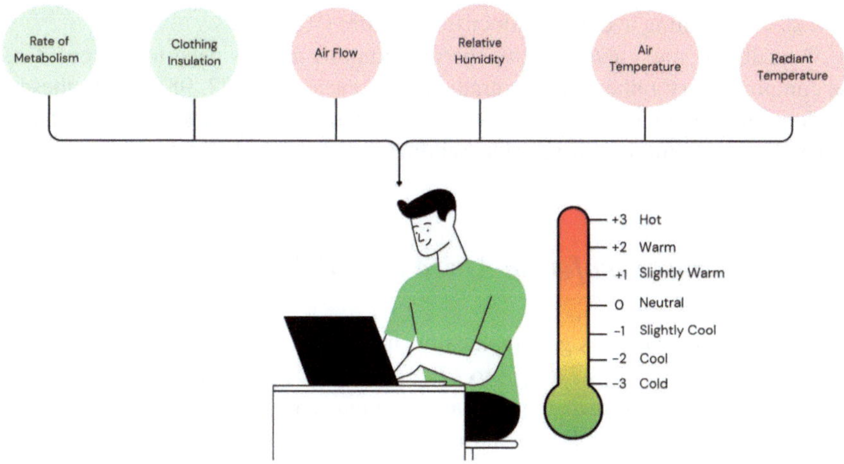

Fig. 4.3 Factors of thermal comfort in the PMV index

The adaptive thermal model, unlike the static model, considers the human as an active player in establishing thermal equilibrium with his environment and also uses data collected from naturally ventilated buildings [12, 56]. Humphrey and Nicol, in the 1970s, proposed the adaptive thermal comfort theory. It is a linear regression model that relates indoor temperature to outdoor environmental parameters [31, 79]. Since the outdoor environmental parameters will vary in different geographic region, this model is contextualized to the prevailing climatic conditions. As a result, Efeoma and Uduku [34] mention that the adaptive model makes provision for a wide range of temperature. Since the human's thermoregulatory system will react in a manner so as to adapt to their environment in order to restore their thermal satisfaction [45]. Brager and de Dear [22] avow that the adaptive model takes into account the people's behavioural changes, physiological and psychological factors. Accordingly, developing a generally feasible model is considered strenuous [35].

Over the years, researchers have developed various adaptive models for different building settings in different climatic regions [3, 14, 33, 78, 79]. Bienvenido-Huertas et al. [21] contend the standards can be adopted for three types of buildings: natural ventilated buildings, air-conditioned buildings and mixed-mode buildings. The adaptive models have been adopted in international standards like the ASHRAE Standard 55 and EN 15251. Kiki et al. [49] contend that the adaptive thermal models offer a solution to soaring energy consumption and the emission of greenhouse gases. According to the ASHRAE Standard 55, the adaptive thermal assessment for naturally ventilated buildings can be computed, using this equation:

$$T_c = 17.80 + 0.31 \times T_{mo} \tag{4.3}$$

where

T_c = Comfort Temperature (°C);

T_{mo} = Prevailing average outdoor temperature (°C);

Overall, the static and adaptive thermal comfort models need to be considered as complimentary rather than contradictory to each other [22, 98]. Using these in tandem can present a more comprehensive thermal comfort assessment, which is an imperative to promoting sustainable buildings.

4.2.1 Thermal Comfort in Africa and Global Standards

The continent of Africa is geographically situated on latitude 37° 21′ N to 34° 51′ 15″ S and longitude 51° 27′ 52″ E to 17° 33′ 22″ W. Among the continents characterized by the warm climate, the African continent is most centred within the tropics (the tropic of Cancer (23.5° N) and tropic of Capricorn (23.5° S)). The region is usually hot and humid due to the direct exposure to the Sun's rays [83]. According to the Köppen-Geiger climatic map, Africa is largely characterized by Tropical Rainforest, Arid, and Temperate regions (see Fig. 4.2) [19]. However, a closer look at Fig. 4.4 shows 13 climatic conditions prevail in the continent [69]. For instance, the warm desert climate is noticeable in northern countries like Egypt, Libya, Algeria, Mauritania, Niger, etc. These regions, characterized by minimal rainfall, freezing nights and hot days, contain the Sahara and Kalahari deserts. Morocco and Tunisia tend to be characterized by warm Mediterranean climate. In contrast, the Equatorial and Tropical Savana climate are stretches from Guinea-Bissau through to some part of Madagascar. In the south the humid subtropical, warm and temperature oceanic climates prevail in South Africa, some part of Madagascar, Botswana, Angola, Kenya, and Zambia. Basically, these regions have two seasons: the raining and dry seasons. During the dry season, temperature in northern part of the region can rise above 40 °C [3]. In the wet season, regions along the coastal belt could record an average of about 5000 mm rainfall [69]. Recognized as an uncomfortable region due to heighten humidity and temperature, Africa is projected to experience significant growth in industrial activities and urbanization [73]. Subsequently, these activities coupled with the need to provide thermal comfortable spaces can significantly contribute to energy consumption, environmental degradation, climate change, and so on. There is therefore necessary to promote sustainable related activities within the region. In previous dispensations, many African countries relied on various traditional practices to achieve thermally conducive buildings. Some of these include the use of rammed earth or hemp blocks, orienting buildings, using grass roofs, etc. [4, 6, 44]. This notwithstanding, in recent times, Mohammadpourkarbasi et al. [59] for instance, argue the prevalence of building designs that are unfit for the harsh climatic condition. Aside failing to preserve the African vernacular architectural culture, these contemporary building concepts rely heavily on mechanical systems to engender thermal comfort among building users [65, 70].

Additionally, the use of inefficient building materials and the unavailability of sustainable country-specific design principles have largely contributed to unsustainable building practices in some parts of the continent [51, 53]. Considering this,

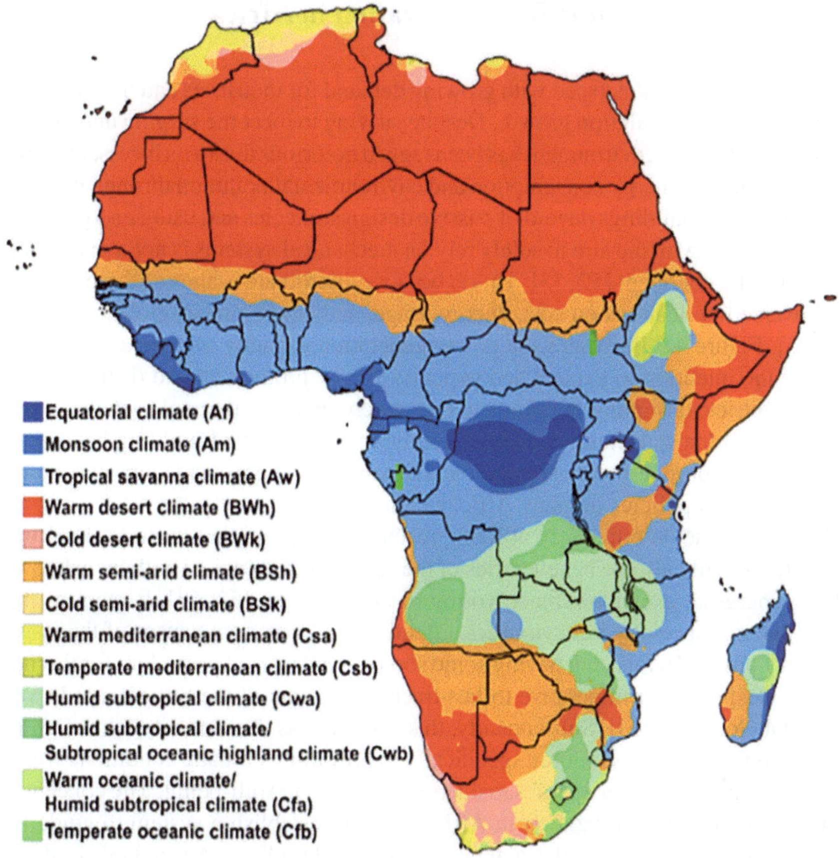

Fig. 4.4 Climate zones in Africa adapted from Nwaerema et al. [69]

studies in the continent have highlighted the imperative of using local materials and passive techniques to achieve thermal comfort in indoor environments [107]. Nwalusi et al. [70] stated that using roof verandas and balconies are effective means of reducing heat within indoor spaces. Kimemia et al. [50] proposed the use of cool coatings on surfaces of building envelopes. Cool coating is a paint based product that have high solar reflective properties [96]. Others mentioned adobe, rammed earth, earthbag, and superadobe as effective materials for buildings [23]. The use of façades, shading devices and vegetation can induce thermal comfort [32, 57]. Concerning standards that are used to evaluate indoor thermal comfort, the ASHRAE Standard 55 adaptive thermal comfort model is predominantly used [57]. This is due to its comprehensive guidelines for assessing thermal comfort based on environmental factors and responses from building users [79]. However, applying this standard has its own limitation due to the diversified climatic, socio-economic, and cultural conditions in the continent.

4.3 Thermal Comfort Standardization in Africa

The African continent is faced with growing demand for industries and urban centres as proponents to population growth. Despite striving to meet the surging demands of the society, the built environment has been tagged notorious due to its incessant depletion of resources, energy consumption and environmental pollution through its activities. Designing buildings devoid of passive design strategies and using energy inefficient materials, with the aim to solely rely on mechanical systems is not a sustainable choice for the continent [52, 53]. This is because, inordinate usages of these systems has adverse implications on cost, energy usage, and environment (global warming) [67, 75]. Figure 4.5 illustrates the power generation capacity of African countries. Despite the intention to expand and expedite energy production and distribution in these countries, designing buildings to rely much on active thermal systems is way to further burden stakeholders. Consequently, this approach of building has gained less appreciation due to the temperamental nature of energy and improper power implementation policies in many African countries [36].

The discourse on efficiently utilizing resources today so that those of the future generations could also be beneficiaries of the same is not complete without acknowledging energy as a major subject worthy of our attention [52, 53]. It requires that stakeholders, especially policymakers, effectively address root causes of the issues of energy utility. To this end, having some sort of standardized procedure for thermal comfort designs is an imperative to fast track the agenda of the Sustainable Developmental Goals [81, 83]. Subsequently, this can serve as the modality for designing net-zero building spaces tailored specifically to prevailing climatic conditions within the continent to engender users' satisfaction, health, and well-being. The concept of thermal comfort standardization in Africa seeks to establish a system to guide the actual thermal and expected thermal performance of buildings. A standard is a document that specifies common rules, guidelines, or certain distinct characteristics for an activity or product in order to achieve an optimal result [106]. The use of a widely accepted standard within the built environment industry would minimize the impact of separation and fragmentation, and promote consistency in building performance assessment and certification in Africa [68].

Despite these benefits of developing a thermal comfort standard, there exist certain hindrances. These includes the following.

4.3.1 Climatic Diversity and Climate Change

Nicol and Humphreys [66] consider climate a significant factor to developing thermal comfort models. According to the Köppen-Geiger climate classification, Africa has 13 different climatic regions (Fig. 4.4) [54]. This classification is based on data on temperature and precipitation gathered over many years. The nature of a climate in a geographic area is influenced by prevailing ecological features and the movement of

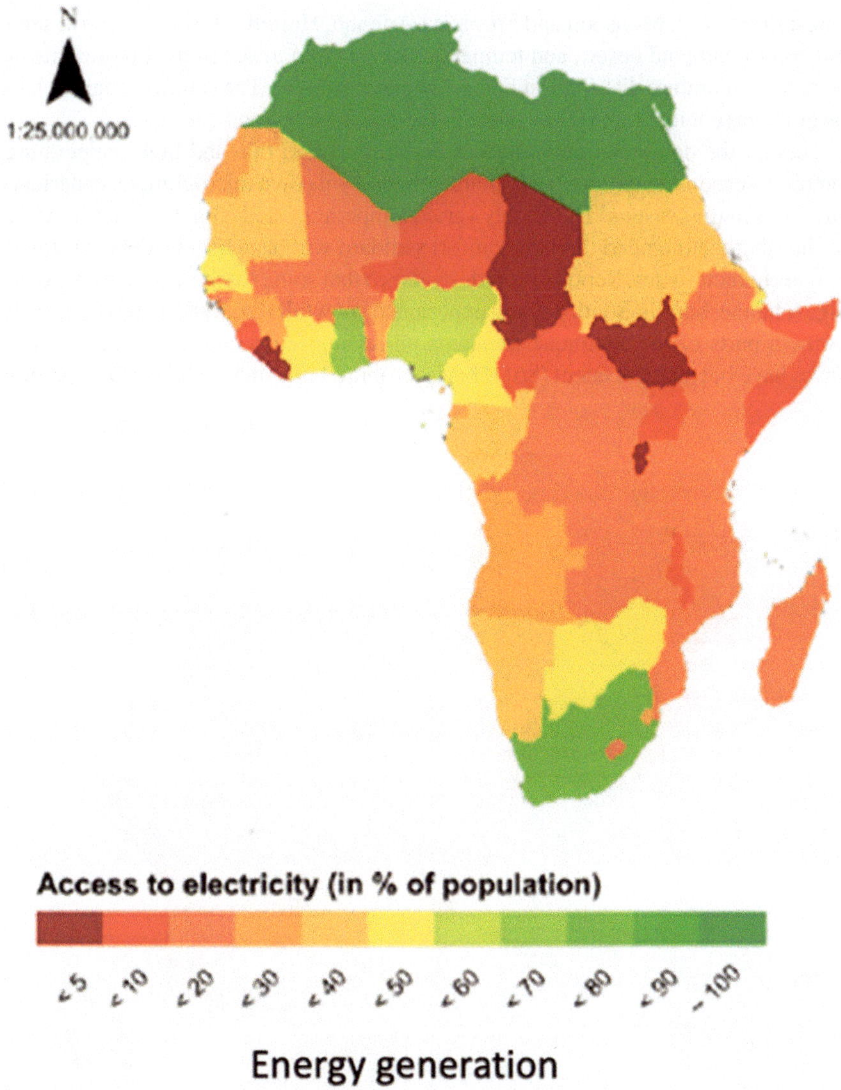

Fig. 4.5 Energy generation in Africa adopted from Emetere et al. [36]

air masses [58]. The northern part of the African continent is characterized by warm Mediterranean climate, warm climate, and cold desert climate. The western part of the continent is characterized by warm semi-arid climate, monsoon climate, and tropical savannah climate. Warm desert climate, warm semi-arid climate, humid subtropical climate, and Temperate Mediterranean climate are prevalent in the Eastern part of the continent. In Central Africa, Equatorial, Monsoon, and Tropical savannah climates prevail. The Southern part of Africa is characterized by a greater number of climatic zones: Equatorial, Monsoon and Tropical savannah, Humid subtropical, warm semi-arid, warm and cold desert, and temperate oceanic climates. Figure 4.6 presents the average temperature distribution in the African continent. The southern regions have lower average temperature than those in the northern regions.

Though the continent generally has dry (little to no rain and high temperature) and rainy seasons, each country is characterized by its own microclimate, experience variations in the seasons. The widely varied temperature and climatic zones in Africa implies that the intended thermal comfort standard will have to consider each zone's unique characteristics. Reports provide evidence that some southern regions of Africa (e.g. Johannesburg in South Africa) experience snowfall [24, 89]. Meanwhile, in the northern parts of the continent, incidents on raising temperatures and intense heat waves have become prevalent [86]. This is attributed to climate change. Our inability

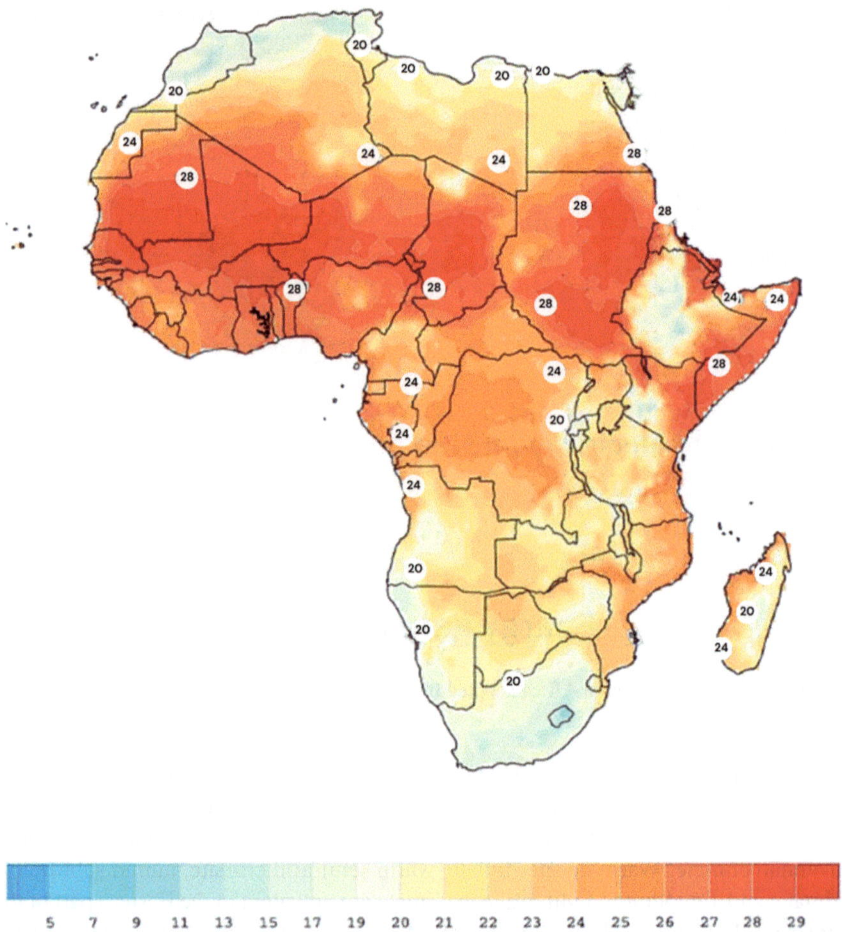

Fig. 4.6 Average temperature in different regions in African adapted from Nwaerema et al. [69]

to accurately predict these changes in the climate could serve as a major limitation to determining the range of comfort temperatures [66].

4.3.2 Cultural and Behavioural Differences

Consequently, the impact of each climate on its inhabitants tend to influence their lifestyles and behaviours. The distinct cultural elements and individuals' cognitive system influences their thermal perception [31]. The belief and behaviour of a people influences the styles or design, methods, and materials that are adopted during the construction of their buildings [99]. In other cases, the culture of a people can be seen in their clothing styles, food and work activities, which can affect their rate of metabolism [105]. Consequently, these do impact their thermal satisfaction to a large extent. Salata et al. [85] contend that using active thermal systems are not adequate during extreme weather conditions, hence the indigenous involuntarily respond by adapting to certain practices (e.g. changing their clothing) in order to find thermal equilibrium with their environment. Yadeta et al. [97] claim adaptation in this regard, can be genetic and acclimatized. Genetic adaptation occurs over a longer period of time whereas, the acclimatization adaptation happens over a few days. Figure 4.7 shows a two part self-regulating system which shows how an individual will act in a manner so as to restore his comfort [67]. Efeoma and Uduku [34] maintain that individuals with diverse socio-cultural backgrounds will by all means have different thermal perception. In addition, other factors like affluence, varied working conditions, age, gender, and health can also affect how people will respond to their environment. Consequently, a pan-African thermal comfort standard will not be adequate without much considerations for these diversified inhabitants.

4.3.3 Lack of Data, Collaboration, and Funding

Rodriguez et al. [81, 83] identified data collection and field work in thermal comfort studies are an arduous task since the accuracy in results is highly dependent on the quality of data gathered from a test sample. Despite the ease in automated weather station installations for this purpose, Caramel [24] pointed out regular maintenance and calibration to be a major setback. This is can be attributed to technological gaps and the lack of capacity building programs in areas bordering on climate, architecture, building technology, and building performance assessment. Africa remains the continent with the least dense observation network [24, 34]. Furthermore, the lack of research funds tends to limit international collaborative activities among researchers. Procuring simple gadgets or sophisticated software, training staff for data collective, organizing forums and so on require some level of investment and financial sponsorship. Having formulated the model, Nicol and Roaf [67] reveal that its accessibility and usage by designers, policymakers, etc. can be a daunting task researchers will

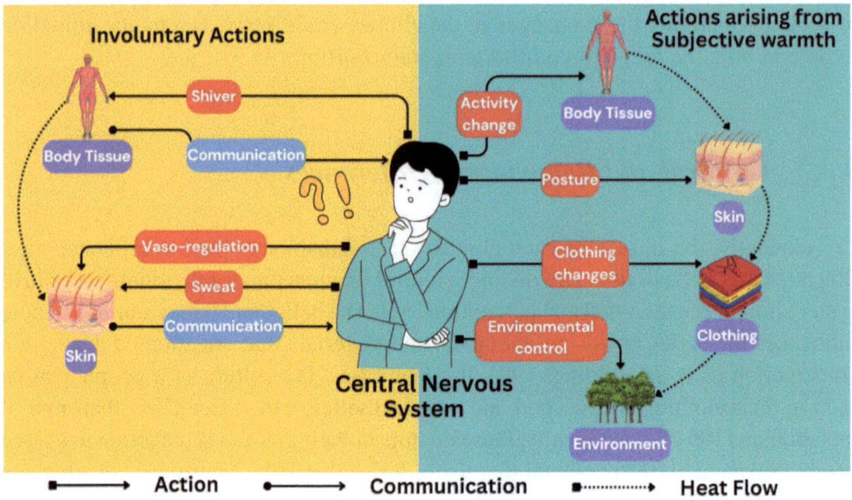

Fig. 4.7 Thermal regulatory system adapted from Nicol and Roaf [67]

have to deal with. Not addressing issues on funding and data acquisition is tantamount to making no progress with getting a pan-African thermal comfort standard.

4.3.4 Insufficiency of the "One-Size-Fits-All" Coverage of Foreign Standards

The blind implementation of existing thermal comfort standards may not adequately support the decarbonization agenda in Africa. This is because, evidence has revealed their inconsistencies in relation to their databases, basic equations, lower and upper temperature level [88]. The adaptive thermal model for instance, uses data from the ASHRAE RP-884 and ASHRAE Global Thermal Comfort Database II [31, 38]. The ASHRAE RP-884 gathered data from 23 field projects, whiles the ASHRAE Global Thermal Comfort Database II analysed data from 42 field projects [51, 53]. The ASHRAE Global Thermal Comfort Database II, for instance uses data taken from Nigeria (Fig. 4.8) out of the 54 countries and 13 distinct climatic zones in Africa. This means that, the thermal model would not be sufficiently suitable for use in different socio-culture and regions [42].

Despite its wide acceptance, Bienvenido-Huertas et al. [21] contend that temperature thresholds in foreign international standards are not always adjusted to meet local thermal expectations of building users due to personality differences, socio-cultural, and socio-economic factors. Even though, the adaptive model in the ASHRAE Standard 55 helps reduce energy consumption, having a region-specific standard is a better option in the quest for decarbonization and energy efficiency since the model

Fig. 4.8 Location of ASHRAE Global Thermal Comfort Database II field studies [38]

would be tailored more specifically to thermal expectations and behaviours of users within Africa [21, 34].

4.4 Issues and Way Forward

Addressing these challenges to developing a pan-African thermal comfort standard require a multifaceted approach that considers both socio-cultural, socio-economic, and technological solutions. These may includes the following.

4.4.1 Community Engagement and Education

Community engagement and education means involving members of the community in planning or implementation of an initiative while teaching them about the issue at hand and the importance of their participation. Engaging the community plays an important part of data collection since it fosters collaboration, increases awareness, and builds capacity. Since developing the standard requires the collection of bioclimatic data of the various regions within the continent, it is imperative to utilize a mixed research method in order to have a better appreciation of the different responses to adapting and maintaining thermal comfort of building occupants [34]. This could serve as an effective way of fine-tuning models. Consequently, making the standard socially acceptable and locally relevant.

4.4.2 Policy Intervention

This refers to the actions taken by policymaking bodies in addressing specific issues or problems. It has been considered as an effective approach in problem solving [26]. Problems related to energy use, health implication of buildings users, and environmental degradation requires that regulatory bodies strategically inculcate guidelines or recommendations from studies into the designing and construction activities of the Architectural Engineering and Construction industry. Furthermore, policies related to education need to be reinforced especially to the societies and nations with lower levels of education [64]. Knowledge about the relevance of a conducive indoor environment would foster the sustainable construction practices even during the design stages of a building. Policies that give grants to researchers, tax incentives to building owners based on adherence to sustainable practices or certification, would serve as a way of advocating for the standard for the continent. Having supports from the government or these agencies makes it easier for international collaboration activities and awareness creation.

4.4.3 Research, Funding, and Collaboration

Financial investment for such extensive research aimed at collecting different thermal preferences across the varied climatic and cultural contexts within the African continent can be considered as a major limitation. The survival of the research lies in the consistency of funds allocated to it. Without funding, it will actually be impossible to get the necessary tools, skills, physical structures needed to carry out this vision. In fact, the absence of funding even implicates human resources (such as technicians, support staff) on the project since this in-depth study into thermal comfort requires an interdisciplinary effort. This explain why there is a need for either international or local collaboration. Collaborating with institutions or individuals at the national or international level having different backgrounds and research interests is an imperative to enhance the different capacities and ensure a robust investigation of the subject matter [62]. This notwithstanding, forming a consortium of universities, research institutions and industry partners, or encouraging public–private research partnerships can be a way to conduct such extensive studies. In this way, partners will be able to leverage on the resources and expertise of each other whiles undertaking localized studies and providing solutions. Moreover, participating institutions can implement standardized protocols when collecting data in order to ensure consistency. International and national bodies can also establish grants purposely for thermal comfort in Africa. In addition, strategic collaboration with institutions in the Global North would facilitate bidirectional knowledge exchange—leveraging their technical expertise while simultaneously uncovering and integrating context-specific insights relevant to the local environment.

4.4.4 Data Accessibility

Data plays an integral role in every industry. Addressing issues on data accessibility can be done by creating a comprehensive data sharing and retrieval platforms. These repositories or platforms should be made easily accessible to researchers, industry players, and policymaking bodies. This will further create a localized system where a wide range of climatic and thermal comfort data across the continent can be accessed. Seeing that, the African continent is a large mass which will require several years to compile thermal data from, the data platform will serve as a means of keeping track of what has been gathered so far. Similarly, having an open data system can inform the creation of technological tools like building energy modelling software specific to the region.

Following these strategies, it is possible to overcome the challenges associated with developing a pan-African thermal comfort standard. Consequently, these will provide means of improving occupants' well-being, reduce energy consumption, and foster the global decarbonization agenda.

4.5 Summary

A thermal environment is influenced by meteorological factors such as temperature, humidity, airflow, and solar radiation. Regions in the warm climates are typically found in the tropical and subtropical climatic zones. Most often, these climatic zones experience prolonged high temperatures and humidity, with distinct wet and dry seasons. Economies in African, Southeast Asia, Central and South America, and the Pacific islands, collectively known as the global south, tend to experience this harsh weather conditions. Consequently, this creates uncomfortable indoor spaces and health implications. To cope, people in the regions adopt various coping mechanisms, such as changing activities, clothing styles, using certain building materials. The built environment is saddled with providing sustainable solutions to some of these problems. The need to create safe indoor environments have led to the development of various thermal models, notably the Predicted Mean Vote based on Fanger's model and the Adaptive thermal model. These have been adopted in international standards like the ASHRAE Standard 55, ISO 7730 and the EN 15251.

Africa's diverse climates, socio-cultural, socio-economic, and personal expectations pose a serious challenge for developing a standardized thermal comfort model. These also render some of the assumptions in international thermal standards sufficiently unfit to achieve thermal satisfaction in some parts of the continent. Meanwhile, thermal comfort standards that are region specific are able to promote climate resilience buildings, energy efficiency and occupants' health, well-being and satisfaction. Developing a pan-African thermal comfort standards requires a holistic approach, combing personal adaptions in various climates and bioclimatic architectural principles within all the major regions of the continent. This model should make provision for naturally ventilated, air-conditioned and mixed designs. In the attempt to develop this comprehensive framework for sustainable building practices in Africa, strategic policies, community engagement and education, data accessibility, funding for research, promoting collaboration among research institutions, NGO's, etc., are essentials.

References

1. Abd El-Hafeez M, El-Mokadem A, Sallam I, Nashaat B (2014) Adapting the green pyramid rating system for assessing zero energy concept on neighborhoods level. Port-Said Eng Res J 18(1):153–160. https://doi.org/10.21608/pserj.2014.46819
2. Abdulgader M, Lashhab F (2021) Energy-efficient thermal comfort control in smart buildings. In: 2021 IEEE 11th annual computing and communication workshop and conference (CCWC), pp 0022–0026. https://doi.org/10.1109/CCWC51732.2021.9376175
3. Adaji MU, Adekunle TO, Watkins R, Adler G (2019) Indoor comfort and adaptation in low-income and middle-income residential buildings in a Nigerian city during a dry season. Build Environ, Elsevier 162:106276
4. Agyekum K, Adinyira E, Oppon JA (2021) Factors limiting the adoption of hemp as an alternative sustainable material for green building delivery in Ghana. Int J Build Pathol Adapt, Emerald Publishing Limited 40(2):202–218. https://doi.org/10.1108/IJBPA-11-2020-0100
5. Agyekum K, Amudjie J (2024) Challenges to implementing circular economy: empirical evidence among built environment firms in Ghana. Int J Constr Manag, Taylor & Francis 24(3):281–297. https://doi.org/10.1080/15623599.2023.2222993
6. Agyekum K, Goodier C, Oppon JA (2021) Key drivers for green building project financing in Ghana. Eng Constr Arch Manag, Emerald Publishing Limited 29(8):3023–3050. https://doi.org/10.1108/ECAM-02-2021-0131
7. Ahmad A, Prakash O, Kumar A, Mozammil Hasnain SM, Verma P, Zare A, Dwivedi G et al (2022) Dynamic analysis of daylight factor, thermal comfort and energy performance under clear sky conditions for building: an experimental validation. Mater Sci Energy Technol 5:52–65. https://doi.org/10.1016/j.mset.2021.11.003
8. Ahmed I, Khan T (2023) Finding post-colonial identity through vernacular and climate responsive adaptations: the architecture of Bangladesh from 1947 to 1971. Open House Int, Emerald Publishing Limited 49(2):384–408. https://doi.org/10.1108/OHI-03-2023-0064
9. Al Horr Y, Arif M, Kaushik AK, Arsalan H, Mazroei A, Rana MQ (2023) Thermal comfort in buildings: scientometric analysis and systematic review. J Arch Eng 29(2):03123001. https://doi.org/10.1061/JAEIED.AEENG-1490
10. Alamirah H, Schweiker M, Azar E (2022) Immersive virtual environments for occupant comfort and adaptive behavior research—a comprehensive review of tools and applications. Build Environ 207:108396. https://doi.org/10.1016/j.buildenv.2021.108396
11. Al-Tamimi N (2022) Passive design strategies for energy efficient buildings in the Arabian desert. Front Built Environ, Frontiers 7:805603. https://doi.org/10.3389/fbuil.2021.805603
12. Amaripadath D, Rahif R, Velickovic M, Attia S (2023) A systematic review on role of humidity as an indoor thermal comfort parameter in humid climates. J Build Eng 68:106039. https://doi.org/10.1016/j.jobe.2023.106039
13. Amos-Abanyie S, Gyimah KA, Adjei EA (2021) Towards climate responsive building design: bio-climatic design features of residential building typologies in the warm-humid climate of Ghana. J Build Constr Plan Res, Scientific Research Publishing 9(2):170–187
14. Aqilah N, Rijal HB, Zaki SA (2022) A review of thermal comfort in residential buildings: comfort threads and energy saving potential. Energies, MDPI 15(23):9012
15. ASHRAE (2017) Thermal environmental conditions for human occupancy

16. ASHRAE (2020) 2020 ASHRAE handbook. Inch-Pound., Atlanta
17. Balabel A, Alwetaishi M (2021) Towards sustainable residential buildings in Saudi Arabia according to the conceptual framework of 'mostadam' rating system and vision 2030. Sustainability 13:793. https://doi.org/10.3390/su13020793
18. Baruti MM, Johansson E, Åstrand J (2019) Review of studies on outdoor thermal comfort in warm humid climates: challenges of informal urban fabric. Int J Biometeorol 63(10):1449–1462. https://doi.org/10.1007/s00484-019-01757-3
19. Beck HE, Zimmermann NE, McVicar TR, Vergopolan N, Berg A, Wood EF (2018) Present and future Köppen-Geiger climate classification maps at 1-km resolution. Sci Data, Nature Publishing Group 5(1):180214. https://doi.org/10.1038/sdata.2018.214
20. Bhandari N, Tadepalli S, Gopalakrishnan P (2023) Influence of non-uniform distribution of fan-induced air on thermal comfort conditions in university classrooms in warm and humid climate, India. Build Environ 238:110373. https://doi.org/10.1016/j.buildenv.2023.110373
21. Bienvenido-Huertas D, Sánchez-García D, Tejedor B, Rubio-Bellido C (2024) Energy savings in buildings applying ASHRAE 55 and regional adaptive thermal comfort models. Urban Clim 55:101892. https://doi.org/10.1016/j.uclim.2024.101892
22. Brager GS, de Dear RJ (1998) Thermal adaptation in the built environment: a literature review. Energy Build 27(1):83–96. https://doi.org/10.1016/S0378-7788(97)00053-4
23. Cabeza LF, Chàfer M (2020) Technological options and strategies towards zero energy buildings contributing to climate change mitigation: a systematic review. Energy Build 219:110009. https://doi.org/10.1016/j.enbuild.2020.110009
24. Caramel (2023) North Africa is used to the heat waves Europe is now experiencing. Le Monde.Fr, Paris
25. Chan IYS, Liu AMM (2018) Effects of neighborhood building density, height, greenspace, and cleanliness on indoor environment and health of building occupants. Build Environ 145:213–222. https://doi.org/10.1016/j.buildenv.2018.06.028
26. Chen Y, Yang Y, Luo L, Ding C (2023) Impacts of BIM policy on the technological progress in the Architecture, Engineering, and Construction (AEC) industry: evidence from China. J Asian Arch Build Eng, Taylor and Francis Ltd. https://doi.org/10.1080/13467581.2023.2278888
27. Cheung T, Schiavon S, Parkinson T, Li P, Brager G (2019) Analysis of the accuracy on PMV–PPD model using the ASHRAE global thermal comfort database II. Build Environ 153:205–217. https://doi.org/10.1016/j.buildenv.2019.01.055
28. Economy C (2023) The circularity gap report 2023. Circle Economy, Amsterdam, pp 1–64
29. Cody B (2014) The role of technology in sustainable architecture. Wolkenkuckucksheim, Internationale Zeitschrift Zur Theorie Der Architektur 19(33):239–249
30. Dadzoe F, Addy M, Duah DYA, Adesi M (2022) Towards a circular economy: a knowledge-attitude gap between demand and supply-side operators on green building construction in Ghana. Smart Sustain Built Environ (ahead-of-print). https://doi.org/10.1108/SASBE-03-2022-0048
31. de Dear R, Brager GS (1998) Developing an adaptive model of thermal comfort and preference
32. Dev G, Saifudeen A (2023) Dynamic facade control systems for optimal daylighting, a case of Kerala. Sustain Anal Model 3:100018. https://doi.org/10.1016/j.samod.2023.100018
33. Djamila H, Chu C-M, Kumaresan S (2013) Field study of thermal comfort in residential buildings in the equatorial hot-humid climate of Malaysia. Build Environ 62:133–142. https://doi.org/10.1016/j.buildenv.2013.01.017
34. Efeoma M, Uduku O (2014) Assessing thermal comfort and energy efficiency in tropical African offices using the adaptive approach. Struct Surv 32:396–412. https://doi.org/10.1108/SS-03-2014-0015
35. Elnabawi MH, Hamza N (2020) Behavioural perspectives of outdoor thermal comfort in urban areas: a critical review. Atmosphere, Multidisciplinary Digital Publishing Institute 11(1):51. https://doi.org/10.3390/atmos11010051
36. Emetere ME, Agubo O, Chikwendu L (2021) Erratic electric power challenges in Africa and the way forward via the adoption of human biogas resources. Energy Explor Exploit, SAGE Publications Ltd STM 39(4):1349–1377. https://doi.org/10.1177/01445987211003678

37. Fanger PO (1973) Assessment of man's thermal comfort in practice. Occup Environ Med, BMJ Publishing Group Ltd 30(4):313–324. https://doi.org/10.1136/oem.30.4.313
38. Földváry Ličina V, Cheung T, Zhang H, De Dear R, Parkinson T, Arens E, Chun C et al (2018) Development of the ASHRAE global thermal comfort database II. Build Environ 142:502–512. https://doi.org/10.1016/j.buildenv.2018.06.022
39. Frontczak M, Wargocki P (2011) Literature survey on how different factors influence human comfort in indoor environments. Build Environ 46(4):922–937. https://doi.org/10.1016/j.buildenv.2010.10.021
40. Gavu EK, Peiser RB (2024) Embodied carbon and the nuances in office-to-residential conversions. Sustainability 16(7):2711. https://doi.org/10.3390/su16072711
41. González-Lezcano RA (2023) Editorial: design of efficient and healthy buildings. Front Built Environ, Frontiers 9:1210956. https://doi.org/10.3389/fbuil.2023.1210956
42. Gou Z, Gamage W, Lau SS-Y, Lau SS-Y (2018) An investigation of thermal comfort and adaptive behaviors in naturally ventilated residential buildings in tropical climates: a pilot study. Buildings, MDPI 8(1):5
43. Gut P, Ackerknecht D (1993) Climate responsive buildings: appropriate building construction in tropical and subtropical regions
44. Hailu H, Gelan E, Girma Y (2021) Indoor thermal comfort analysis: a case study of modern and traditional buildings in hot-arid climatic region of Ethiopia. Urban Sci, MDPI 5(3):53
45. Humphreys M, Nicol F (1998) Understanding the adaptive approach to thermal comfort. ASHRAE Trans 104:991–1004
46. Jain N, Burman E, Robertson C, Stamp S, Shrubsole C, Aletta F, Barrett E et al (2020) Building performance evaluation: balancing energy and indoor environmental quality in a UK school building. Build Serv Eng Res Technol, SAGE Publications Ltd STM 41(3):343–360. https://doi.org/10.1177/0143624419897397
47. Kanna K, AIT Lachguer K, Yaagoubi R (2022) MyComfort: an integration of BIM-IoT-machine learning for optimizing indoor thermal comfort based on user experience. Energy Build 277:112547. https://doi.org/10.1016/j.enbuild.2022.112547
48. Khalil A-A, Fikry M, Abdeaal W (2018) High technology or low technology for buildings envelopes in residential buildings in Egypt. Alex Eng J 57(4):3779–3792. https://doi.org/10.1016/j.aej.2018.11.001
49. Kiki G, Kouchadé C, Houngan A, Zannou-Tchoko SJ, André P (2020) Evaluation of thermal comfort in an office building in the humid tropical climate of Benin. Build Environ, Elsevier 185:107277
50. Kimemia D, Van Niekerk A, Annegarn H, Seedat M (2020) Passive cooling for thermal comfort in informal housing. J Energy South Afr, University of Cape Town 31(1):28–39
51. Koranteng C (2010) Energy performance of office buildings in Ghana. J Sci Technol (Ghana) 30.https://doi.org/10.4314/just.v30i2.60542
52. Koranteng C, Simons B, Abrokwa Gyimah K, Nkrumah J (2021) Ghana's green building assessment journey: an appraisal of the thermal performance of an office building in Accra. J Eng Des Technol, Emerald Publishing Limited 21(1):188–205. https://doi.org/10.1108/JEDT-02-2021-0109
53. Koranteng C, Simons B, Gyimah KA (2021) Potential measures towards the reduction of cooling loads of office buildings in Ghana. Front Eng Built Environ, Emerald Publishing Limited 1(2):161–172. https://doi.org/10.1108/FEBE-03-2021-0016
54. Kottek M, Grieser J, Beck C, Rudolf B, Rubel F (2006) World map of the Köppen-Geiger climate classification updated. Meteorol Z 15(3):259–263. https://doi.org/10.1127/0941-2948/2006/0130
55. Kueh M-T, Lin C-Y, Chuang Y-J, Sheng Y-F, Chien Y-Y (2017) Climate variability of heat waves and their associated diurnal temperature range variations in Taiwan. Environ Res Lett, IOP Publishing 12(7):074017. https://doi.org/10.1088/1748-9326/aa70d9
56. Lamsal P, Bajracharya SB, Rijal HB (2023) A review on adaptive thermal comfort of office building for energy-saving building design. Energies, Multidisciplinary Digital Publishing Institute 16(3):1524. https://doi.org/10.3390/en16031524

57. Li J, Sun R, Chen L (2023) A review of thermal perception and adaptation strategies across global climate zones. Urban Clim 49:101559. https://doi.org/10.1016/j.uclim.2023.101559
58. Lionello P (ed) (2012) "Foreword", The Climate of the Mediterranean Region. Elsevier, Oxford, pp xxi–xxiii. https://doi.org/10.1016/B978-0-12-416042-2.00013-6
59. Mohammadpourkarbasi H, Jackson I, Nukpezah D, Appeaning Addo I, Assasie Oppong R (2022) Evaluation of thermal comfort in library buildings in the tropical climate of Kumasi, Ghana. Energy Build 268:112210. https://doi.org/10.1016/j.enbuild.2022.112210
60. Mujan I, Anđelković AS, Munćan V, Kljajić M, Ružić D (2019) Influence of indoor environmental quality on human health and productivity—a review. J Clean Prod 217:646–657. https://doi.org/10.1016/j.jclepro.2019.01.307
61. Munaro MR, Tavares SF, Bragança L (2020) Towards circular and more sustainable buildings: a systematic literature review on the circular economy in the built environment. J Clean Prod 260:121134. https://doi.org/10.1016/j.jclepro.2020.121134
62. Naziz A (2019) Collaboration for transition between TVET and university: a proposal. Int J Sustain High Educ, Emerald Publishing Limited 20(8):1428–1443. https://doi.org/10.1108/IJSHE-10-2018-0197
63. Neves L (2006) Architecture integrated to nature: the use of timber by Severiano Porto in Brazilian Amazon
64. Neves SA, Marques AC (2022) Drivers and barriers in the transition from a linear economy to a circular economy. J Clean Prod 341:130865. https://doi.org/10.1016/j.jclepro.2022.130865
65. Neya I, Yamegueu D, Messan A, Coulibaly Y, Ouedraogo ALS-N, Ayite YMXD (2023) Effect of cement and geopolymer stabilization on the thermal comfort: case study of an earthen building in Burkina Faso. Int J Build Pathol Adapt (ahead-of-print). https://doi.org/10.1108/IJBPA-05-2022-0069
66. Nicol F, Humphreys M (2002) Adaptive thermal comfort and sustainable thermal standards for buildings. Energy Build 34:563–572. https://doi.org/10.1016/S0378-7788(02)00006-3
67. Nicol JF, Roaf S (2017) Rethinking thermal comfort. Build Res Inf, Routledge 45(7):711–716. https://doi.org/10.1080/09613218.2017.1301698
68. Nizam Akbar AR, Mohammad M, Ahmad N, Maisyam M (2018) Standardization in construction environment: adopting standard method of measurements. Asian J Behav Stud 3:147. https://doi.org/10.21834/ajbes.v3i12.131
69. Nwaerema P, Diagi B, Edokpa D, Ajiere S (2019) Population variability and heat bias prediction of Africa from 2019 to 2049: an approach to sustainable continental heat management
70. Nwalusi DM, Obi NI, Chendo IG, Okeke FO (2022) Climate responsive design strategies for contemporary low-rise residential buildings in tropical environment of Enugu, Nigeria. In: IOP conference series: earth and environmental science, vol 1054, no 1, p 012052. https://doi.org/10.1088/1755-1315/1054/1/012052
71. Odim OO (2008) Experimental studies of comfort levels of east-west and north-south solar-oriented buildings in a warm-humid climate. Arch Sci Rev, Taylor & Francis 51(4):403–406. https://doi.org/10.3763/asre.2008.5144
72. Ohene E, Krarti M, Chan APC, Hsu S-C, Ansah MK (2024) Optimal design guidelines for net zero energy residential buildings in cooling-dominated climates: case study of Ghana. Build Environ 260:111685. https://doi.org/10.1016/j.buildenv.2024.111685
73. Okeke FO, Ezema EC, Nnaemeka-Okeke RC, Okosun AE, Okeke CA (2023) Architectural design response to population issue in sub-Saharan cities. In: Tursunov O (ed) E3S web of conferences, vol 434, p 02005. https://doi.org/10.1051/e3sconf/202343402005
74. Olesen BW (1995) International standards and the ergonomics of the thermal environment. Appl Ergon 26(4):293–302. https://doi.org/10.1016/0003-6870(95)00033-9
75. Opoku R, Edwin IA, Agyarko KA (2019) Energy efficiency and cost saving opportunities in public and commercial buildings in developing countries—the case of air-conditioners in Ghana. J Clean Prod 230:937–944. https://doi.org/10.1016/j.jclepro.2019.05.067
76. Özdamar Seitablaiev M, Umaroğulları F (2018) Thermal comfort and indoor air quality. Int J Sci Res Innov Technol 5:90–109

77. Pradeep Kumar B, Anusha BN, Raghu Babu K, Padma Sree P (2023) Identification of climate change impact and thermal comfort zones in semi-arid regions of AP, India using LST and NDBI techniques. J Clean Prod 407:137175. https://doi.org/10.1016/j.jclepro.2023.137175
78. Rijal HB, Humphreys MA, Nicol JF (2019) Adaptive model and the adaptive mechanisms for thermal comfort in Japanese dwellings. Energy Build 202:109371. https://doi.org/10.1016/j.enbuild.2019.109371
79. Rijal HB, Yoshida K, Humphreys MA, Nicol JF (2021) Development of an adaptive thermal comfort model for energy-saving building design in Japan. Arch Sci Rev 64(1–2):109–122. https://doi.org/10.1080/00038628.2020.1747045
80. Rînjea C, Chivu OR, Darabont D-C, Feier AI, Borda C, Gheorghe M, Nitoi DF (2022) Influence of the thermal environment on occupational health and safety in automotive industry: a case study. Int J Environ Res Public Health 19(14):8572. https://doi.org/10.3390/ijerph19148572
81. Rodriguez C, Coronado M, D'Alessandro M, Medina J (2019) The importance of standardised data-collection methods in the improvement of thermal comfort assessment models for developing countries in the tropics. Sustainability, Multidisciplinary Digital Publishing Institute 11(15):4180. https://doi.org/10.3390/su11154180
82. Rodriguez C, D'Alessandro M (2019) Indoor thermal comfort in the tropics, pp 9–15
83. Rodriguez C, Medina J, Coronado M, D'Alessandro M (2019) The development of data-collection methods for thermal comfort assessment in tropical countries. In: IOP conference series: materials science and engineering, vol 603, p 052001. https://doi.org/10.1088/1757-899X/603/5/052001
84. Royal Scientific Society (2020) Jordan national building codes
85. Salata F, Golasi I, Ciancio V, Rosso F (2018) Dressed for the season: clothing and outdoor thermal comfort in the Mediterranean population. Build Environ 146:50–63. https://doi.org/10.1016/j.buildenv.2018.09.041
86. Sippy P (2024) Mali heatwave: ice becomes a hot commodity as temperatures soar. British Broadcasting Corporation, 27 April
87. Sitorus A, Sembiring D, Abdillah W, Harisdani D (2023) Thermal comfort in humid tropical climate areas (case study on open spaces and shaded spaces in Medan City). Int J Arch Urban 7:294–318. https://doi.org/10.32734/ijau.v7i2.13512
88. Taleghani M, Tenpierik M, Kurvers S, van den Dobbelsteen A (2013) A review into thermal comfort in buildings. Renew Sustain Energy Rev 26:201–215. https://doi.org/10.1016/j.rser.2013.05.050
89. Taylor DB (2023) Johannesburg sees first snowfall in over a decade. Residents flock to the flakes. The New York Times, 12 July
90. TERI (2021) Thermal comfort prescription for cooling dominated Indian residential buildings. The Energy and Resources Institute, New Delhi
91. Tham KW (2016) Indoor air quality and its effects on humans—a review of challenges and developments in the last 30 years. Energy Build 130:637–650. https://doi.org/10.1016/j.enbuild.2016.08.071
92. Tham KW, Willem HC (2010) Room air temperature affects occupants' physiology, perceptions and mental alertness. Build Environ 45(1):40–44. https://doi.org/10.1016/j.buildenv.2009.04.002
93. Toyin JO, Mewomo MC (2023) Overview of BIM contributions in the construction phase: review and bibliometric analysis. J Inf Technol Constr, International Council for Research and Innovation in Building and Construction 28:500–514. https://doi.org/10.36680/j.itcon.2023.025
94. Trebilcock M, Soto-Muñoz J, Piggot-Navarrete J (2020) Evaluation of thermal comfort standards in office buildings of Chile: thermal sensation and preference assessment. Build Environ 183:107158. https://doi.org/10.1016/j.buildenv.2020.107158
95. Wright CY, Wernecke B, Kapwata T, Kunene Z, Mathee A, Vande Hey J, Theron L (2022) Perceptions of thermal comfort and coping mechanisms related to indoor and outdoor temperatures among participants living in rural villages in Limpopo province, South Africa. South Afr J Psychol, SAGE Publications 52(4):449–459. https://doi.org/10.1177/00812463221129363

96. Xu B, Chen X, Fei Y, Gan W, Pei G (2023) Optimizing the applicability of cool paint through phase change material according to the energy consumption characteristics in different regions. Renew Energy 212:953–971. https://doi.org/10.1016/j.renene.2023.05.107
97. Yadeta C, Indraganti M, Alemayehu E, Tucho GT (2022) An investigation of human thermal comfort and adaptation in naturally ventilated residential buildings and its implication for energy use in tropical climates of Ethiopia. Sci Technol Built Environ, Taylor & Francis 28(7):896–915. https://doi.org/10.1080/23744731.2022.2048574
98. Yao R, Zhang S, Du C, Schweiker M, Hodder S, Olesen BW, Toftum J et al (2022) Evolution and performance analysis of adaptive thermal comfort models—a comprehensive literature review. Build Environ 217:109020. https://doi.org/10.1016/j.buildenv.2022.109020
99. Yemoh OO, Opoku R, Takyi G, Adomako EK, Uba F, Obeng G (2024) Green building adaptation in hot-humid climates: assessment of coconut and corn husk fiber composite bricks as energy-efficient building envelopes. Int J Build Pathol Adapt (ahead-of-print). https://doi.org/10.1108/IJBPA-11-2022-0178
100. Zafarmandi S, Matzarakis A, Norford L (2024) Effects of clothing's thermal insulation on outdoor thermal comfort and thermal sensation: a case study in Tehran, Iran. Sustain Cities Soc, Elsevier BV. https://doi.org/10.1016/j.scs.2023.104988
101. Zhang J, You Q, Ren G, Ullah S, Normatov I, Chen D (2023) Inequality of global thermal comfort conditions changes in a warmer world. Earth's Future 11(2):e2022EF003109. https://doi.org/10.1029/2022EF003109
102. Zhang Y (2024) Application of ecological architecture in architectural design based on information technology integration. Appl Math Nonlinear Sci Sciendo 9(1). https://doi.org/10.2478/amns.2023.1.00327
103. Zhang Y, Chen J, Liu H, Chen Y, Xiao B, Li H (2024) Recent advancements of human-centered design in building engineering: a comprehensive review. J Build Eng 84:108529. https://doi.org/10.1016/j.jobe.2024.108529
104. Zhao Q, Lian Z, Lai D (2021) Thermal comfort models and their developments: a review. Energy Built Environ 2(1):21–33. https://doi.org/10.1016/j.enbenv.2020.05.007
105. Zheng P, Yao R, Toftum J, Liu S, Kalmár F, Li B (2024) Evaluation methods and factors influencing the thermal comfort of children—a comprehensive literature review. J Build Eng 95:110063. https://doi.org/10.1016/j.jobe.2024.110063
106. Zhou G-D, Yi T-H, Li W-J, Zhong J-W, Zhang G-H (2020) Standardization construction and development trend of bridge health monitoring systems in China. Adv Bridge Eng 1(1):13. https://doi.org/10.1186/s43251-020-00016-5
107. Zhu J, Feng J, Lu J, Chen Y, Li W, Lian P, Zhao X (2023) A review of the influence of courtyard geometry and orientation on microclimate. Build Environ 236:110269. https://doi.org/10.1016/j.buildenv.2023.110269

Chapter 5
Daylighting

5.1 Introduction

Daylighting is a passive design strategy which utilizes natural lighting to enhance visual comfort whereas reducing the energy consumption in buildings. It supplements or replaces natural lighting to create a good indoor environmental quality and reduce energy consumption for lighting in buildings. This promotes sustainable practices through the reduction of the reliance of energy produced by fossil fuels to power luminaries and enhance occupant comfort and environmental health [47]. Strategic utilization of daylighting in architectural designs promotes energy efficiency and reduce cost to energy consumption and increased productivity. However, these can be harnessed through effective consideration of factors such as building orientation, positioning of fenestrations, window to wall ratio (WWR), and the local climate of the building [7].

Beyond the energy saving potentials of daylighting, it has a profound impact on the physiological and psychological well-being of occupants. Exposure to daylighting diminishes headache, eyestrain and stress while enhancing the overall health and productivity [57]. In offices, daylighting reduces eyestrain and depression of occupants and improves productivity [61]. It further enhances cognitive performance, improved mood and better circadian rhythm that is the natural oscillation which regulates the sleep–wake cycle.

Daylight performance can be measured both qualitatively and qualitatively. However, the quantitative measurement is widely used due to their objectivity and compatibility with daylight simulation models. The quantitative metrics conform to standards for daylight performance evaluation such as BREEM and LEED. Common among these metrics are the Daylight Factor (DF), Daylight Autonomy (DA), Useful Daylight Illuminance (UDI), and Annual Sunlight Exposure (ASE) [30]. Daylight Factor is a conventional approach for measuring daylight. It is often measured under overcast sky conditions as the ratio of indoor illuminance at a point to the simultaneous outdoor illuminance [66]. This approach is static in nature. That is, it does not

M. N. Addy et al., *Decarbonization of Building Energy in Developing Countries*, Green Energy and Technology, https://doi.org/10.1007/978-3-032-09113-0_5

account for daylight availability in an entire year but only a specific point in time [41]. This limitation was address by DA which measures the percentage of occupied hours during which minimum threshold for illuminance can be met by daylighting alone under a continuous overcast sky condition. It is dynamic and can estimate energy consumption by lighting systems depending on weather conditions [1]. Since DA does not specify maximum threshold, Useful Daylight Illuminance further refines DA by defining illuminance range at which are considered beneficial for occupants [46].

In developing countries, the integration of daylighting in architectural designs presents a promising approach to harness the importance of daylighting. Most African countries have enormous daylight availability for potential daylight harvesting. Despite this potential, buildings in this region heavily rely on artificial lighting even during daytime. Also, cultural and socio-economic factors, urban morphology and construction practices has influenced the neglect of daylighting models in local standards. This chapter aims to explore the role of daylighting in building decarbonization while addressing challenges affecting the integration of daylighting in building designs and propose strategies to improve it particularly in developing countries.

5.2 Daylighting and Shading in Warm Climates

Designing for daylighting is essential to enhance energy efficiency and occupants' health and well-being. However, uncontrolled solar radiations can lead to glare, thermal discomfort, and overheating. Regions in the warm-humid climates are characterized by excessive solar radiations due to increased sunshine and high ambient temperatures. These results in intense sunlight over these regions. Due to high solar radiations in this region, buildings are susceptible to overheating which increase the energy consumption for cooling, thermal discomfort, and glare as a result of the increased exposure to the solar radiations. Despite these challenges, elevated daylight intensities can be controlled through some passive design strategies. They play a crucial role in enhancing energy efficiency and improving thermal discomfort and glare. As discussed in chapter two some of the climate-responsive design strategies include building orientation, positioning of windows, window to wall ratio (WWR), and shading devices [25].

In the warm climate, it is essential to properly orient buildings and correct window positioning. The position of windows should be determined to minimize solar gain while maintaining daylight consistency, with orientation strategies adapted to the building's latitude. For example, in tropical regions, east- and west-facing windows are more prone to glare and heat gain from low morning and afternoon sun angles and thus north and south facing windows are more preferred [8]. Orientation is considered significant if the WWR is less than 60% hence, it is essential for the WWR and the aspect ratio of the building to be considered. Altaf and Hill [6] recommended the WWR of a building should be between 13 and 25% for optimal daylight performance. However, Priya et al. [51] proposed a window to wall ratio of 25% should be suitable for optimal daylight performance across several building orientations. The WWR, the height, and size can minimize direct sunlight exposure while increasing deeper daylight penetration. Amidst all, shading devices is the most effective strategy in reducing the direct gain of solar radiations especially during peak hours. It provides a trade-off between visual comfort and thermal comfort however offers a balanced requirement for visual and thermal demands [11]. Effective shading devices regulates adequate daylight entry, prevents glare, reduces overheating, and promotes optimal thermal comfort [5].

Shading devices can broadly be categorized into static and dynamic shading. Static shading devices are fixed and unresponsive to changes in the environment. It includes overhangs, fins, brise-soleils, deep window, etc. It is usually preferred due to its optimal cost of construction and maintenance. They further provide significant energy saving potential. According to a study by De Luca et al. [17], exterior static shadings reduced energy use by 29% in Northern Europe. In Taiwan, a substantial impact on energy consumptions was recorded as a result of the static shading devices [55]. In Africa, several studies have highlighted the importance of static shading as an efficient energy saving strategy. In Algeria, Elouadjeri et al. [21] assessed the impact of fixed shading on thermal comfort, energy usage and daylighting performance. The study concluded that, static and external shading has a high energy saving potential and reduces glare and any visual discomfort especially during the day and summer times. However, optimal daylight performance during winter is reduced and hence increases energy consumption again. In Malaysia, Lau et al. [34] emphasized that, integrating shading devices into buildings lead to significant energy saving and enhances sustainability. Likewise in Ghana and Nigeria, several studies have reiterated the energy saving potential of integrating static shading in building design in the warm climate. A case in point is a study by Ohene et al. [48] that showed that passive design strategies such as shading, minimizes heat gains and thereby reducing cooling load. Alegbe et al. [4] concluded that, cooling energy reduced by 23% and overall energy demand reduced by 16% when shading was provided for classrooms in Nigeria.

Dynamic shading responds to the varying environmental conditions such and daylight availability to provide optimal indoor environmental performance. Various configurations such as angles, heights and opacity of the shading devices can be varied either through manual occupant control, sensors and automated algorithms to regulate the level of penetration of daylight due to solar gains and glare. This helps to reduce energy consumption for cooling and lighting while improving thermal and visual comfort.

Dynamic shading systems have gained attention over the past years due to the shift to net-zero buildings and the need for climate-responsive buildings. Most common of these shading systems are the venetian blinds, roller shades, and louvres and usually integrated with technologies such as electrochromic glazing, automated louvres, and motorization of blinds. Dynamic shading outperforms static shading and provides more uniform performances across diverse climatic conditions. In the US climate it reduced energy consumption by about 39% [32]. In addition to the energy saving, it prevents visual discomfort more than 85% of the active occupants' hours and improves daylight levels [33].

In the warm-humid climates, dynamic shading systems have the potential to significantly provide optimal solar gains while improving visual and thermal comfort. A study by Sorooshnia et al. [56] assessed how dynamic shading design can improve indoor environmental comfort. The study showed that, there was an annual recorded reduction in energy by 71.77% for cooling while improving visual performance by a reduction of 72.2% in glare and 71.43% in sunlight exposure. Budaiwi and Fasi [14] further showed that, electrochromic glazing maintains visual comfort throughout the building and controls glare. Energy consumption is reduced by 23% when electrochromic glazing is adopted.

Despite the potential of dynamic shading in the warm-humid climate, studies are more focused on static and passive shadings such as overhangs and fixed fins. This is mostly associated to the lack of region specific and implementation framework for integration into the industry. Also, high initial cost and limited technological infrastructure impedes the implementation of dynamic shading. Kabanshi et al. [29] explored the passive shading however, integrated with PV modules which produces energy to be used by the building. It maximizes the solar radiations penetrating into the building and converts it into energy for the building use.

5.3 Standardization of Daylight Metrics and Guides for Africa

Daylight metrics are essential tools for assessing the daylight performance of buildings, visual comfort and potential energy savings. Historically, Daylight Factor (DF) has been a dominant and widely used metric to assess the daylighting performance in buildings. It is calculated as the ratio of internal illuminance at a given point to the unobstructed illuminance under a standard overcast sky [65]. It is expressed as;

$$\mathrm{DF} = \frac{E_i}{E_o} \times 100\%$$

where E_i = illuminance due to daylight at a point on a working plane

E_o = simultaneous outdoor illuminance on a horizontal plane from an unobstructed overcast sky.

It is easy to calculate and hence easy to interpret results from the calculations. DF is typically independent of time and location hence can serve as a uniform benchmark to assess the daylight performance of different building designs. Due to its ease of use, it has been enshrined in global standards such as the UK's Building Research Establishment (BRE) later BREEAM, British Standard 8206-2 [64] and the US Green Building council's LEED. Currently, DF has a threshold of 2% for kitchens and 1% for bedrooms. However, it has limitations in accounting for the dynamic nature of climate. It is static and measures daylight at a particular point in time without considering any climate variations. The results from DF do not factor the visual discomfort levels such as glare and thermal discomfort performance such as overheating. When region has fairly clear skies, DF does not indicate the true daylight levels [16].

Due to these limitations, climate-based daylight metrics are preferred to overcome the shortcomings of DF. These include Daylight Autonomy (DA), Useful Daylight Illuminance, and Spatial Daylight Autonomy (sDA). They adopt annual climatic data and hourly simulation tools to predict the daylight performance of a building [36, 37].

5.4 Daylight Autonomy

Daylight Autonomy (DA) quantifies the natural light utilization and potential energy savings in a building. It accounts for sun positioning, local weather conditions, geometry of the building and how daylight availability varies throughout the year [1]. Its impact has been outlined in the study of Gómez et al. [26] to assess how Daylight Autonomy varies in a space. Liu et al. [38] compared daylight performance across varying climatic zones in Turkey. While the DF remained constant, the DA ranged from within 40 to 70% which confirms the importance of local weather conditions in the assessment of daylight performance in a building. DA can predict daylight in all regions regardless of weather conditions and sky paths [36, 37].

In the Africa region, some studies have adopted DA to assess the daylight performance of buildings and propose some passive design strategies for optimal building performance. For instance, the daylight performance of aerogel glazing in school buildings was compared to the conventional glazing. Adopting dynamic daylight models, the study showed that the aerogel glazing performed better than the conventional glazing type. This is because, the aerogel glazing attained an increased number of occupied hours with lux levels above 300. The measured DA for the aerogel glazing was higher than that of the conventional types hence, improved visual comfort and thermal performance [44]. Similarly, Mehrvarz et al. [43] used DA to assess the daylight performance of some passive design strategies specifically, responsive façades, orientation and positioning of fenestration in East Africa. The study achieved an average DA level above 60%. Moreover, this was further reiterated by Wanas et al. [59]. The study showed the importance of adaptive façade system to increase the DA in hot arid climates. The study adopted simulation tools to assess the illuminance level of a carefully programmed kinetic louvres. This resulted in a well day-lit zone of about 63% of the total area of the building hence emphasizing the use of DA for daylight performance simulations. In educational facilities in Ghana, DA levels remained below 40% even with optimized window types and orientation hence indicating poor classroom designs in this region [3]. Contrary to that, Hassan et al. [27] using simulation tools to assess the daylight performance of passive design strategies in Nigeria, the study concluded that, an optimal WWR of about 20% achieved adequate DA and higher thermal performance. The study further indicated that, higher

WWR increased the DA however, it also resulted in overheating, reduced occupancy comfort, and increased energy consumption.

Even though DA was generally used, this metric had some limitations. Key among them is, DA is a point-based metric which quantifies daylight availability at a specific location but does not give a clear indication on how daylight is distributed across a given area. Hence not suitable for daylight measurement in deeper zones [36, 37]. Also, DA is one-tailed and do not differentiate between comfortable daylighting and excessive daylighting leading to visual and thermal discomfort. DA assess the minimum required daylight availability, however, does not account for a maximum threshold for daylighting in a space [35]. Hence, cannot clearly indicate occupant comfort levels and glare.

5.5 Spatial Daylight Autonomy

Spatial Daylight Autonomy (sDA) was developed to improve on the DA by measuring the percentage of floor area which receives adequate daylight through the year. This overcomes the spatial limitation of DA, where daylight performance was measured at only specific points. They are commonly used for analysis of office spaces, classrooms, meeting rooms, lobbies, etc. [28]. The accuracy of this method was validated by Reinhart and Walkenhorst [63] and Mardaljevic et al. [42]. According to these studies, sDA is predicted to be more realistic in measuring daylight availability, actual light savings and occupancy comforts. It has been integrated into standards such as the America's IES LM-83 and Europe's EN 17037. Typically, a building should receive 300 lx for 50% of the time occupied according to sDA.

In Africa where daylight availability contrasts with poor daylight performance, sDA has been adopted by some studies to understand daylight performance of these buildings. Musa et al. [45] highlighted poor daylight penetration in deeper floors of an office building in Nigeria. The sDA values were typically below 50% which were below average. According to the study, the well-lit areas were areas closer to the façade with WWR above 10% in order to attain adequate Daylight Autonomy hence highlighting the importance of room depth and window to wall ratio in office building designs. However, in South Africa, a more favourable result were attained by Overen et al. [50]. They adopted simulations tools to assess the daylight performance of an educational facility. The measured sDA values satisfied the global threshold of sDA 300/50%, confirming adequate daylight performance and acceptable visual comfort levels. Factors leading to high sDA values were associated to the WWR, the use of shading materials and good building orientation. Contrary to this, education buildings underperformed in the hot dry regions of Ghana. According to Akubah et al. [3], all classrooms tested through the climate-based daylight modelling in IES VE did not meet the threshold for sDA even with high WWR and optimal fenestration types. This disparity can be associated with the current school building designs in Ghana and the need to improve positioning of fenestration and building orientation. Zoure and Genovese [62] further highlights the importance of passive design strategies in attaining adequate sDA. The study showed that, the orientation of the building, shape of the building, the glazing type and a WWR of 30% were adequate to attain an sDA of about 45%. Mehrvarz et al. [43] also emphasized that, the use of operable louvres and adaptive façade system is essential to improve the daylight availability and enhance visual comfort. Besbas et al. [10] assessed the daylight performance of a hospital building in Algeria. From the study, it showed that, orienting a building to the north receives excessive illuminance due to the high exposure to solar radiations. It resulted in visual discomfort, glare and overheating. However, applying multi-objective optimisation to enhance daylight performance and reduce energy consumption led to an increase in sDA, while maintaining a glare-free indoor environment. This approach also improved thermal comfort and further reduced overall energy use.

sDA is not without limitations. sDA fails to account for excessive daylight illumination which results in visual and thermal discomfort and increased cooling load. It mainly determines if illumination levels exceed the minimum threshold of 300 lx. In regions that experience high solar radiations, this metric usually cannot provide a true indication of illuminance levels to achieve visual comfort. Therefore, resulting in overheating, glare and increased cooling loads. To improve upon the weakness of the sDA, the Useful Daylight Illuminance (UDI) was developed.

5.6 Useful Daylight Illuminance

UDI quantifies the amount of daylight that contributes positively to visual comfort and enhances energy efficiency. It offers a more realistic approach and it ranges from 100–3000 lx [13, 46]. A UDI below 100 lx is considered dim hence artificial lighting is required. When it is between 100 and 3000 lx, then it is considered useful by occupants and no other artificial lighting is required. A UDI greater than 3000 is deemed excessively too bright which leads to visual discomfort, glare and an increase in cooling load [46]. UDI has been tested and evaluated in the developed countries and even developing countries. This shows the accuracy of UDI for daylight assessment. In UK, Mardaljevic et al. [42] showed that in 75% of the total floor area of schools in UK, a good UDI of 50% of active occupants time can be achieved. In the hot humid climate and sub-Saharan Africa, due to the intense solar radiations, varying climatic conditions and resource-limited architectural designs, the use of UDI has been low but increasing steadily. It offers a good benchmark for assessing visual comfort and potential energy savings. Zoure and Genovese [62] applied UDI in office buildings in Burkina Faso and achieved a useful illuminance of 45% of active occupants' hours. Overen et al. [50] assessed the daylight performance of heritage and office buildings in South Africa using UDI. They adopted simulation tools such as Radiance and Daysim to run simulations based on different orientations of the building and assessed the daylight performance. With 300 and 2000 lx as the minimum and maximum threshold for UDI, respectively, approximately 49.6% UDI was reordered when the building was oriented towards the East and about 67.3% when it is facing the West. Elakkad and Ismaeel [20] integrated daylight metrics such as UDI with prescriptive designs to develop a unified framework for daylighting designs. The study adopted UDI to compare the daylight performance of a LEED-certified building and a conventional building using the simulation tools such as Radians and Daysim. The study concluded that, the LEED-certified buildings performed better than the conventional buildings. Higher values of UDI were attained for the LEED-certified building. Also, WWR between 20 and 40% yielded improved UDI results. The study recommended the integration of UDI into the building code to enhance early decision support for building designs in Egypt. Hence, daylight performance variability is largely influenced by the room orientation and the window size. Similarly, based on the study of Akubah et al. [3], they suggested the need for guidelines in the selection of windows especially due to the climatic conditions. This is due to the poor lighting conditions in schools after long term assessment. The UDI values attained throughout the entire year were below globally accepted range with lux values averagely below 45 lx hence necessitating the need for artificial lighting to supplement the natural lighting. In Ghana, Addy et al. [2] applied UDI to assess the impact of passive design strategies on the thermal and visual performance of buildings. The study further proposed recommendations on some passive design strategies such as WWR and orientation after a multi-objective optimization to increase the daylight performance and minimize energy performance.

The gradual evolution of daylight metrics from the static Daylight Factor (DF) to more robust and dynamic climate-responsive indices such as Daylight Autonomy, Spatial Daylight Autonomy, and Useful Daylight Illuminance has resulted in more effective tools for assessing the daylight performance of buildings. In regions such as Africa, these metrics consistently highlight a significant gap between the high potential for daylighting and the actual performance achieved, a disparity largely attributed to poor architectural design and limited resource.

5.7 Developmental Issues and Way Forward

Climate-Based Daylight Metrics (CBDM) integrate the varying sky conditions and climate data for daylight performance assessment. Although climate-based daylight metrics are globally adopted, its implementation in the developing countries has faced significant challenges.

First of all, assess to robust climate data is a crucial for the implementation of accurate CBDM. A reliable CBDM assessment requires high-resolution Typical Meteorological Year (TMY) data, which contains hourly weather records representing average yearly conditions and reflecting local daylight patterns. In many developing countries, however, such datasets are absent, incomplete, or outdated [18, 31]. Many daylight performance metrics, such as Spatial Daylight Autonomy (sDA) and Annual Sunlight Exposure (ASE), were developed and validated primarily in mid-latitude regions. Specifically, the thresholds in IES LM-83 were derived from studies conducted between approximately 37° and 48° N. As such, caution and local calibration are advisable when applying them in warm-humid or tropical climates. For example, in Brazil, Santos et al. [54] applied sDA 300/50% in a tropical savanna context and recommended adapting such standards to better suit local climatic conditions. In warm-humid climates, achieving acceptable daylight levels can increase the risk of overheating and elevate cooling energy demand, making it essential to balance visual and thermal comfort [40]. While Li et al. [36, 37] showed that CIE Standard Skies can approximate certain CBDM outcomes, notable discrepancies remain in locations where local sky conditions deviate significantly from the standard models. The accuracy of CBDM outputs is also influenced by the choice and quality of weather files. For instance, Bellia et al. [9] demonstrated that results can vary depending on the data source, underscoring the importance of developing accurate, location-specific weather datasets. Also, CBDM utilizes building performance simulations and hence require advanced hardware and software simulation tools like Radiance/Daysim and workflow such as Honeybee. Furthermore, a significant amount of time and money is spent on developing models for simulations. Reinhart [52] posits that, computational modelling and simulation is expensive and time consuming as simulation is conducted for the entire year. The complexity of some building complicates the entire building simulation [12]. Moreover, beyond hardware and software availability, widespread adoption is often constrained by limited practitioner expertise and training and by cost/time overheads.

CBDMs have been integrated into global standards however, countries in sub-Saharan Africa and developing countries lack well defined policies and frameworks for the integration into the local climate. There are no modern guidelines to aid in the adoption of these metrics. There is lack of institutional and regulatory support for adoption of climate specific daylight metrics [24]. Developed nations typically show higher levels of policy implementation and technology adoption than developing countries [53]. A case in point is Ghana, where building legislation provides prescriptive natural light provisions (e.g. minimum opening areas) but lacks widely adopted, climate-based daylight performance criteria and detailed technical guidance, limiting the integration of advanced daylighting practice in design and compliance. Zoure and Genovese [62] used CBDM in Burkina Faso to assess strategies in implementing daylight strategies in buildings. The study outlined that, the high initial cost to acquire simulation software, computers and capacity building is a crucial factor for the slow implementation in developing countries. Firms in these regions operates with tight budget hence always unwilling to invest in these technologies without economic incentives hence, CBDMs are often overlooked. Mahame et al. [39] emphasized that, high initial cost and inadequate market supply is a major hurdle for implementation of CBDM. Notwithstanding these challenges, CBDMs are yet essential to address building energy demands, improved occupants' health, comfort and productivity. Hence, developing countries and sub-Saharan regions should underline strategies to promote the adoption of these metrics.

Infrastructure on local climate data has to be developed, improved or updated in these regions. Countries should prioritize developing metrics peculiar to their local climate conditions. There is a need for Africa to develop a climate data tool to enhance quality of climate data [19]. Without robust weather data, results from daylight and thermal performance simulation are sometimes unreliable. Hence, high quality and resolution climate data file is requested to improve the accuracy of the findings [62]. An exhaustive climate data file increases correlation and reduces biasness and develop real-time accessibility to climate data in sub-Saharan Africa [23]. Also, CBDMs parameters should be tailored to suit climate, culture and use of building. That is, the various indicators of these metrics should be contextualized to suit the local climate.

Increasing the adoption of CBDM requires strengthening local capacity through targeted training programmes, workshops, and other professional development initiatives focused on climate-based daylight modelling and the effective use of computational tools for daylight simulation. Fasehun [22] further proposes the establishment of a dedicated technical capacity-building facility to enhance the expertise of local professionals and improve their ability to adopt CBDM practices. For example, a green building school in Iran allowed architects to identify and correct daylighting errors using CBDM, hence, promoting early design decisions at the design stages [58]. This improves energy consumption, thermal performance and optimized shading devices [15]. These initiatives should be focused on promoting knowledge sharing and development of skills [60]. With economic barriers being a major hinder to adoption of innovation, developing countries thereby are more exposed to slow adaptation due to significant financial constraints [49]. Diverse and sustainable financing should be implemented such as green financing to promote the adoption of these metrics. Incentives for private sector can enhance the investment into the developing and utilization of the daylight metrics to enhance energy saving and thermal comfort.

5.8 Summary

This chapter highlights the role of daylighting in building decarbonization especially in developing countries. It outlines how daylight can be leveraged as a passive design strategy to reduce energy consumption, enhance visual and thermal comfort and promote the general well-being and productivity of occupants. It also contributes to sustainable practices by reducing the dependence on energy consumption for lighting. It further outlines the health-related benefits associated with the exposure to daylighting including improved mood, enhancing the regulation of the circadian rhythm and boost productivity and performance. Daylighting is measured with metrics such as Daylight Factor, Daylight Autonomy, Spatial Daylight Autonomy, and Useful Daylight Illuminance. Although DF is commonly and generally used, it does not account for difference in climate data due to its static nature. However, DA, sDA and UDI provide a more holistic metric for assessing the performance of daylighting by factoring difference in climate data and thresholds for occupant comfort.

In the warm-humid climate, the high exposure to solar radiations causes visual discomfort, overheating and consequently, thermal discomfort. Hence, the need for passive design strategies such as building orientation, WWR, and shading devices is necessitated. Although all the passive designs are essential, shading devices have the high potential to control the level of solar radiations received in to a room. A shading device could be static or dynamic. Static shading is preferred due to its low cost however; dynamic shading offers the ability to regulate the level of daylight at any point in time. The adoption of the CBDM is warm-humid climate is impeded by factors such as the unavailability of a robust local climate data, lack of skilled personnel, high initial cost for software for simulation, and little or no policy regulations. To overcome these challenges, it is recommended that, a high-resolution climate infrastructure has to be developed. Also, daylight metrics should be tailored to the local climate conditions as these CBDM were developed for countries in the temperature region. Also, capacity building of local professionals should be encouraged to fill the skill gap and financial incentives should be provided to encourage adoption.

References

1. Acosta I, Campano MÁ, Domínguez S, Fernández-Agüera J (2019) Minimum daylight autonomy: a new concept to link daylight dynamic metrics with daylight factors. LEUKOS J Illum Eng Soc North Am 15:251–269
2. Addy MN, Amos-Abanyie S, Tetteh BK, Kwofie TE, Aigbavboa C (2024) Design optimisation of fenestration and orientation for daylighting and energy performance in Ghana using genetic algorithm. In: Adinyira E, Amoako C, Kwofie TE, Aigbavboa C, Agyekum K, Addy M (eds) Sustainable and resilient infrastructure development in Africa's changing climate. ICIDA 2023. Springer, Cham. https://doi.org/10.1007/978-3-031-69606-0_22
3. Akubah JT, Amos-Abanyie S, Simmons B (2023) Simulation-based exploration of daylighting strategies for a public basic school in a hot-dry region of Ghana. In: Aigbavboa C et al (eds) Sustainable education and development—sustainable industrialization and innovation. ARCA 2022. Springer, Cham. https://doi.org/10.1007/978-3-031-25998-2_17
4. Alegbe M, Chukwuemeka L, Kalu JL, Eke-Nwachukwu A (2023) Building optimisation Vis-À-Vis solar shading for improved comfort and energy efficiency in classrooms. Dimensi 50:53–68
5. Al-Masrani SM, Al-Obaidi KM (2019) Dynamic shading systems: a review of design parameters, platforms and evaluation strategies. Autom Constr 102:195–216
6. Altaf M, Hill F (2019) Optimizing window to wall ratio for conserving energy in office buildings for cooling dominant climates with and without daylight utilization. J Green Build 14:129–142
7. Ayoub M (2019) 100 Years of daylighting: a chronological review of daylight prediction and calculation methods. Sol Energy 194:360–390
8. Basnet B, Uprety S (2022) A pilot study on the effect of daylighting and orientation in the office building: a case of Kathmandu. Adv Eng Technol Int J 2:33–45
9. Bellia L, Pedace A, Fragliasso F (2015) Dynamic daylight simulations: impact of weather file's choice. Sol Energy 117:224–235
10. Besbas S, Nocera F, Zemmouri N, Khadraoui MA, Besbas A (2022) Parametric-based multi-objective optimization workflow: daylight and energy performance study of hospital building in Algeria. Sustainability 14:12652
11. Bisegna F, Aureli C (2020) Shading devices in building design. In: CRC Press eBooks, pp 383–393
12. Brembilla E, Chi DA, Hopfe CJ, Mardaljevic J (2019) Evaluation of climate-based daylighting techniques for complex fenestration and shading systems. Energy Build 203:109454

13. Brzezicki M (2021) An evaluation of useful daylight illuminance in an office room with a light shelf and translucent ceiling at 51° N. Buildings 11:494
14. Budaiwi IM, Fasi MA (2023) Assessing the energy-saving potential and visual comfort of electrochromic smart windows in office buildings: a case study in Dhahran, Saudi Arabia. Sustainability 15:9632
15. Costanzo V, Evola G, Marletta L, Nascone FP (2018) Application of climate based daylight modelling to the refurbishment of a school building in Sicily. Sustainability 10:2653
16. De Luca F, Kiil M, Simson R, Kurnitski J, Murula R (2020) Evaluating daylight factor standard through climate based daylight simulations and overheating regulations in Estonia. In: Building simulation conference proceedings
17. De Luca F, Sepúlveda A, Varjas T (2022) Multi-performance optimization of static shading devices for glare, daylight, view and energy consideration. Build Environ 217:109110
18. Dinku T (2019) Challenges with availability and quality of climate data in Africa. In: Elsevier eBooks, pp 71–80
19. Dinku T, Faniriantsoa R, Islam S, Nsengiyumva G, Grossi A (2022) The climate data tool: enhancing climate services across Africa. Front Clim 3
20. Elakkad N, Ismaeel WSE (2021) Coupling performance-prescriptive based daylighting principles for office buildings: case study from Egypt. Ain Shams Eng J 12:3263–3273
21. Elouadjeri SM, Boussoualim A, Haddou HA (2021) Evaluating the effect of external horizontal fixed shading devices' geometry on internal air temperature, daylighting and energy demand in hot dry climate. Case study of Ghardaïa, Algeria. Buildings 11:348
22. Fasehun AO (2015) The war on climate change: ushering in sustainable development to VDCs through a technical capacity-building facility. Environ Claims J 27:196–225
23. Gebrechorkos SH, Hülsmann S, Bernhofer C (2018) Evaluation of multiple climate data sources for managing environmental resources in East Africa. Hydrol Earth Syst Sci 22:4547–4564
24. Genovese PV, Zoure AN (2023) Architecture trends and challenges in sub-Saharan Africa's construction industry: a theoretical guideline of a bioclimatic architecture evolution based on the multi-scale approach and circular economy. Renew Sustain Energy Rev 184:113593
25. Gil-Ozoudeh NI, Iwuanyanwu NO, Okwandu NAC, Ike NCS (2022) The role of passive design strategies in enhancing energy efficiency in green buildings. Eng Sci Technol J 3:71–91
26. Gómez AJ, Ortiz AN, Correa EP, Díaz LA, Trujillo JHS (2021) Influence of zoned use and occupancy on daylight autonomy in high-rise mass housing in Colombia. In: Building simulation conference proceedings
27. Hassan FHF, Ali KaY, Ahmed SaM (2023) Biomimicry as an approach to improve daylighting performance in office buildings in Assiut City, Egypt. J Daylighting 10:1–16
28. Heschong L, Wymelenberg VD, Andersen M, Digert N, Fernandes L, Keller A, Loveland J, McKay H, Mistrick R, Mosher B, Reinhart C, Rogers Z, Tanteri M (2012) Approved method: IES spatial daylight autonomy (sDA) and annual sunlight exposure (ASE). Environ Sci Eng
29. Kabanshi A, Choonya G, Ameen A, Liu W, Mulenga E (2023) Windows of opportunities: orientation, sizing and PV-Shading of the glazed area to reduce cooling energy demand in Sub-Sahara Africa. Energies 16:3834
30. Kalaimathy K, Priya RS, Rajagopal P, Pradeepa C (2023) A review of daylighting calculation metrics, variables and methods in evaluating the daylight performance of a residential building. In: AIP conference proceedings, vol 2813, p 020016
31. Kaspar F, Andersson A, Ziese M, Hollmann R (2022) Contributions to the improvement of climate data availability and quality for Sub-Saharan Africa. Front Clim 3
32. Krarti M (2022) A comparative energy analysis of dynamic external shadings for office buildings. ASME J Eng Sustain Build Cities 3
33. Kunwar N, Cetin KS, Passe U, Zhou X, Li Y (2020) Energy savings and daylighting evaluation of dynamic venetian blinds and lighting through full-scale experimental testing. Energy 197:117190
34. Lau AKK, Salleh E, Lim CH, Sulaiman MY (2016) Potential of shading devices and glazing configurations on cooling energy savings for high-rise office buildings in hot-humid climates: the case of Malaysia. Int J Sustain Built Environ 5:387–399

35. Lee J, Boubekri M, Liang F (2019) Impact of building design parameters on daylighting metrics using an analysis, prediction, and optimization approach based on statistical learning technique. Sustainability 11:1474
36. Li S, Li D, Chen W, Tsang E, Lou S, Wang Z (2023) Determination of climate-based daylight metrics under 15 CIE (International Commission on Illumination) standard skies and three representative skies. Buildings 13:2523
37. Li S, Li DHW, Chen W, Lou S, Tsang EKW (2023) Simple mathematical models to link climate-based daylight metrics with daylight factor metrics and daylighting design implications. Heliyon 9:e15786
38. Liu B, Liu Y, Deng Q, Hu K (2023) A study on daylighting metrics related to the subjective evaluation of daylight and visual comfort of students in China. Energy Build 287:113001
39. Mahame C, Kikwasi GJ, Baruti MM (2024) Barriers to the effective selection of sustainable materials for residential building projects: a qualitative study. Sustainability 16:9526
40. Mahmoud MAA, Mashaly IA, Rashed YM, Nassar K (2016) A new dynamic climate-based daylight metric for sustainable building design in hot climates. Mod Environ Sci Eng 2:358–364
41. Mangione A, Mattoni B, Bisegna F, Iatauro D, Zinzi M (2018) On the validity of daylight factor for evaluating the energy performance of building. In: 2022 IEEE international conference on environment and electrical engineering and 2022 IEEE industrial and commercial power systems Europe (EEEIC/I&CPS Europe), pp 1–4
42. Mardaljevic J, Heschong L, Lee E (2009) Daylight metrics and energy savings. Light Res Technol 41:261–283
43. Mehrvarz F, Bemanian M, Nasr T, Mansoori R, Al-Kazee MF, Khudhayer WA, Mahdavinejad M (2024) Designerly approach to design responsive façade for occupant visual comfort in different latitudes. J Daylighting 11:149–164
44. Mohamed AF, Amir AA, Ragab A (2024) The effect of aerogel glazing on daylight and heat gain in school buildings in hot and dry climate. Environ Dev Sustain
45. Musa MA, Salisu AS, Tukur RB, Stanley AAM (2019) Assessment of daylight of mid-rise office buildings in tropical wet and dry climate of Nigeria
46. Nabil A, Mardaljevic J (2005) Useful daylight illuminance: a new paradigm for assessing daylight in buildings. Light Res Technol 37:41–57
47. Nasrollahi N, Shokri E (2016) Daylight illuminance in urban environments for visual comfort and energy performance. Renew Sustain Energy Rev 66:861–874
48. Ohene E, Hsu S-C, Chan APC (2022) Feasibility and retrofit guidelines towards net-zero energy buildings in tropical climates: a case of Ghana. Energy Build 269:112252
49. Okisatari M, Lu W, Kandpal R, Morita K (2023) Closing the gap on city climate finance in developing countries
50. Overen OK, Meyer EL, Makaka G (2023) Daylighting assessment of a heritage place of instruction and office building in Alice, South Africa. Buildings 13:1932
51. Priya PV, Priya MG, Mahesh E, Sai KN, Fareez D, Chandrika TN (2024) Daylight performance of an office building. In: E3S web of conferences, vol 591, p 09013
52. Reinhart CF, Jones L (2017) Speedup potential of climate-based daylight modelling on GPUs. In: Building simulation conference proceedings
53. Rossi A (2024) Evaluation of climate change mitigation policies in developed vs developing countries. Int J Clim Stud 3:46–56
54. Santos IGD, Auer T, De Souza RVG (2017) Optimized indoor daylight for tropical dense urban environments. Ambiente Construído 17:87–102
55. Sari DP, Chiou Y-S (2018) Energy saving potential from shading design for residential house in rural area. In: MATEC web of conferences, vol 164, p 01007
56. Sorooshnia E, Rahnamayiezekavat P, Rashidi M, Sadeghi M, Samali B (2023) Passive intelligent kinetic external dynamic shade design for improving indoor comfort and minimizing energy consumption. Buildings 13:1090
57. Tabadkani A, Roetzel A, Li HX, Tsangrassoulis A (2021) Daylight in buildings and visual comfort evaluation: the advantages and limitations. J Daylighting 8:181–203

58. Vaisi S, Gorji HM, Shafei N (2024) Application of simulation-based metrics to improve the daylight performance of a secondary school, an approach for green building designers and architects. Adv Environ Eng Res 05:1–15
59. Wanas A, Aly S, Farghal A, El-Dabaa R (2015) Use of kinetic facades to enhance daylight performance in office buildings with emphasis on Egypt climate. J Eng Appl Sci 62(1):339–361
60. Wieszczeczynska KA, Tollin N, Spaliviero M (2024) Capacity building within urban climate resilience in the Global South—a literature review. Front Sustain Cities 6
61. Woo M, MacNaughton P, Lee J, Tinianov B, Satish U, Boubekri M (2021) Access to daylight and views improves physical and emotional wellbeing of office workers: a crossover study. Front Sustain Cities 3
62. Zoure AN, Genovese PV (2023) Implementing natural ventilation and daylighting strategies for thermal comfort and energy efficiency in office buildings in Burkina Faso. Energy Rep 9:3319–3342
63. Reinhart CF, Walkenhorst, O (2001) Validation of dynamic RADIANCE-based daylight simulations for a test office with external blinds. Energy Build 33(7):683-697
64. BSI (2008) British Standard 8206-2: lighting for buildings – Part 2: code of practice for daylighting. British Standards Institution, London
65. Li A, Xiao F, Fan C, Hu M (2021) Development of an ANN-based building energy model for information-poor buildings using transfer learning. Build Simul 14(1):89–101. Tsinghua University Press, Beijing
66. Li DH, Li S, Chen W, Lou S (2022) Simple correlations between point daylight factor, average daylight factor and vertical daylight factor under all sky conditions and building design implications. Indoor Built Environ 31(6):1700–1714

Chapter 6
Building Occupancy and Management

6.1 Introduction

Buildings are designed to keep people safe and comfortable by shielding them from external weather conditions [37]. To achieve this, building elements like artificial lighting and heating, ventilation, and air-conditioning (HVAC) systems are integrated to control the indoor environmental parameters such as temperature, air quality, illuminance, and so forth. However, these active systems, despite consuming a substantial amount of energy [37], often fall short in meeting building users' expectations [35, 76]. This is largely due to the varied individuals' comfort preferences, satisfaction levels, and perceptions of the indoor environment, which are basically shaped by a combination of physiological factors (e.g., gender and age), psychological influences, and cultural contexts [25, 66, 67].

Existing studies underscore that space utilization patterns, operational factors, along with the behaviours of building occupants, significantly influence the variability in a building's energy use [15, 32, 36]. The unpredictable nature of human behaviour, in particular, is considered as a key factor in a building's energy performance. According to existing studies, actions of building users create a significant disparity between a building's expected and actual energy consumption throughout its operational life cycle [15, 37]. Chen et al. [14] and Jia et al. [36], for instance, note that inefficient occupant behaviours can result in an energy waste of at least 30% vis-à-vis a building's designed performance.

Since conventional techniques like total volume conditioning have proven inadequate in addressing the varied comfort needs of building users [76], more personalized approaches are being explored. Systems like the Occupancy-Based Building Management Systems (Ob-BMS) are being developed to optimize energy use by detecting real-time occupancy and adjusting building operations accordingly, often using AI-driven using sensors and AI-driven automation [9, 18, 23].

M. N. Addy et al., *Decarbonization of Building Energy in Developing Countries*, Green Energy and Technology, https://doi.org/10.1007/978-3-032-09113-0_6

6.2 Overview of Building Occupancy Usage and Building Management in Africa

Jin et al. [37] define Building Occupancy as the presence state, occupant number, and trajectory, which indicate whether the occupant is in a certain space in the building, the number of occupants, and how the occupants are moving, as shown in Fig. 6.1. Fluctuations in occupancy patterns, significantly impact indoor environmental conditions, and influence energy consumption.

Occupants are a primary focus of building services; thus, the design, construction, and operation of buildings should consider them as a priority [29, 37]. Occupants' interaction with their indoor environment influences the indoor conditions. For instance, personal factors such as gender, anthropometry, clothing, demographics, and other cognitive variables influence occupants' perception of indoor thermal conditions [24]. Building users are not passive agents, instead, they are active participants in their thermal environment, constantly adjusting to maintain comfort [33]. This process of adjusting and adapting is known as thermal adaptation. Turning on a fan, opening or closing windows, putting on a sweater or removing a shirt, or adjusting curtains are all examples of occupants' adaptive behaviours. These behaviours are influenced by a wide range of factors, including the season, time of day, indoor and outdoor temperature, building orientation, and a person's mood, personality, and cultural background. Consequently, this combination of factors and individual choices makes it difficult to predict how a group of building users will behave. As a result, several studies, driven by the need to enhance comfort and reduce energy consumption, have focused on comprehending human adaptive behaviours.

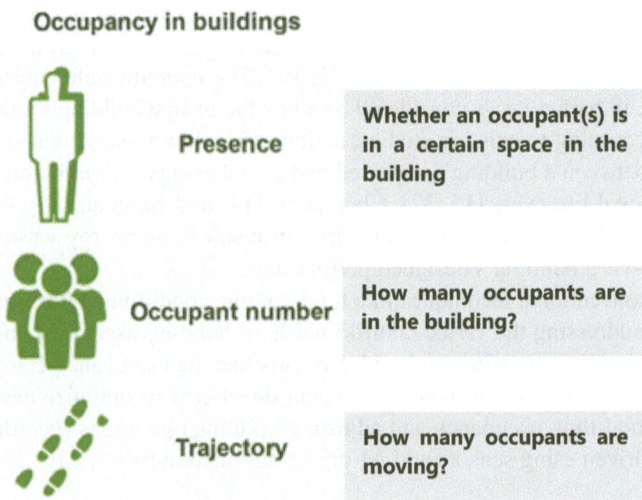

Fig. 6.1 Definition of building occupancy [37]

According to Jin et al. [37], this area of emphasis is flourishing mainly because of advancement in technologies, like Building Information Modelling (BIM), Virtual Reality (VR), Digital Twin, Artificial Intelligence (AI), used for detecting and analysing the presence of people [8, 73]. Earlier research studied how using data on where people are can improve the smart control of lighting and HVAC systems, which handle heating, cooling, and air. These studies found that using occupancy information can make HVAC scheduling more efficient. This efficiency leads to better system performance and smarter energy use [3, 45].

Making accurate models to show how buildings are occupied is challenging. It requires a lot of effort, time, and resources, and mistakes are usually inevitable [18]. This is particularly crucial in Net Zero Energy buildings. In these buildings, just optimizing the systems is not enough to achieve energy neutrality. Understanding Building Occupancy patterns is essential to achieve this goal. The building might not meet its Net Zero Energy target if its occupants do not operate it as predicted. While ASHRAE 90.1 focuses mainly on building components and prescriptive/system performance requirements, occupant behaviour is typically represented through simplified, standard schedules rather than explicit behaviour models. Jia et al. [36] noted this gap, although recent 90.1 editions have taken limited steps (e.g., occupancy diversity assumptions) to better reflect occupancy effects.

6.2.1 Building Management Systems in Africa

Building Management Systems (BMS), also referred to as Building Control Systems or Building Automation Systems, consist of physical and computational equipment designed to make buildings more comfortable, safe, and energy-efficient [6, 69]. The primary purpose of a BMS is to conserve energy while ensuring the comfort of the occupants. It is responsible for managing essential functions such as controlling heating, ventilation, and air-conditioning (HVAC), adjusting lighting, regulating hot water, and keeping track of electricity use. These strategies help keep energy costs low while ensuring residents stay comfortable [56]. To enhance service quality, safety, and cost-effectiveness, BMS employ modern technologies. These include data analytics, data gathering, data storage and visualization. Efficient operation requires ongoing adjustments to performance settings to adapt to changing needs. Various technologies are utilized such as high-precision imaging devices like optical and thermal cameras, and specialized environmental sensors like carbon dioxide and passive infrared sensors. These tools are essential for tracking real-time occupancy and optimizing energy efficiency. Designing BMS comes with uncertainty, even though models and model-based techniques play a crucial role. This uncertainty makes accurate system modelling through offline simulations difficult. Eini et al. [19] introduced a smart BMS that operates in real time. This system blends model-based control with machine learning to enhance aspects like cost-effectiveness, occupant comfort, security, and energy efficiency. They addressed research gaps by combining the

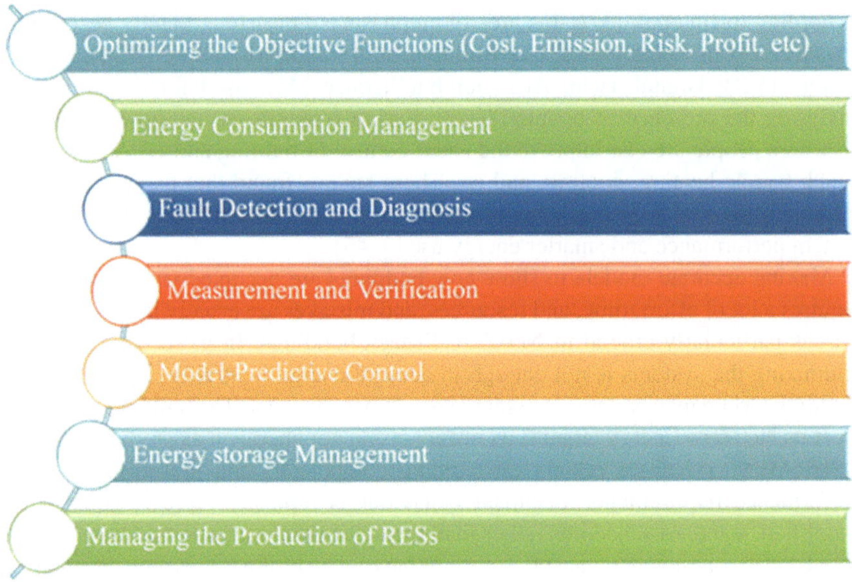

Fig. 6.2 Key features of building management systems

management of multiple building subsystems with real-time learning about subjective factors. Their study showed that a unified management framework that adapts to how people act and environmental shifts can ensure optimal building performance. In the same year, Vahidinasab and his team identified seven key features of a BMS, depicted in Fig. 6.2.

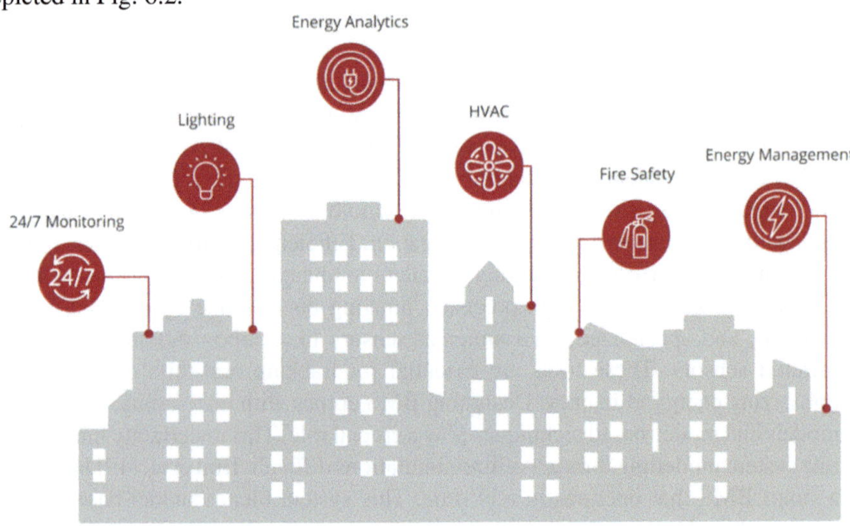

6.3 Challenges to Building Management for Energy Decarbonization

The pursuit of decarbonization goals necessitates the employment of strategic ways to address energy use. This has pushed for the adoption of several innovative ways in various sectors towards energy management, with buildings as one of the critical areas of the energy sector that need attention [31]. System innovation and advanced technologies are essential for the transition to sustainable buildings. Accordingly, the integration of buildings with automated systems and controls to aid in energy management is a burgeoning area of discussion. A key development in this field is the use of Digital Twins, which are virtual models of physical buildings that collect real-time data through inbuilt monitoring systems. This technology allows for performance validation, optimization, and the updating of control strategies, thereby enhancing a building's role in decarbonization [73]. The adoption of advanced technologies like BMS has also gained recognition in discourse regarding net-zero buildings, even though, it faces significant hindrances in Africa. Addressing these barriers is crucial for improving the use of such systems and achieving decarbonization goals across the continent.

6.3.1 Cultural Practices

Individual and organizational actions, as bounded by routine and habit, may aid or encroach upon the adoption of energy efficiency technologies. In their review, Sovacool and Griffiths [86] found that people's ideology and social disposition hinder their adoption of climate-responsive solutions and digital technologies. Ultimately, these practices make people resistant to change. Following Dabaieh and Johansson [87], people either resist or slowly embrace change when cost implications of adopting sustainable practices are relatively higher than conventional practices. As a consequence, individuals overlook cost-saving potentials of energy efficiency measures, even if their advantages are clear [52]. People have a propensity to justify past choices by highlighting the advantages of the chosen course of action and the drawbacks of the option that was not selected. Organizational culture directly influences the values a people, and can act as a significant barrier to the implementation of energy-efficient technologies [13, 40]. Most African countries express strong environmental advocacy and claim to be zealous to undertake and promote sustainable environmental activities. The challenge, however, is translating such attitudes and claims into daily actions and modus operandi. Research by Amoah and Addoah [7] and Berndt and Gikonyo [10], for example, underscores a significant disconnect between individuals' professed commitment to environmental awareness and their actual behaviour.

According to Abdullahi and Lee [1], people are more likely to accept solutions that respect and are integrated into their existing norms. Ignoring indigenous knowledge and social practices, like traditional courtyard house or passive cooling design strategies with earthen materials, can result in a relatively low utilization of automated building systems. Furthermore, the wide language diversity and varying literary levels across the African continent can serve as a great limitation [86]. When little consideration is given to designing a system with these factors in mind, it can lead to confusion, user error, and ultimately hinder the successful adoption and effective use of a new technology. Addressing these barriers requires promoting education, raising awareness, implementing supportive policies, and adapting automations management systems to reflect local conditions.

6.3.2 Limited Availability of Energy-Efficient Building Technologies

The limited availability of energy-efficient technologies in the majority of African countries is a huge limitation to their massive deployment and utilization [40]. Underdevelopment of the manufacturing sector and weak supply chains are among the leading causes of this limitation. The majority of African countries rely on international imports for energy-efficient building equipment, sensors, and smart control systems. This reliance, coupled with access constraints due to the high duty duties, fluctuating currency exchange rates, and transportation expenses, leads to a significant escalation in the cost for the same technology that is being used in a developed country with a more productive manufacturing sector and supply channels. As a consequence, developers and building managers face considerable difficulty in justifying the financial costs of integrating such sophisticated systems in buildings, ultimately restricting the adoption of such technology in the continent. Another challenge is the lack of information and technical know-how regarding the technology [40]. Studies in Ghana and similar contexts report limited professional awareness and skills for implementing smart building systems (including BAS/BMS), alongside gaps in procedures, tools, and integration—factors that constrain demand and slow wider adoption [26]. Due to the lack of training and certification programs, even where these systems are available, there are not enough skilled people to install and operate them effectively. This lack of knowledge reinforces the perception that BMS are complex, not feasible, or too costly to maintain, further inhibiting their uptake. This technological deficit as well as knowledge gap needs to be tackled to enhance energy efficiency in the African building sector.

6.3.3 Climate Variability

Indoor climate control is a major energy demand everywhere. Design and operation of buildings and HVAC systems crucially depend on climate data and real-time meteorological conditions [72]. Energy-efficient buildings also contribute to reducing air pollution and climate change in urban areas as well as regionally and globally. Climate is the basic determinant for the design and operation of buildings. Climatic factors like temperature, solar irradiation, wind humidity impact the energy balance of buildings. This is because the building exchanges energy with its surroundings. The control of this exchange complicates BMS strategies to optimize energy and comfort [72]. Climate change has led to erratic weather patterns, extreme temperatures, unpredictable rainfall, and an increased frequency of natural disasters. These factors not only complicate energy management strategies but also hinder the widespread implementation of BMS in the region. Climate variability results in unpredictable temperature extremes, increasing the demand for heating, ventilation, and air-conditioning (HVAC) systems in buildings. Many African nations, particularly those in sub-Saharan Africa, are experiencing a growing demand for energy due to increasing urbanization and rising living standards [50]. Notwithstanding, a lot of buildings are not built to save energy. A lack of proper design strategy and advanced HVAC systems is a significant challenge in African buildings. According to Leal Filho et al. [43], the continent continues to struggle with energy poverty—a condition describing the lack of access to affordable and sustainable energy. The region is home to 44% of the 1.2 billon people who lack access to electricity worldwide [11], and even those with access often experience unstable power due to deficient grid systems [58]. This makes the implementation of automated energy management systems, such a BMS particularly difficult. For instance, in Nigeria, where frequent power outages are endemic [21], a steady power supply is a basic necessity effective function of BMS. Furthermore, countries facing resource limitations struggle with acquiring adequate and reliable systems for climate monitoring, computational resources, and integrating real-time weather information into the BMS platforms [68]. The absence of such equipment can significantly hinder the adoption of advanced building solutions in Africa. Each region within the continent is distinguished by different bioclimatic materials and techniques, which needs to be captured within the automated systems. Lastly, the use of foreign building solutions that are not adapted to the various African climates, can ultimately reduce the effectiveness and sustainability of BMS [27].

6.3.4 Paucity of Research

There is a dearth on studies on energy management specific to Africa, creating a knowledge gap. This gap makes it hard to develop, implement and improve BMS throughout the continent [4]. An indirect implication of limited studies is the failure to provide continent-level solutions attuned to the peculiar energy demands of Africa, limited infrastructure within areas, as well as meteorological conditions. Research is crucial for creating evidence-based policies to help adopt BMS. Notwithstanding, without enough data on building energy use, energy efficiency, and BMS cost-effectiveness, it will be difficult for governments to make effective policies and regulations [42]. Policymakers need details on energy performance in various building types, the financial payoff from adopting BMS, and the effects of different incentive programs [49]. Without this information, it's difficult to draft policies that encourage building owners, developers, and policymakers to focus on BMS. Many BMS technologies come from developed regions like Europe and North America, where energy systems are established and electricity is consistent. These solutions often do not work well in most parts Africa, where electricity grids are unstable, blackouts are recurrent, and reliance on off-grid options like solar energy presents unique problems. Without enough research, it is tough to create energy management solutions that handle these challenges well. The lack of accurate data makes it hard to set national energy efficiency goals and create benchmarks, which slows the progress towards sustainable building practices.

The research gap in BMS also leads to the lack of local technical skills for architects, engineers, and energy professionals. Education programs and training courses within most Developing Countries do not cover recent energy management technology, and therefore, there is the gap in applying and sustaining it. A study by Malatji [47] revealed the meagre presence of specialist courses and certification in energy-efficient building technology among African institutions, thus lowering the ability of experts to install BMS effectively into building schemes. With inadequate research, there is also limited awareness within industry stakeholders, and hence creating misconceptions on the viability, affordability, and advantages of BMS in Africa. There is a lack of studies on the economic viability and long-term savings of BMS that discourages public and private sector investments [40]. In the absence of thorough studies establishing the return on investment and energy savings, policymakers and building owners are likely reluctant to invest funds in these technologies.

6.4 Strategies to Improve Occupancy Usage and Energy Decarbonization

In the pursuit of an energy-efficient building, occupancy is a critical factor that cannot be overlooked [79]. Governments worldwide have been strengthening policies for energy efficiency to drive the decarbonization of buildings. To this end, the operational energy of a building, in particular, has been identified as a significant contributor to overall energy use and greenhouse gas emissions throughout a building's life cycle. Although lighting and HVAC systems are primary services in any building, they nonetheless constitute key energy-consuming elements, which account for a significant portion of a building's operational energy use [57, 85].

Beyond their inherent reliance on energy, the operation of these systems is basically centred on the building occupants who use them. This connection has led to a growing dominance of research focused on using occupancy data to optimize energy consumption.

6.4.1 Integration of Occupancy Sensors into Buildings

The development and integration of occupancy sensors into buildings facilitates the efficient operation of lighting and HVAC systems through automated controls. The energy saving potential of occupancy sensors is seen in their ability to coordinate the building mechanical system with occupancy levels and behaviours [78]. In 2018, the United States Energy Information Administration (EIA) conducted a Commercial Buildings Energy Consumption Survey (CBECS) which indicated that about 17% of commercial buildings in the U.S. were integrated with an occupancy sensing system for energy management [57]. Consequently, there was a 26% increase in energy efficiency as compared to six years before. Nagy et al. (2015) conducted a six-week case study in 10 offices where they achieved 37.9% energy savings as compared to a standard setting control baseline scenario using set-points for switching artificial lights on and off, based on statistical analysis of occupancy sensor data. A simulation study by Zhang et al. (2013), of a large-sized office in a Commercial Prototype Building Models (CPBM), developed by the U.S. Department of Energy (DOE), revealed a 5.9% energy saving of lighting and HVAC energy consumption. Energy saving by occupancy sensors is also confirmed by Kong et al. [84] in their side-by-side testbed study, revealing up to 35% energy savings in an office setting.

6.4.2 Post-Occupancy Evaluation Studies

Post-occupancy Evaluation (POE) is the systematic assessment of the actual performance of a building, as compared to its performance criteria specified in the design phase after it has been occupied for some time [16, 20]. Creating a conducive environment for end-users is generally an envisioned performance criterion and prerequisite for the design of buildings as the artificial environment created in the building directly impacts the health and safety of occupants. Subject to the performance of the building, it can encounter some challenges as shown in Table 6.1. Practitioners have long either failed to undertake a POE or been inconsistent in their practice. This inconsistency stems partly from the wide variety of POE implementation strategies reported in literature and practice [65]. Consequently, valuable chances to reduce unnecessary energy use, lessen reliance on resources, and prevent material waste are commonly lost. Meanwhile, this discord harms investment returns on finances and diminishes the level of satisfaction from occupants. One study attributes such resistance to using POE as a response to a focus on short-term rewards and passive

Table 6.1 Common challenges of buildings

Challenges	Security issues
	Leakage problems
	Inadequate signage and navigation problems
	Poor air circulation and temperature control
	Accessibility issues for individuals with disabilities
	Insufficient or lack of storage
	Lack of privacy
	Blocked hallways
	Aesthetic problems
	Entry door issues, including wind interference and dirt accumulation
	Insufficiently designed spaces for equipment
	Maintainability of glass surfaces (e.g. skywalks or inaccessible skylights)

Source Preiser et al. [63]

commitment to sustainability as causing monstrous walls for stakeholders such as designers, contractors, and clients. However, higher public and political pressures for increasingly greener buildings in terms of their sustainability might have the result of altering such rooted negative behaviours.

POE is regarded as one of various techniques designed to understand design criteria, predict the effectiveness of new designs, assess completed designs, aid in building activation and facilities management, and link user feedback to building performance. POE is progressing towards assessments of capital asset management, programming, and planning that are increasingly process-oriented.

POE is employed to ascertain whether choices made by the design and construction professionals have satisfied the intended needs of end-users and the development commissioners [65]. POE focuses on the needs of building occupants (including their health, safety, security, psychological comfort, aesthetic quality, and satisfaction), offering valuable insights into the impacts of previous design choices and the subsequent performance of the building. The evaluation helps to provide feedback on the difference between the actual performance and the stated performance, and most importantly, occupant satisfaction for the purposes of identifying problems present in the building, testing new building prototypes, improving the life cycle cost of future buildings and developing design guidance for the planning, programming, and design processes for future buildings. POE is useful, especially for organizations with recurring construction programmes.

POE encompasses a range of advantages including the transfer of knowledge to inform future building designs, enabling the continuous enhancement of an existing facility's performance and providing a reliable method for comparing building performance across multiple facilities, especially within the same estate. Some advantages of POEs in the short-term, medium-term, and long-term are summarized in Table 6.2.

Table 6.2 Advantages of post-occupancy evaluations

Short-term advantages	• Identifying problems and their appropriate solutions • Proactive facility management • Improved space utilization • Improved attitude of building occupants • Understanding of the performance implications of changes dictated by budget cuts • Informed decision making • Better understanding of the consequences of design
Medium-term advantages	• Facility adaptation to organizational change and growth over time • Improvement in life cycle cost • Accountability for building performance
Long-term advantages	• Long-term improvements in building performance • Improved design databases and criteria

POEs are conventionally carried out using questionnaires, interviews, site visits, and observation of building users. Information obtained are based on the commissioning and construction phases, perspectives from end-users on quality of finishes and functional performance, technical assessments of building system performance, and a strategic analysis integrating data from the above evaluation stages. Throughout the years, evolving stakeholder needs and budget considerations have driven the development of more specialized processes, diverse survey tiers, and innovative technologies. Researchers and evaluators can now get accurate and valid information in a relatively shorter period due to shortcut techniques that have been developed. How POEs are conducted and the data gathered are analysed may change dramatically with the advent of the Web and other technologies. POE was easier in the past where the evolution of building types was slow. Lessons learned from buildings could easily be passed on to another generation. In modern times, where there is increasing evolution, innovation and client preferences, such informal techniques will fail. The complexity is compounded by the fact that the building delivery process now involves diverse professionals and must comply with more stringent technical codes and requirements; hence, critical decisions affecting the output are made by a group or committee and not individuals. The need to comply with and balance all these requirements necessitates the clear definition and documentation of facility performance expectations, typically in the form of a detailed facility program. The facility program is usually documented in the form of a functional program and shared with all parties involved in the building delivery process.

6.5 Pathways to Building Management for Energy Decarbonization in Africa

Future population growth and urbanization will result in an enormous demand for new buildings, placing an immense strain on the natural environment and its finite resources [48]. While architectural design has traditionally centred on passive or low-energy design strategies to curtail building energy consumption, it seems obvious that a more expansive, whole-life attitude is required [83]. BMS can mitigate these effects, improve building energy efficiency, and balance the imbalance between energy supply and demand. BMS is integrated with sophisticated techniques and systems that are able to monitor and control the energy requirements of operational buildings to improve energy efficiency and conservation. Changing how people think and spreading information is important to build an energy-smart culture and overcome challenges to using BMS.

6.5.1 Behaviour Change and Public Awareness Campaigns

There is a need for more education within the building industry since most people within the sector are not as knowledgeable in new systems as is required. The successful adoption of BMS in Africa relies not only on technological advancements but also on changing public perceptions and behaviours regarding energy use [46]. A big hindrance to the widespread uptake of BMS in Africa is that many people are not aware of energy management solutions and their benefits. Changing how people think and spreading information is important to instigate an energy-conscious culture and overcome informational barriers to BMS adoption. In fact, many building owners, facility managers, and residents either do not know about these systems or doubt their cost-effectiveness [26]. Public awareness campaigns can teach people and businesses about BMS, showing how they work, how they save money over time and their role in reducing energy use and carbon footprints. These campaigns can use social media, workshops, and community programs to correct misunderstandings and help more people accept BMS technology. More so, it can help individuals adopt more efficient energy-use practices with emphasis on how BMS can increase comfort and cut expenses. In Africa, changing traditional habits can be tough due to cultural norms [50, 55]. Many people prefer manual control over electrical devices and embrace building designs that might not be compatible with modern energy systems. Engaging local leaders and influencers can aid these campaigns in encouraging communities to gradually adopt new energy saving habits and technologies respectfully. Community-based education and training programs play a crucial role in this process. These programs can clarify how BMS operate and illustrate their long-term benefits, making them more acceptable to the public [26]. This includes training workshops, awareness drives, and demonstration projects to showcase real-world examples of how BMS can optimize energy use, reduce costs, and contribute to environmental sustainability.

6.5.2 Strategic Policy Directives

Policies need to be put in place to drive building sustainability transitions, which evolve incrementally and have significant roles to play in addressing environmental problems. They are meant to influence the behaviour of people; to get them to do things they would not otherwise do without the policy and to get them to do them as desired by the policymaker. The lack of supportive policies and regulations further complicates the situation. In the majority of African countries, building codes and energy requirements have fallen behind the call for sustainability. They are obsolete in some and poorly enforced in others, allowing developers to disregard energy-efficient alternatives. Without stringent requirements or incentives, there is little motivation to prioritize sustainability in construction, and many buildings become less efficient and their long-term energy costs go up. Policy formulation for innovative systems, like BMS, is designed to guide the direction of action towards specific goal and encourage the adoption of energy-efficient technologies by organizations and individuals. For instance, Finland, a country that ranks among the top three for achieving significant milestones in energy efficiency policies, implemented a series of 30% improvement targets for new buildings between 2003 and 2012 [41]. Kivimaa et al. [41] added that such targeted policies are a crucial to successful transitions, as evidenced by the fact voluntary energy conservation agreements in 2010 catered for about 80% of Finland's total energy consumption. The tools enabling such evolution are of diverse character and have been referred to with different terms in the literature. Bemelmans-Videc et al. [82] frame the discussion regarding such policy tools in terms of "carrots", "sticks", and "sermons". Sticks consist of regulations that the recipient is compelled to adhere to. Carrots are economic tools that render an action more or less costly. Sermons, lastly, are information-disseminating tools that seek to influence the addressee by persuading or presenting facts on a topic. The most potent method of influencing energy consumption, nonetheless, is the use of the three categories of tools in tandem. Governments should have strong political will and determination to introduce necessary mandates and policies to set priorities for the initiative of building efficiency improvement [22]. It is urgently needed to convince the government to establish policies to enforce the uptake of BMS [5].

6.5.3 Research

In Africa, there are several challenges in adopting modern BMS. These include traditional building methods, limited awareness, weak policies, and a lack of local expertise. These issues can be addressed with better, sustainable ways to manage buildings by investing in research [4]. By funding research, we can advance BMS in Africa by providing crucial information, encouraging innovation, and offering policy advice to improve energy efficiency and sustainability. Research can offer reliable data about energy usage in buildings, their efficiency, and their environmental impact, leading to evidence-based solutions. By exploring new materials, smart energy systems, and automation technologies through academic and industrial research, we can enhance energy use in buildings. This approach encourages technological innovation in managing buildings. For example, researching IoT-based energy systems, smart grids, and passive cooling can help develop efficient and affordable BMS solutions in Africa. Collaboration between local universities and foreign institutions is key to adapting these technologies to African conditions, ensuring they are suitable and scalable. Through research, we can develop new materials, smart energy systems, and automation technologies, optimizing energy use in buildings and driving progress in building management.

Governments will have to have suitable policies and building codes to promote sustainable construction of buildings and energy conservation that they will implement. Governments can make policy through analysis of the impact of energy efficiency building codes, tax credits, and subsidy regulations to inform them on what will stimulate the adoption of BMS. Evidence-based policies also ensure greater enforcement of building codes, and therefore greater compliance. Capacity and skills development can be enabled through research. Without proper research, policy decisions may lack accuracy or fail to address the actual needs of the sector. The governments need stringent policies and building codes to promote sustainable construction practices and energy management. Research on the efficacy of energy efficiency codes, tax credits, and subsidy policies can help governments create policies that promote the adoption of BMS while keeping economic feasibility in consideration. Research-backed policies can also ensure better enforcement of building regulations, leading to improved compliance rates.

Among the primary issues confronting effective building management in Africa is the shortage of competent professionals in the area of building energy management and intelligent building technologies [26]. Research institutions can customize training programs, course offerings, and certification programs to equip engineers, architects, and facility managers with the expertise to effectively utilize BMS. By doing this, an effective pool of labour is established that has the potential to innovate and build the efficiency of buildings across the continent. Consequently, research will contribute to capacity building and skills development. Research can further guide investment and financial strategies to support sustainable building initiatives [40]. High initial expenses and unclear Return on Investment (ROI) discourage most investors and developers from utilizing BMS [2]. Economic research and case studies of the long-term financial benefits of energy efficiency can provide conclusive evidence that the use of such systems leads to cost savings, appreciation in property value, and reduced environmental impact. Research can catalyse further investment in energy-efficient building technology in the public and private sectors by illustrating compelling financial models.

6.5.4 Upgrading Building Code and Regulations

Codes and standards are broadly considered to be drivers of change (Diabat and Govindan 2011). Enhancing building codes and regulations is a strategic opportunity to advance the uptake of BMS in Africa. In Africa, quite a number of countries continue to use outdated or weakly implemented building regulations that do not prioritize energy efficiency to any significant level, therefore preventing BMS from becoming a commonplace practice within the construction sector. Through updating building codes, governments can provide an enabling environment for energy-efficient technologies and sustainable building practices. Most African governments have not set standards explicitly for energy-efficient buildings, and as a result, there is huge energy consumption in commercial and residential buildings. The building codes need to be revised to have obligatory energy efficiency standards for new and existing buildings to mandate the fitting of energy management systems to make energy management the norm and not an afterthought or optional feature in new buildings. When building codes enforce energy efficiency measures, construction companies, architects, and developers will be compelled to integrate BMS and related systems into their projects. This not only increases the adoption of these technologies but also stimulates local markets for energy-efficient building materials and services. As a result, local industries and suppliers benefit from increased demand, leading to economic growth and job creation in the energy management sector. Enhancing building regulations and codes can also trigger market demand for energy efficiency technologies. When building codes enforce energy efficiency measures, developers, architects, and builders will be compelled to integrate BMS and related systems in their projects. This will not only increase the adoption of such technologies but also stimulate local markets for energy-efficient building materials and services.

Upgraded regulations also play a critical role in reducing the life cycle cost of buildings. Poor energy management leads to excessive electricity use, higher operational costs, and increased strain on national power grids. By enforcing stricter building codes that require BMS integration, governments can help property owners reduce long-term energy expenses, making buildings more affordable and sustainable in the long run. More so, inefficient energy management results in high electricity usage, expensive operations, and additional stress on national grids. Governments can assist building owners in lowering the cost of energy in the long term to enable buildings to be more sustainable and cheaper in the long run by implementing stronger building codes that mandate the incorporation of BMS. Yet, institutional capacity and enforcement are needed for building code enhancement. Most African nations suffer from poor regulatory enforcement, where even building codes in place are not adhered to due to corruption, poor monitoring, or low compliance levels. To counter this, governments need to reinforce inspection agencies, train construction practitioners, and impose fines for non-compliance. Digital platforms may also be leveraged to make permit approvals more efficient as well as monitor compliance more effectively, with transparency and efficiency in the regulatory process.

6.6 Summary

The main objective of buildings is to shield their occupants from the external environment. To achieve this goal, building services are provided through the installation of systems including the lighting and HVAC system. The operation of these systems, however, also results in an undesired outcome, which is the high consumption of energy. This leads to the building sector being a significant contributor to energy consumption globally, especially during its operational phase. For this reason, identifying strategies to mitigate energy consumption in buildings is an effective way of spear-heading towards decarbonization. Since, buildings are primarily constructed for use by people, the occupancy of the building largely determines its energy consumption. Changes in occupancy patterns affect the internal environmental conditions of a building and its energy consumption. The presence, number, and behaviour of occupants in a building are key variables of occupancy that are used in energy and comfort efficiency. More occupants in a building will result in higher energy consumption due to the metabolism and activities of the occupants within the building. Thus, occupancy of a building has a direct influence on the energy consumption of a building and is a critical driver of energy efficiency in buildings when it is managed. Several factors can influence the behaviour of building occupants: temperature conditions, season, mood personality, cultural background, etc. It is however difficult to forecast their behaviour in a specific scenario. Decarbonization requires strategic energy management, particularly in buildings, which are important energy consumers. Smart technologies coordinated by BMSs are a strategic way to achieve a sustainable building. BMS focuses on maximizing energy efficiency and ensuring occupant comfort simultaneously. It can monitor and control the building's mechanical equipment to ensure optimum energy and occupant comfort. Its key features include energy consumption management, fault detection and diagnosis, measurement and verification, model-predictive control and construction operation and building information exchange [30]. In spite of the benefits BMS offers, it is associated with some challenges including data transferability, unpredictable nature of occupants, labour-intensive development of models and lack of efficient data visualization capabilities. The implementation of BMS is hindered by factors including cultural practices, limited availability of energy-efficient building technologies, climate variability and paucity of research. However, through strategies such as behaviour change and public awareness campaigns, strategic policy directives, research and upgrade of building code and regulations, the adoption of BMS in Africa can be improved to foster energy efficiency on the continent.

References

1. Abdullahi AL, Lee A (2020) The influence of culture, beliefs and experience on the sustainable end-of-life management of buildings in Nigeria. J Build Constr Manag. https://journals.co.za/doi/pdf/10.10520/EJC-8505ef9d2

2. Adebisi OD, Wahab AB (2023) Factors influencing adoption of building management system in commercial buildings in Lagos State, Nigeria. Afr J Hous Sustain Dev 4(1)
3. Agarwal Y, Balaji B, Gupta R, Lyles J, Wei M, Weng T (2010) Occupancy-driven energy management for smart building automation. In: Proceedings of the 2nd ACM workshop on embedded sensing systems for energy-efficiency in building, pp 1–6
4. Agbajor FD, Mewomo MC, Umoh VB, Makanjuola SA (2023) Building energy management system research in South Africa—a decade overview. Energy Rep 9:33–39
5. Alam M, Zou PX, Stewart RA, Bertone E, Sahin O, Buntine C, Marshall C (2019) Government championed strategies to overcome the barriers to public building energy efficiency retrofit projects. Sustain Cities Soc 44:56–69
6. Alghassab MA (2024) Fuzzy-based smart energy management system for residential buildings in Saudi Arabia: A comparative study. Energy Rep 11:1212–1224
7. Amoah A, Addoah T (2021) Does environmental knowledge drive pro-environmental behaviour in developing countries? Evidence from households in Ghana. Environ Dev Sustain 23(2):2719–2738
8. Arowoiya VA, Moehler RC, Fang Y (2024) Digital twin technology for thermal comfort and energy efficiency in buildings: A state-of-the-art and future directions. Energy Built Environ 5(5):641–656
9. Batra N, Arjunan P, Singh A, Singh P (2013) Experiences with occupancy-based building management systems. In: 2013 IEEE eighth international conference on intelligent sensors, sensor networks and information processing. IEEE, pp 153–158
10. Berndt A, Gikonyo L (2012) Environmental concern behaviours in Africa: An exploratory study. J Mgmt & Sustainability 2:1
11. Blimpo MP, Cosgrove-Davies M (2019) Electricity access in Sub-Saharan Africa: Uptake, reliability, and complementary factors for economic impact. World Bank Publications
12. Brooks DJ, Coole M, Haskell-Dowland P (2020) Intelligent building systems: security and facility professionals' understanding of system threats, vulnerabilities and mitigation practice. Secur J 33(2):244–265
13. Carlander J, Thollander P (2023) Barriers to implementation of energy-efficient technologies in building construction projects—Results from a Swedish case study. Resour, Environ Sustain 11:100097
14. Chen Z, Jiang C, Xie L (2018) Building occupancy estimation and detection: A review. Energy Build 169:260–270
15. D'Oca S, Hong T, Langevin J (2018) The human dimensions of energy use in buildings: A review. Renew Sustain Energy Rev 81:731–742
16. Dam-Krogh EP, Rupp RF, Clausen G, Toftum J (2024) Scoping review of post occupancy evaluation of office buildings with focus on indoor environmental quality and productivity. J Build Eng:108911
17. De Carli M, Olesen BW, Zarrella A, Zecchin R (2007) People's clothing behaviour according to external weather and indoor environment. Build Environ 42(12):3965–3973
18. Dong J, Winstead C, Nutaro J, Kuruganti T (2018) Occupancy-based HVAC control with short-term occupancy prediction algorithms for energy-efficient buildings. Energies 11(9):2427
19. Eini R, Linkous L, Zohrabi N, Abdelwahed S (2021) Smart building management system: Performance specifications and design requirements. J Build Eng 39:102222
20. Elsayed M, Pelsmakers S, Pistore L, Castaño-Rosa R, Romagnoni P (2023) Post-occupancy evaluation in residential buildings: A systematic literature review of current practices in the EU. Build Environ 236:110307
21. Emovon I, Samuel OD, Mgbemena CO, Adeyeri MK (2018) Electric Power generation crisis in Nigeria: A Review of causes and solutions. Int J Integr Eng 10(1)
22. Enshassi A, Ayash A, Mohamed S (2018) Factors driving contractors to implement energy management strategies in construction projects. J Financ Manag Prop Constr 23(3):295–311
23. Erickson VL, Carreira-Perpinán MÁ, Cerpa AE (2014) Occupancy modeling and prediction for building energy management. ACM Trans Sens Netw (TOSN) 10(3):1–28

24. Erlandson T, Cena K, De Dear R, Havenith G (2003) Environmental and human factors influencing thermal comfort of office occupants in hot—humid and hot—arid climates. Ergonomics 46(6):616–628
25. Feige A, Wallbaum H, Janser M, Windlinger L (2013) Impact of sustainable office buildings on occupant's comfort and productivity. J Corp R Estate 15(1):7–34
26. Ghansah FA, Owusu-Manu DG, Ayarkwa J, Edwards DJ, Hosseini MR (2022) Assessing the level of awareness of smart building technologies (SBTs) in the developing countries. J Eng, Des Technol 20(3):696–711
27. Grynning S, Gradeci K, Gaarder JE, Time B, Lohne J, Kvande T (2020) Climate adaptation in maintenance operation and management of buildings. Buildings 10(6):107
28. Gunay HB, O'Brien W, Beausoleil-Morrison I (2013) A critical review of observation studies, modeling, and simulation of adaptive occupant behaviors in offices. Build Environ 70:31–47
29. Gunay HB, O'Brien W, Beausoleil-Morrison I (2015) Development of an occupancy learning algorithm for terminal heating and cooling units. Build Environ 93:71–85
30. Hannan MA, Faisal M, Ker PJ, Mun LH, Parvin K, Mahlia TMI, Blaabjerg F (2018) A review of internet of energy based building energy management systems: Issues and recommendations. IEEE Access 6:38997–39014
31. Harputlugil T, de Wilde P (2021) The interaction between humans and buildings for energy efficiency: A critical review. Energy Res Soc Sci 71:101828
32. Hoes P, Hensen JL, Loomans MG, de Vries B, Bourgeois D (2009) User behavior in whole building simulation. Energy Build 41(3):295–302
33. Humphreys MA, Fergus Nicol J (2018) Principles of adaptive thermal comfort. In: Sustainable houses and living in the hot-humid climates of Asia. Springer Singapore, Singapore, pp 103–113
34. Indraganti M, Rao KD (2010) Effect of age, gender, economic group and tenure on thermal comfort: A field study in residential buildings in hot and dry climate with seasonal variations. Energy Build 42(3):273–281
35. Jazizadeh F, Ghahramani A, Becerik-Gerber B, Kichkaylo T, Orosz M (2014) Human-building interaction framework for personalized thermal comfort-driven systems in office buildings. J Comput Civ Eng 28(1):2–16
36. Jia M, Srinivasan RS, Raheem AA (2017) From occupancy to occupant behavior: An analytical survey of data acquisition technologies, modeling methodologies and simulation coupling mechanisms for building energy efficiency. Renew Sustain Energy Rev 68:525–540
37. Jin Y, Yan D, Chong A, Dong B, An J (2021) Building occupancy forecasting: A systematical and critical review. Energy Build 251:111345
38. Karima M, Altan H (2017) Interactive building environments: A case study university building in UAE. Procedia Eng 180:1355–1362
39. Karjalainen S (2007) Gender differences in thermal comfort and use of thermostats in everyday thermal environments. Build Environ 42(4):1594–1603
40. Khafiso T, Aigbavboa C, Adekunle SA (2024) Barriers to the adoption of energy management systems in residential buildings. Facilities 42(15/16):107–125
41. Kivimaa P, Kangas HL, Lazarevic D, Lukkarinen J, Åkerman M, Halonen M, Nieminen M (2019) Transition towards zero energy buildings
42. Klass AB, Wilson EJ (2014) Energy consumption data: They key to improved energy efficiency. San Diego J Climate & Energy L 6:69
43. Leal Filho W, Gatto A, Sharifi A, Salvia AL, Guevara Z, Awoniyi S, Mang-Benza C, Nwedu CN, Surroop D, Teddy KO, Muhammad U (2024) Energy poverty in African countries: An assessment of trends and policies. Energy Res Soc Sci 117:103664
44. Lee S, Karava P, Tzempelikos A, Bilionis I (2020) A smart and less intrusive feedback request algorithm towards human-centered HVAC operation. Build Environ 184:107190
45. Lo LJ, Novoselac A (2010) Localized air-conditioning with occupancy control in an open office. Energy Build 42(7):1120–1128
46. Lowry G (2002) Factors affecting the success of building management system installations. Build Serv Eng Res Technol 23(1):57–66

47. Malatji EM (2017) The development of a smart campus-African universities point of view. In: 2017 8th International Renewable Energy Congress (IREC). IEEE, pp 1–5
48. Miatto A, Dawson D, Nguyen PD, Kanaoka KS, Tanikawa H (2021) The urbanisation-environment conflict: Insights from material stock and productivity of transport infrastructure in Hanoi, Vietnam. J Environ Manag 294:113007
49. Morshed ASM, Manjur KA (2024) Optimizing energy efficiency: A comprehensive analysis of building design parameters. Available at SSRN 5079816
50. Mosner-Ansong KF, Duah D, Amos-Abanyie S, Baffoe-Ashun AI (2024) Energy efficiency awareness and assertiveness in Sub-Saharan Africa households: The case of Ghana. Energy Sustain Dev 79:101415
51. Nguyen TA, Aiello M (2013) Energy intelligent buildings based on user activity: A survey. Energy Build 56:244–257
52. Niesing G, Lotriet R, Grobler LJ (2012) Integrating change management for/towards sustainable energy efficient management—a case study. In: 2012 Proceedings of the 9th industrial and commercial use of energy conference. IEEE, pp 1–5
53. Nurminen M, Lindstedt A, Saari M, Rantanen P (2021) The requirements and challenges of visualizing building data. In: 2021 44th international convention on information, communication and electronic technology (MIPRO). IEEE, pp 968–972
54. Oguntona OA, Ngobeni F, Akinradewo O, Ogunbayo B, Aigbavboa C (2023) Barriers to emerging smart solutions adoption for energy efficiency in the construction industry. In: Proceedings the 11th world construction symposium, p 270
55. Ouedraogo NS (2019) Opportunities, barriers and issues with renewable energy development in Africa: a comprehensible review. Curr Sustain/Renew Energy Rep 6(2):52–60
56. Page J, Robinson D, Morel N, Scartezzini JL (2008) A generalised stochastic model for the simulation of occupant presence. Energy Build 40(2):83–98
57. Pang Z, O'Neill Z, Chen Y, Zhang J, Cheng H, Dong B (2023) Adopting occupancy-based HVAC controls in commercial building energy codes: Analysis of cost-effectiveness and decarbonization potential. Appl Energy 349:121594
58. Panteli M, Crossley PA, Kirschen DS, Sobajic DJ (2013) Assessing the impact of insufficient situation awareness on power system operation. IEEE Trans Power Syst 28(3):2967–2977
59. Parsons KC (2002) The effects of gender, acclimation state, the opportunity to adjust clothing and physical disability on requirements for thermal comfort. Energy Build 34(6):593–599
60. Parvin K, Lipu MH, Hannan MA, Abdullah MA, Jern KP, Begum RA, Mansur M, Muttaqi KM, Mahlia TI, Dong ZY (2021) Intelligent controllers and optimization algorithms for building energy management towards achieving sustainable development: challenges and prospects. IEEE Access 9:41577–41602
61. Peng Y, Rysanek A, Nagy Z, Schlüter A (2018) Using machine learning techniques for occupancy-prediction-based cooling control in office buildings. Appl Energy 211:1343–1358
62. Pinto G, Wang Z, Roy A, Hong T, Capozzoli A (2022) Transfer learning for smart buildings: A critical review of algorithms, applications, and future perspectives. Adv Appl Energy 5:100084
63. Preiser WF (1995) Post-occupancy evaluation: how to make buildings work better. Facilities 13(11):19–28
64. Preiser WF, White E, Rabinowitz H (2015) Post-occupancy evaluation (Routledge revivals). Routledge
65. Roberts CJ, Edwards DJ, Hosseini MR, Mateo-Garcia M, Owusu-Manu DG (2019) Post-occupancy evaluation: a review of literature. Eng Constr Archit Manag 26(9):2084–2106
66. Schakib-Ekbatan K, Cakıcı FZ, Schweiker M, Wagner A (2015) Does the occupant behavior match the energy concept of the building? Analysis of a German naturally ventilated office building. Build Environ 84:142–150
67. Schweiker M, Wagner A (2016) The effect of occupancy on perceived control, neutral temperature, and behavioral patterns. Energy and Buildings 117:246–259
68. Selvam AP, Al-Humairi SNS (2023) The impact of IoT and sensor integration on real-time weather monitoring systems: A systematic review

69. Shaikh PH, Nor NBM, Nallagownden P, Elamvazuthi I, Ibrahim T (2014) A review on optimized control systems for building energy and comfort management of smart sustainable buildings. Renew Sustain Energy Rev 34:409–429
70. Soleimanijavid A, Konstantzos I, Liu X (2024) Challenges and opportunities of occupant-centric building controls in real-world implementation: A critical review. Energy and Buildings 308:113958
71. Soonsawad N, Martinez RM, Schandl H (2022) Material demand, and environmental and climate implications of Australia's building stock: Current status and outlook to 2060. Resour Conserv Recycl 180:106143
72. Taesler R (1991) Climate and building energy management. Energy and Buildings 16(1–2):599–608
73. Vahidinasab V, Ardalan C, Mohammadi-Ivatloo B, Giaouris D, Walker SL (2021) Active building as an energy system: Concept, challenges, and outlook. IEEE Access 9:58009–58024
74. van Roosmale S, Audenaert A, Meysman J (2024) Understanding the opportunities and challenges of building automation and control systems to support facility management–an extensive literature review. Facilities 42(7/8):677–693
75. Verbeke S, Laffont K, Rua D (2022) Interoperability as a driver or barrier of smart building technologies? In CLIMA 2022 conference
76. Veselý M, Zeiler W (2014) Personalized conditioning and its impact on thermal comfort and energy performance–a review. Renew Sustain Energy Rev 34:401–408
77. Williams J, Lellouch B, Stein S, Vanderwel C, Gauthier S (2022) Low-carbon comfort management for smart buildings. In: 2022 IEEE International Smart Cities Conference (ISC2), pp. 1–5. IEEE
78. Yang T, Bandyopadhyay A, O'Neill Z, Wen J, Dong B (2022) From occupants to occupants: A review of the occupant information understanding for building HVAC occupant-centric control. In: Building simulation, vol 15, no 6, pp 913–932. Tsinghua University Press, Beijing
79. Yoshino H, Hong T, Nord N (2017) IEA EBC annex 53: Total energy use in buildings—analysis and evaluation methods. Energy and Buildings 152:124–136
80. Yu L, Qin S, Zhang M, Shen C, Jiang T, Guan X (2021) A review of deep reinforcement learning for smart building energy management. IEEE Internet Things J 8(15):12046–12063
81. Zou P, Alam M, Phung VM, Wagle D, Stewart R, Bertone E, Sahin O, Buntine C (2017) Achieving energy efficiency in government buildings through mandatory policy and program enforcement
82. Bemelmans-Videc ML, Rist RC, Vedung EO (eds) (2011) Carrots, sticks, and sermons: policy instruments and their evaluation, Vol 1. Transaction Publishers
83. Pomponi F, D'amico, B (2020) Low energy architecture and low carbon cities: exploring links, scales, and environmental impacts. Sustainability 12(21):9189
84. Kong M, Dong B, Zhang R, O'Neill Z (2022) HVAC energy savings, thermal comfort and air quality for occupant-centric control through a side-by-side experimental study. Appl Energy 306:117987
85. Janda KB (2011) Buildings don't use energy: people do. Archit Sci Rev 54(1):15–22
86. Sovacool BK, Griffiths S (2020) The cultural barriers to a low-carbon future: a review of six mobility and energy transitions across 28 countries. Renew Sustain Energy Rev 119:109569
87. Dabaieh M, Johansson E (2018) Building performance and post occupancy evaluation for an off-grid low carbon and solar PV plus-energy powered building: a case from the Western Desert in Egypt. J Build Eng 18:418–428

Chapter 7
Local Sustainable Materials and Circularity

7.1 Introduction

The activities of the construction industry over the years have amounted to grave challenges, including socio-economic and environmental degradation. They contribute enormously to global warming especially, with up to 39% carbon emissions [1–3]. Many developing countries are undergoing rapid urban development and economic growth. This has led to the continual rise in carbon emissions in the environment [4]. Despite this trend, reductions in carbon emissions from the construction sector remain a challenge [5]. A survey by the World Green Building Council (WGBC 2019) has shown that 93% of Built Environment Professionals (BEPs) believe sustainability should be a core part of their work. However, this high level of recognition of the significance of sustainability is not reflected in construction practices. One of the major causes of global warming in construction is the heavy reliance of construction on conventional construction materials, leading to high greenhouse gas (GHG) emissions [6, 7]. A significant portion of these emissions can be attributed to the materials used in construction [5]. The industry is beset with high levels of waste generation, inefficient use of resources and carbon emissions [67]. Promoting sustainability goals requires intentional material sourcing, prioritizing locally available sustainable materials which have minimal energy demands for production and transportation. Studies have shown that GHG emissions can be reduced by replacing traditional construction materials with more sustainable, low-carbon materials. These include the use of waste and local sustainable materials, which aligns with the principle of circularity. However, despite growing awareness of the importance of decarbonizing the construction industry in developing countries, little attention is given to these local sustainable materials and circular economy practices. Extant studies show that employing local construction materials is one of the effective low-carbon emission building techniques to reduce the carbon footprint. The construction sector is currently facing significant challenges regarding environmental sustainability and waste reduction. There's mounting urgency to not just minimize waste, but to rethink

M. N. Addy et al., *Decarbonization of Building Energy in Developing Countries*, Green Energy and Technology, https://doi.org/10.1007/978-3-032-09113-0_7

how design, build, and demolish is done [68]. The industry can make real progress toward reducing its environmental footprint and fostering more sustainable built environments [72]. Recently, there has been more attention given by researchers toward the development, marketing, and use of renewable materials and other local resources that have a low-carbon footprint. Advocates have promoted the adoption of integrating recycled materials with renewable sources of energy in order to mitigate carbon emissions from construction activities [71]. Other studies indicate that the employment of indigenous construction materials increases environmental sustainability and promotes the sustainability of the local construction industry over time [5]. Awareness of circular economy practices within the construction industry exists, though it remains somewhat limited [8].

7.2 Overview of Local Sustainable Materials and Circularity

7.2.1 Sustainable Materials

Materials used in production impact the environment negatively in several ways, from their extraction to disposal, including waste generation, landscape degradation, decreased water quality, and ecosystem pollution [9–11]. The excess energy consumption throughout this cycle cannot be ignored. One cause of some of these adversities is direct emissions from chemical processes during their manufacturing process. For instance, cement production accounts for emissions of CO_2 due to calcination of limestone into lime during the process, resulting in about 50% of emissions during cement production, with the remaining percentage resulting from fuel and electricity consumption [9]. During the usage phase, the adverse effects posed to the environment are a result of fuel and electricity consumption. Also, when materials are disposed of, resources are lost. In cases where materials are incinerated, greenhouse gases are emitted into the atmosphere. In the case of polymers, improper disposal may lead to enduring microparticles that survive and accumulate in the food chain [9].

Local sustainable materials are those that are renewable, low-carbon-emitting, energy-efficient, and locally sourced [12]. They are found in the environment as raw materials. These materials are important to the circular economy. The use of sustainable materials is a general concern across industries, including fashion, packaging, automotive, manufacturing, agriculture, and construction. Within the context of sustainable building strategies, the concept of localizing materials stands out as a significant approach, extending its influence from the construction phase through to ongoing operations. The construction industry continues to depend heavily on industrial cement, concrete, and sprawling international supply chains, largely to drive down labour costs. Yet, localizing materials and practices presents a compelling alternative. By drawing on regional resources and traditional know-how, such strategies can reduce environmental harm and confront the social challenges that mainstream, cost-focused approaches often ignore [13, 14]. The construction industry accounts for the consumption of roughly 60% of global natural resources each year, including materials like sand, gravel, and water, in order to meet the ever-growing demand for cement and steel [15]. Notably, about 70% of this resource usage is attributed to cement production, while steel accounts for the remaining 30%. This underscores the substantial impact the sector has on natural resource depletion worldwide [16, 17]. Efforts to promote sustainable building practices now focus strongly on introducing alternative materials that can either replace industrial cement or enhance the performance of building components [15].

7.2.2 Circularity

Kazmi and Chakraborty [8] define circularity as a system that "keeps resources in the loop for the longest possible, eliminating waste from the system". It basically deals with the continual use of indigenous materials for the longest possible time while eliminating waste. Benachio et al. [18] describe the circular economy as a restorative process. Unlike the linear economy's wasteful approach, the circular economy promotes a closed-loop system to keep resources in use and reduce waste generation [19]. It is aimed at transforming the existing "take-make-dispose" production and consumption pattern [70]. They contend that it aims to close material loops by treating what would traditionally be waste as a resource to be retained. For instance, the construction and demolition waste, often viewed as disposable, contains numerous building elements with varying lifespans that could be repurposed or reused if managed responsibly. Reprocessing of materials includes recycling of steel scrap and iron ore at the end of their life cycle, which demonstrates 73% savings in carbon emission [20], and production of clinker from industrial waste, with a 38.62% reduction in environmental impact [21]. The construction industry, however, has historically operated on a linear model, that is, take, make, consume, and dispose, leading to the generation of nearly 35% of all global residues [15, 18]. To transition from linearity to circularity, the well-established 3R framework of Reduce, Reuse, Recycle has evolved into a more nuanced 10R model—Refuse, Rethink, Reduce, Reuse, Repair, Refurbish, Remanufacture, Repurpose, Recycle, and Recover [69]. This hierarchy promotes multiple levels of circular strategy aimed at extending material use and reducing waste throughout a building's lifecycle. However, despite increasing academic and policy interest over the past decade, implementation remains limited. Benachio et al. [18], Munaro et al. [22] and Anastasiades et al. [68] highlight the lack of industry-wide tools, business models, and practical know-how. The construction sector's adoption of circular principles remains constrained by regulatory gaps, insufficient financial incentives, and lack of stakeholder engagement. Quantitative measures indicate that the circularity of the global economy stands at approximately 6%, a clear sign that circularity is mainly a concept, not yet mainstream practice.

Construction, as one of the most resource-intensive sectors, plays a significant role in this low percentage.

Shifting from a linear to a circular construction model is increasingly recognized as crucial for achieving sustainability within the industry [23, 24]. Circularity principles have become central in construction practices, directly supporting the realization of Sustainable Development Goals (SDGs) [25]. The European Union, for instance, actively promotes the adoption of reuse and recycling strategies to improve the circularity of building materials and manage construction waste more efficiently [26]. Methods such as repairing, reclaiming, and repurposing materials are designed to optimize resource efficiency by preserving product value, while also minimizing waste generation, energy consumption, and the need for new raw materials [27, 66]. Reusing building components up to three times can result in around 60% saving [65]. Effective reuse of building materials hinges on strategies like selective disassembly and selective demolition. Selective disassembly involves carefully reclaiming and maintaining components throughout the building's life, so they can be reused rather than discarded [28]. Selective demolition, meanwhile, focuses on sorting construction waste directly on site, making it easier to recover valuable materials [29]. Recent research highlights that circular economy practices offer notable economic and environmental advantages, particularly by reducing resource depletion and minimizing pollution [30].

Olivetti and Cullen [9] proposed five ways a sustainable materials system can be improved, hence, circularity (Table 7.1).

Production of construction materials consumes more than 80% of the total energy used in building construction (Zhang et al. 2023). Sustainable materials play a pivotal role within the construction industry, primarily aiming to mitigate natural resource depletion and minimize waste generation. The construction sector has been known as

Table 7.1 Five ways for improving sustainable material system

	Strategies	Description
1	Resource optimization	Reducing material usage through smarter designs, lightweight structures, and finding ways to achieve the same function with fewer resources
2	Lifetime extension	Designing products for durability, repairability, and modularity to ensure they remain functional for a longer period. This will help reduce the demand for conventional materials for the production of physical infrastructure [16]
3	Manufacturing efficiency	Streamlining production processes to minimize waste, energy consumption, and emissions, ensuring that resources are used as efficiently as possible
4	Substitution	Replacing traditional materials with more sustainable, renewable, or lower-impact alternatives to reduce the environmental footprint
5	Recovery	Implementing recycling, reuse, and circular economy practices to reclaim materials at the end of their lifecycle, reducing reliance on virgin resources

Source Olivetti and Cullen [9]

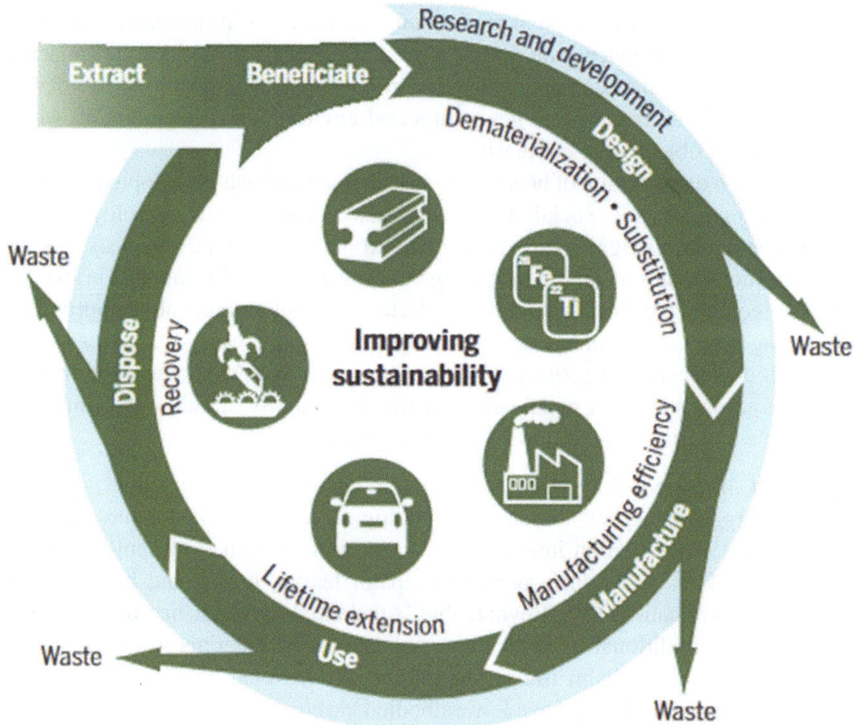

Fig. 7.1 Concept of sustainable materials system by Olivetti and Cullen [9]

one of the three sectors with high potential to implement circular economy strategies. The adoption of circular economy principles in the construction industry promotes the use of sustainable materials, maximizing material recovery, and avoiding unnecessary waste generation and waste disposed of to landfill. In Africa, a growing wave of research highlights how using locally–sourced sustainable materials can reshape the construction sector, promoting lower embodied carbon, cutting costs, and revitalizing local economies through techniques rooted in earth, agriculture, and traditional knowledge (Fig. 7.1).

7.3 Local Sustainable Materials

The construction industry greatly relies on conventional construction materials for projects, resulting in a high carbon footprint. In the context of developing countries, carbon footprint continues to increase due to rapid urban development. While in abundance, indigenous and waste materials are often disregarded and unutilised. Local sustainable materials are more advantageous, having less environmental impact, than the conventional materials used in production. The use of local sustainable materials can minimize the dependence on energy-intensive materials and reduce the impact on the ecosystem. Local sustainable materials are mostly biodegradable or recyclable, which will help mitigate waste generation and environmental degradation, which aligns with the objective of circularity.

A thorough examination of how local construction materials are adopted is crucial for advancing economic, social, and environmental development. Employing local sustainable materials significantly reduces environmental impact, enhances recyclability, and contributes to a more insulating building envelope. Sustainable bricks, for instance, constructed from sustainable materials, present an innovative approach to addressing both waste management challenges and the increasing expenses associated with construction supplies [31]. A variety of indigenous materials have been identified across Africa for sustainable construction initiatives. Below are some local sustainable materials common in some developing countries.

Bio-Based Materials: Bio-based materials include agricultural residues, animals, and microorganisms. They are renewable biological resources. A notable milestone occurred in 1986, when an international symposium convened to explore the use of plant-based materials such as papyrus, palm leaves, husks, and bamboo reed in building applications [6]. Beyond that, substantial research has been devoted to substituting traditional reinforcement and aggregate materials in concrete with natural fibres such as Palm fibres, Diss fibres, Alfa, Kenaf and Cork. Bio-based materials are renewable and have low embodied carbon.

Rammed Earth: Rammed earth construction involves the compaction and compression of locally sourced soil to form a material for building. The rammed earth has commonly been used as building material for load-bearing posts. This technique has a long-standing history, with notable examples found in ancient sites such as Jericho, Nemrik, and Aswad in the Middle East, where partially buried structures were constructed using this method. Although rammed earth construction has ancient roots, its broader adoption in Africa did not occur until between 264 and 241 BC. This construction technique dates back to antiquity. Its adoption in Africa, however, started between 264 and 241 BC. The use of rammed earth and adobe became the most used techniques in several villages. The structures had good resilience against the harsh climatic conditions of the time. Since the 1960s, in the Middle East and North Africa, rammed earth has been used to build several structures. For instance, rammed earth was used for the walls of the Luxor House Construction in Egypt. Earth-based structures have been known to have good thermal performance and insulation.

Gypsum: According to Sonebi et al. [6], Gypsum is one of the greatly mined materials worldwide due to their unique properties such as quick setting, smooth surface, fire resistance and recyclability, and vast availability in the environment. Tunisia is one of the countries with high levels of gypsum production. Morocco has 10 million national reserves of gypsum. Gypsum was common in the past for the use as construction material. In Egypt, it was commonly used for vessels and plaster. Algeria was well-known for using gypsum for quick setting and as binder in cementitious mixtures. Gypsum was used to produce gypsum mortar for the external walls of a floor building in a research laboratory of the ministry of housing in Algeria (Centre National d'Études et de Recherches Intégrées du Bâtiment, CNERIB), in 1995, for experimental purposes. The walls of this building have, reportedly, had no major defects. In 1988, an experimental house was constructed in Rabat, Morocco, using locally sourced gypsum, with the aim of exploring affordable housing options. The mix design for the gypsum mortar, referred to as "béton banché" in French, was developed by the rsearcher Sonebi at the Laboratoire Public des Essais et Etudes (LPEE) in Casablanca. He also oversaw quality control for the materials used in both the walls and roof. Notably, this villa has remained intact for over thirty years, from its completion in 1989 through to 2022, demonstrating the durability and potential of gypsum-based construction in this context.

Pulverized Fuel Ash or Fly Ash: Fly ash is an industrial waste obtained when pulverized coal is burnt in thermal plants. Singh et al. (2024) demonstrated in their research that fly ash used as a concrete material exhibits a reduction in carbon emissions. Sizirici et al. [32] also revealed that fly ash, when used in concrete instead of cement, effectively improves the physical and mechanical properties—compressive strength, durability and lightweight—of concrete with reduced carbon emissions. Cheah et al. [33] combined fly ash with high calcium wood ash to produce blockwork. Their research showed that the hybrid of the two materials produced a block work with the optimal flexural and compressive strength.

Recycled Aggregates (RA): RA consists of crushed and reprocessed concrete structural elements that have been previously used. RA may contain materials such as bricks, metals, tiles, and miscellaneous materials like wood, plastic, and glass, which were crushed together with concrete [34]. The quality of concrete produced with RA in terms of its durability and mechanical properties is inferior to concrete produced with natural aggregates. However, concrete produced with RA with the addition of industrial waste such as fly ash or blast slag produces an optimal quality of concrete while mitigating the carbon footprint [32]. Construction projects continue to rise rapidly in developing countries due to urbanization. The Construction and Demolition Waste (CDW) generated from these projects are often dumped improperly in inappropriate places, posing environmental challenges. Alternatively, these generated wastes can be reprocessed and applied as recycled aggregates in place of natural aggregates. A study by Wagih et al. [35] demonstrates that, reprocessing of concrete rubble into RA could be useful for structural concrete applications. Hassanieh et al. [36] also demonstrated that RA in asphalt mixes can replace 30% of fine aggregate in hot mix asphalt. More so, it was demonstrated by a study in Saudi Arabia that RA can substitute natural aggregates in precast concrete manufacturing, construction of manholes, and non-loading bearing walls. Raini et al. [37] also contend that up to 15% of sand can be substituted by CDW while retaining the mechanical properties of the mortar. It is seen that CDW can be recycled and reused in construction, which will reduce the generation of waste as amounting from the construction industry. However, utilization of the CDW remains largely underexplored in many developing countries.

7.4 Importance of Local Sustainable Materials and Circularity

The utilization of local sustainable materials and the principles of circularity have become increasingly central to discussions surrounding environmental sustainability and socio-economic resilience. Their application in fields such as construction, manufacturing, and urban planning represents a pragmatic approach to reducing reliance on external resource flows. As climate action and sustainable development gain urgency, the integration of locally sourced materials and circular systems is widely recognized as a critical strategy for enhancing both environmental outcomes and economic stability at the community level. Below are some advantages of their adoption.

7.4.1 Resource Conservation and Optimization

Embracing a circular economy fosters efficient resource management in the primary sector. It promotes practices like sustainable farming, responsible forestry, and efficient mining methods, cutting waste and protecting natural resources. Circular economy principles help the secondary sector optimize the use of materials and energy. This involves strategies such as remanufacturing, recycling, and adopting closed-loop production systems, which reduce dependence on virgin resources and minimize waste generation. In addition, the "Product as a Service" model shifts the focus from selling physical goods to providing ongoing services. For instance, rather than simply selling lighting equipment, companies may offer lighting as a service, thereby retaining responsibility for maintenance and upgrades. This structure encourages companies to prioritize product longevity and supports circular economic principles by extending product lifespans and minimizing resource consumption.

7.4.2 Waste Reduction

Reductions in waste output are achievable in the primary sector (which deals mainly with extraction and collection of raw materials) through the adoption of circular practices [19]. Such methods as repurposing agricultural by-products, reusing irrigation water and implementing sustainable land management are effectual to promoting avenues for minimizing resource depletion in developing countries. In Spain and the Czech Republic, agricultural waste such as olive residues and wheat straw have successfully been transformed to biofuels [38]. The secondary sector often generates considerable amounts of waste such as production scraps, surplus goods, and excess packaging. The framework of a circular economy encourages a reprocessing of these materials. Rather than viewing them merely as refuse, this perspective treats such by-products as potential assets within the production cycle. Waste accumulation in developing countries not only leads to resource depletion but continues to render the environment degraded and unattractive. Reprocessing waste to make them reusable is a great way of mitigating environmental pollution while reducing loss of resources.

7.4.3 Economic Opportunities

Circular economy practices present opportunities for innovation and employment growth, particularly within sectors such as recycling, remanufacturing, and emerging service-oriented models [63]. When focus is shifted toward optimizing material usage and minimizing waste in the developing countries, organizations in secondary sectors can achieve notable cost reductions, both in terms of raw material procurement and waste disposal expenses. Adopting circular strategies, like product-as-a-service or leasing arrangements, not only encourages extended product lifespans but also supports greater profitability through sustainable solutions. This approach catalyses advancements in product design and manufacturing, emphasizing principles such as modularity, design for disassembly, and enhanced recyclability [39]. Standardization of components facilitates efficient material recovery, while advanced technologies, such as additive manufacturing (3D printing), enable the production of complex items with minimal waste.

7.4.4 Reduced Environmental Impact

The circular economy plays a significant role in mitigating climate change and enhancing environmental protection through the reduction of waste and pollution, promotion of the use of renewable energy, and reduction of resource consumption. Within the construction sector, circular economy principles directly address a range of environmental issues by implementing energy-efficient practices, integrating cleaner technologies, and prioritizing recycling. Local sustainable materials like bio-based materials and recycled aggregates have low embodied carbon. Cross-Laminated Timber (CLT) which is a bio-based material, as highlighted by Shin et al. [40], for instance, is a good substitute for concrete and steel. In their study, CLT was applied to the exterior walls of 52 apartment building types. It was discovered that there was 44.6% savings in greenhouse gas emissions and 49.3% in energy demand. In the research by Singh et al. (2024), fly ash used in concrete production exhibited lower carbon footprint with a reduction of 54%. Fly ash is also used to enhance the physical and mechanical properties—strength, durability, and lightweight—of concrete while minimizing its environmental impact [32]. These initiatives play a significant role in promoting environmental sustainability in developing countries and align with global objectives aimed at environmental protection.

7.4.5 *Resilience and Security*

Circularity promotes resilience against scarce resources, fragile supply chain, and energy shock [41, 42]. A circular economy fundamentally reduces reliance on limited resources, thereby enhancing economic resilience in the face of price fluctuations and disruptions to supply chains [41]. It also enhances supply chain by looping products back into production through reuse, recycling and repair. This approach marks a transition toward a more sustainable and regenerative economic framework, one designed to generate beneficial outcomes for the environment, society, and the broader economy. Adopting circularity not only mitigates resource scarcity but also supports long-term stability and societal well-being.

7.5 Impediments to the Use of Local Sustainable Materials and Circularity

The concept of the circular economy represents a significant departure from the conventional linear economic model, emphasizing strategies such as reuse, remanufacturing, repair, and recycling of materials and products. While the theoretical benefits of CE are widely recognized, its practical implementation within the building sector encounters persistent challenges. These include entrenched industry practices, regulatory complexities, and limited stakeholder engagement, all of which hinder widespread adoption.

7.5.1 *Lack of Policy and Regulation*

The absence of effective policy and regulatory frameworks hinders the adoption of local sustainable materials and circular economy practices in developing countries. There is noticeable hesitation to adopt circular economy practices in the construction industry. The absence of dedicated legislation, end-of-waste definitions, and missing technical standards leaves stakeholders in a sort of regulatory limbo. Without these foundational elements, investing in or utilizing reused and recycled materials feels risky and uncertain [43]. It is difficult to expect widespread commitment to circularity when regulatory frameworks are absent. Empirical evidence from Nigeria lines up with these observations. Akindele et al. [44] highlight insufficient enforcement, institutional corruption, and a lack of incentives as primary obstacles to sustainable construction. Similarly, research by Oyewobi et al. [45] suggest that, despite existing procurement and environmental legislation, the absence of targeted policy mechanisms, such as public procurement requirements, tax incentives, or material certification, undermines the competitiveness of sustainable materials and hampers the progression of circular market initiatives.

7.5.2 *Influx of Foreign Materials*

The construction sector in developing countries continue to adhere to conventional methods. As a result, a lot of foreign materials including cement clinker, steel, and glass products continue to be imported. This impedes the integration of local sustainable materials into construction practice. More so, because some of the imported materials are relatively cheaper and more readily available than local sustainable materials, import of foreign materials undermines the adoption of local materials. The sustained importation of construction materials has had the corollary effect of reshaping market preferences, gradually displacing the demand for locally sourced alternatives. This shift has fostered a growing dependency on foreign products, creating both a structural and perceptual bias in favour of imported goods. The absence of robust quality assurance systems, combined with limited domestic production capacity, has further reinforced this reliance. Within the context of globalization, these factors collectively entrench the dominance of foreign construction materials across the sub-region.

7.5.3 *Lack of Capacity and Expertise*

In many developing countries, persistent gaps in technical expertise, infrastructure, institutional support, and financial resources continue to limit the adoption of locally sourced sustainable materials and circular practices. Taking Nigeria as an example, while promising alternatives such as rice husk ash and coconut-fibre composites have been identified through research, their broader application remains restricted. This is primarily due to insufficient research funding, a lack of industrial-scale production facilities, and limited access to markets. These systemic challenges significantly impede progress toward more sustainable construction practices. In Ghana, as in other developing countries of Africa, the construction industry faces significant challenges in adopting circular economy practices. New and emerging digital tools like BIM and IoT are hardly in use, and the regulatory environment is either underdeveloped or inconsistent [46]. As a result, efforts toward circularity in developing countries remain slow and fragmented, leaving the sector trailing behind global advancements.

7.5.4 *Lack of Metric Tools*

It is quite problematic to evaluate the circularity of buildings without adequate analytical tools. The current lack of standardized indicators or performance metrics means there's no consistent way to assess the environmental, social, or economic impacts of construction projects. For deconstruction to gain real traction as a common practice, there is a pressing need to develop effective methodologies and instruments that ensure the safe and economically viable removal of structural components. In essence, these tools are foundational and without them, the dismantling process becomes inefficient and potentially unsafe, undermining both sustainability goals and practical feasibility. Benites et al. [47] argue that widely recognized frameworks such as LEED, Green Star, and BREEAM fall short when it comes to genuinely advancing circular and regenerative approaches within the built environment. In their view, these tools simply do not provide sufficient support for meaningful progress in these areas.

7.5.5 *Lack of Awareness*

A significant barrier frequently mentioned is the widespread lack of understanding regarding circular economy concepts, underlying principles, and their potential advantages among both stakeholders and policymakers. Shooshtarian et al. [64] argued that the level of awareness among stakeholders plays a pivotal role, in that, the more informed they are, the more inclined they are to embrace circular practices. However, knowledge of circularity remains limited among clients, designers, and subcontractors [48]. Without a solid grasp of these fundamentals, meaningful engagement or support is understandably limited. With the level of awareness in developing countries, promoting awareness on local sustainable materials and circularity is crucial to encourage stakeholders to engage in circular practices.

7.6 Pathways to Promoting Local Sustainable Materials and Circularity

Advancing local sustainable materials and fostering circularity is not a one-dimensional task. It calls for a comprehensive strategy which includes combining policy development, educational outreach, innovative solutions, adequate financial support, and active community involvement. Together, these elements create the framework necessary to shift away from linear, resource-heavy models and toward circular, regenerative systems that are both environmentally responsible and tailored to local realities. Some pathways necessary to promote local sustainable materials and circularity are discussed in the next section.

7.6.1 Enforcing Standards, Tax Incentives, and Procurement Mandates

The benefits of circularity are undeniable and cannot be overlooked. Installing policy frameworks is able to widely encourage behaviours that promote waste reduction, recycling and reuse of materials, and key practices that are geared toward making the circular economy a culture. A policy framework focused on the circular economy has the capacity to establish clear benchmarks across building design, construction, and ongoing operations. This includes mandating the use of recycled or reclaimed materials, enforcing robust energy efficiency standards, and implementing strategies to minimize waste, essentially weaving circular principles into the very fabric of industry practices. Scholars such as Akinade et al. [49] have observed that strict regulatory frameworks within construction tend to compel professionals to align with government-mandated targets. Bilal et al. [50] highlight the dual necessity of penalties for non-compliance and incentives for proactive adherence, emphasizing the value of comprehensive government support. When rigorous policies and consistent guidelines are set by authorities, a unified approach across the sector can be fostered [51]. This will create a regulatory environment that not only encourages but actively drives the modernization of engineering and construction practices in line with circular economy ideals.

7.6.2 Investment in Education, Training, and Circular Curricula

Research centred on circularity principles plays a pivotal role in shaping effective regulations and ensuring meaningful implementation within the building sector in developing countries. To strengthen the initiatives surrounding circularity in developing countries, it is imperative to engage in methods to train and educate stakeholders. Educational initiatives, whether through structured training programs, workshops, or accessible informational resources, help equip professionals with the practical skills and foundational knowledge necessary for circularity integration. New technologies and strategies can be developed through academic research for the adoption of local sustainable materials and circularity.

7.6.3 Enabling Recycling Facilities, Material-Tracking Systems

A marketplace dedicated to reused materials functions as a centralized platform where businesses can exchange second-hand products, components, and raw materials. In essence, it fosters the collection, recovery, and resale of building materials that might otherwise become waste. The establishment of such circular marketplaces can dramatically expand both the availability and demand for reused and recycled materials. This increased accessibility enables stakeholders to source building materials at lower costs, which, in turn, has the potential to significantly reduce overall construction expenses. Atta et al. [52] aver that material passports are digital records that document the composition, origin, and carbon footprint of materials. Incorporating material passports can further enhance the efficiency of such marketplaces. By enabling accurate tracking and assessment of material reuse potential, material passports support carbon-efficient decision-making throughout the building life cycle. Together, circular marketplaces and material passports not only promote sustainability but also deliver tangible economic and environmental benefits for the construction sector.

7.6.4 Raising Awareness and Engaging Communities

Undertaking awareness campaigns to advance circular practices can take numerous forms, including workshops, seminars, conferences, and outreach through both social and traditional media channels. Such campaigns serve as a critical foundation, fostering a clearer understanding of circularity principles and their practical advantages. This, in turn, can motivate stakeholders to consider and ultimately adopt circular strategies, delivering tangible benefits for both business operations and the broader environment [50, 53]. Nirmani [54] demonstrates that community participation through dialogue, workshops and stakeholder co-design is essential for encouraging circular practices. Also, in Indonesia, Ecobrick programs are held frequently, which raises environmental awareness and promotes circular practices [55–57]. Campaigns to promote the adoption of sustainable materials will generally encourage more building professionals to begin to practice circularity.

7.7 Summary

This chapter critically examines how the integration of local, sustainable materials and the adoption of circularity could transform the construction sector, particularly within developing nations. It underscores the construction industry's substantial environmental impact, responsible for nearly 40% of global carbon emissions, while highlighting the persistent reliance on conventional materials such as cement and steel, despite a broad consensus regarding the need for more sustainable approaches. The core argument centres on the environmental and socio-economic benefits of shifting toward renewable, low-carbon, and energy-efficient local materials. These alternatives, including bio-based composites, rammed earth, gypsum, fly ash, and recycled aggregates, not only reduce embodied carbon but also promote resource conservation and local economic development. The chapter points out that employing such materials does not merely benefit the environment; it also enhances community resilience and can offer cost advantages. A central concept explored is circularity, the continual reuse and repurposing of construction materials to minimize waste and close resource loops. This stands in contrast to the entrenched linear model of "take, make, dispose". Despite the potential benefits, the chapter identifies significant barriers to the adoption of sustainable and circular construction practices. These include inadequate policy frameworks, insufficient technical expertise, underdeveloped recycling infrastructure, a preference for inexpensive imported materials, lack of standardised metrics, and low stakeholder awareness. Particularly problematic is the absence of enforceable regulations and incentives, which stifles investment in more sustainable approaches. To address these challenges, the chapter proposes a multifaceted strategy which includes implementing clear policy mandates and incentives, investing in education and workforce training, developing robust recycling infrastructure and material-tracking systems, and fostering community engagement through awareness campaigns and participatory projects. These recommendations aim to facilitate a systemic shift from linear to circular models of construction that are contextually grounded and socially inclusive.

References

1. Chen L, Huang L, Hua J, Chen Z, Wei L, Osman AI, Fawzy S, Rooney DW, Dong L, Yap PS (2023) Green construction for low-carbon cities: a review. Environ Chem Lett 21(3):1627–1657
2. Pomponi F, Moncaster A (2016) Embodied carbon mitigation and reduction in the built environment–What does the evidence say? J Environ Manage 181:687–700
3. Bourbia S, Kazeoui H, Belarbi R (2023) A review on recent research on bio-based building materials and their applications. Mater Renew Sustain Energ 12(2):117–139
4. Jahanger A, Usman M, Balsalobre-Lorente D (2022) Autocracy, democracy, globalization, and environmental pollution in developing world: fresh evidence from STIRPAT model. J Public Aff 22(4):e2753
5. Mensah S, Ameyaw C, Yeboah WA (2024) Localizing urban buildings carbon emissions reduction: causes and strategies. Int J Build Pathol Adapt, (Ahead-Of-Print)
6. Sonebi M, Abdalqader A, Amziane S, Dvorkin L, Ghorbel E, Kenai S, Khatib J, Lushnikova N, Perrot A (2022) Trends and opportunities of using local sustainable building materials in the Middle East and North Africa region. RILEM Tech Lett 7:127–138
7. Cheah CB, Part WK, Ramli M (2015) The hybridizations of coal fly ash and wood ash for the fabrication of low alkalinity geopolymer load bearing block cured at ambient temperature. Constr Build Mater 88:41–55
8. Kazmi R, Chakraborty M (2023) Identification of parameters and indicators for implementing circularity in the construction industry. J Eng Appl Sci 70(1):77
9. Olivetti EA, Cullen JM (2018) Toward a sustainable materials system. Science 360(6396):1396–1398
10. Chen L, Yang M, Chen Z, Xie Z, Huang L, Osman AI, Farghali M, Sandanayake M, Liu E, Ahn YH, Al-Muhtaseb AAH (2024) Conversion of waste into sustainable construction materials: A review of recent developments and prospects. Mater Today Sustain:100930
11. Li J (2021) Municipal solid waste incineration ash-incorporated concrete: One step towards environmental justice. Buildings 11(11):495

12. Firoozi AA, Firoozi AA, Oyejobi DO, Avudaiappan S, Flores ES (2024) Emerging trends in sustainable building materials: Technological innovations, enhanced performance, and future directions. Results Eng:103521
13. Cheng B, Lu K, Li J, Chen H, Luo X, Shafique M (2022) Comprehensive assessment of embodied environmental impacts of buildings using normalized environmental impact factors. J Clean Prod 334:130083
14. Geissdoerfer M, Savaget P, Bocken NM, Hultink EJ (2017) The Circular Economy–A new sustainability paradigm? J Clean Prod 143:757–768
15. Kouvara A, Priavolou C, Ott D, Scherer P, van Zyl-Bulitta VH (2023) Circular, local, open: a recipe for sustainable building construction. Buildings 13(10):2493
16. Mohamad N, Muthusamy K, Embong R, Kusbiantoro A, Hashim MH (2022) Environmental impact of cement production and Solutions: A review. Mater Today: Proc 48:741–746
17. Llatas C, Bizcocho N, Soust-Verdaguer B, Montes MV, Quiñones R (2021) An LCA-based model for assessing prevention versus non-prevention of construction waste in buildings. Waste Manage 126:608–622
18. Benachio GLF, Freitas MDCD, Tavares SF (2020) Circular economy in the construction industry: A systematic literature review. J Clean Prod 260:121046
19. Aithal S, Aithal PS (2023) Importance of circular economy for resource optimization in various industry sectors–A review-based opportunity analysis. Int J Appl Eng Manag Lett (IJAEML) 7(2):191–215
20. Broadbent C (2016) Steel's recyclability: demonstrating the benefits of recycling steel to achieve a circular economy. Int J Life Cycle Assess 21:1658–1665
21. Ren C, Wang W, Mao Y, Yuan X, Song Z, Sun J, Zhao X (2017) Comparative life cycle assessment of sulfoaluminate clinker production derived from industrial solid wastes and conventional raw materials. J Clean Prod 167:1314–1324
22. Munaro MR, Tavares SF, Bragança L (2020) Towards circular and more sustainable buildings: A systematic literature review on the circular economy in the built environment. J Clean Prod 260:121134
23. Yang Y, Guan J, Nwaogu JM, Chan AP, Chi HL, Luk CW (2022) Attaining higher levels of circularity in construction: Scientometric review and cross-industry exploration. J Clean Prod 375:133934
24. Corona B, Shen L, Reike D, Carreón JR, Worrell E (2019) Towards sustainable development through the circular economy—a review and critical assessment on current circularity metrics. Resour Conserv Recycl 151:104498
25. Ghufran M, Khan KIA, Ullah F, Nasir AR, Al Alahmadi AA, Alzaed AN, Alwetaishi M (2022) Circular economy in the construction industry: A step towards sustainable development. Buildings 12(7):1004
26. Murray A, Skene K, Haynes K (2017) The circular economy: an interdisciplinary exploration of the concept and application in a global context. J Bus Ethics 140:369–380
27. Baker-Brown D (2024) The re-use atlas: A designer's guide towards a circular economy. Routledge
28. Andersen R, Ravn AS, Ryberg MW (2022) Environmental benefits of applying selective demolition to buildings: A case study of the reuse of façade steel cladding. Resour Conserv Recycl 184:106430
29. Morseletto P (2020) Targets for a circular economy. Resour Conserv Recycl 153:104553
30. Hammad AW, Akbarnezhad A, Wu P, Wang X, Haddad A (2019) Building information modelling-based framework to contrast conventional and modular construction methods through selected sustainability factors. J Clean Prod 228:1264–1281
31. Murali G, Azab M (2023) Recent research in utilization of phosphogypsum as building materials. J Market Res 25:960–987
32. Sizirici B, Fseha Y, Cho CS, Yildiz I, Byon YJ (2021) A review of carbon footprint reduction in construction industry, from design to operation. Materials 14(20):6094
33. Cheah CB, Part WK, Ramli M (2017) The long term engineering properties of cementless building block work containing large volume of wood ash and coal fly ash. Constr Build Mater 143:522–536

34. Kisku N, Joshi H, Ansari M, Panda SK, Nayak S, Dutta SC (2017) A critical review and assessment for usage of recycled aggregate as sustainable construction material. Constr Build Mater 131:721–740
35. Wagih AM, El-Karmoty HZ, Ebid M, Okba SH (2013) Recycled construction and demolition concrete waste as aggregate for structural concrete. HBRC Journal 9(3):193–200
36. Hassanieh DZA, Chehab GR, Srour I, Kassem HA (2016) Recycling cementitious constituents of construction demolition waste in asphalt mixes: the case of Lebanon. Int J Sustain Soc 8(2):109–125
37. Raini I, Jabrane R, Mesrar L, Akdim M (2020) Evaluation of mortar properties by combining concrete and brick wastes as fine aggregate. Case Stud Constr Mater 13:e00434
38. Duque-Acevedo M, Ulloa-Murillo LM, Belmonte-Ureña LJ, Camacho-Ferre F, Mercl F, Tlustoš P (2023) Sustainable and circular agro-environmental practices: A review of the management of agricultural waste biomass in Spain and the Czech Republic. Waste Manage Res 41(5):955–969
39. Tian G, Sheng H, Zhang L, Zhang H, Fathollahi-Fard AM, Zhang X, Feng Y (2024) Enhancing end-of-life product recyclability through modular design and social engineering optimiser. Int J Prod Res:1–19
40. Shin B, Chang SJ, Wi S, Kim S (2023) Estimation of energy demand and greenhouse gas emission reduction effect of cross-laminated timber (CLT) hybrid wall using life cycle assessment for urban residential planning. Renew Sustain Energy Rev 185:113604
41. Di Stefano C, Elia S, Garrone P, Piscitello L (2023) The Circular Economy as a new production paradigm to enhance resilience of MNEs and the economic system. AIB Insights 23(3):1–7
42. Tvaronavičienė M, Razminienė K, Adekola AF (2024) Transition towards a circular economy to build societal resilience to energy shocks. J Infrastruct, Policy Dev 8(15):1–32
43. Timm JFG, Maciel VG, Passuello A (2023) Towards sustainable construction: a systematic review of circular economy strategies and ecodesign in the built environment. Buildings 13(8):2059
44. Akindele OE, Ajayi S, Toriola-Coker L, Oyegoke AS, Alaka H, Zulu SL (2023) Sustainable construction practice in Nigeria: barriers and strategies for improvement. Built Environ Proj Asset Manag 13(4):590–609
45. Oyewobi LO, Jimoh RA (2022) Barriers to adoption of sustainable procurement in the Nigerian public construction sector. Sustainability 14(22):14832
46. Pittri H, Godawatte GAGR, Esangbedo OP, Antwi-Afari P, Bao Z (2025) Exploring barriers to the adoption of digital technologies for circular economy practices in the construction industry in developing countries: A case of Ghana. Buildings 15(7):1090
47. Sala Benites H, Osmond P, Prasad D (2023) Inquiry on perceptions and practices of built environment professionals regarding regenerative and circular approaches. Buildings 13(1):63
48. Adams KT, Osmani M, Thorpe T, Thornback J (2017) Circular economy in construction: current awareness, challenges and enablers. In: Proceedings of the institution of civil engineers-waste and resource management, vol 170, no 1, pp 15–24. Thomas Telford Ltd
49. Akinade O, Oyedele L, Oyedele A, Davila Delgado JM, Bilal M, Akanbi L, Ajayi A, Owolabi H (2020) Design for deconstruction using a circular economy approach: barriers and strategies for improvement. Prod Plan & Control 31(10):829–840
50. Bilal M, Khan KIA, Thaheem MJ, Nasir AR (2020) Current state and barriers to the circular economy in the building sector: Towards a mitigation framework. J Clean Prod 276:123250
51. Ansere B, Ayarkwa J, Addy MN, Osei-Asibey D, Acheampong A (2025) Environmental sustainability and procurement law compliance in Ghana's university goods procurement. J Public Procurement
52. Atta I, Bakhoum ES, Marzouk MM (2021) Digitizing material passport for sustainable construction projects using BIM. J Build Eng 43:103233
53. Cruz Rios F, Grau D, Bilec M (2021) Barriers and enablers to circular building design in the US: An empirical study. J Constr Eng Manag 147(10):04021117
54. Nirmani IAP (2024) Understanding the role of community participation in enhancing circular economy outcomes in developing regions

55. Suci AC, Syaefudin A, Nafisah DAM, Hanifah L, Mufrodah N, Nalim N (2024) Utilizing plastic waste as a source of creativity in ecobrick training in Purana Village. Dharmahita: J Community Serv Dev:110–116
56. Masluha S, Hendra J, Rahmansyah AI (2023) Ecobrick innovation for economic empowerment and sustainable plastic waste management. Int J Community Serv Implement, 1(3)
57. Agustang A, Oruh S, Agustang ADMP (2022) Building environmental awareness through the makassar eco-brick community social movement in plastic waste management. In: SHS web of conferences, vol 149, p 02005. EDP Sciences
58. Çimen Ö (2021) Construction and built environment in circular economy: A comprehensive literature review. J Clean Prod 305:127180
59. Crowther P (2018) Re-valuing construction materials and components through design for disassembly. In: Unmaking waste in production and consumption: Towards the circular economy, pp 309–321. Emerald Publishing Limited
60. Popescu C, Dissanayake H, Mansi E, Stancu A (2024) Eco Breakthroughs: Sustainable materials transforming the future of our planet. Sustainability 16(23):10790
61. Singh GB, Prasad VD (2024) Environmental impact of concrete containing high volume fly ash and ground granulated blast furnace slag. J Clean Prod 448:141729
62. Zhang Y, Yan D, Hu S, Guo S (2019) Modelling of energy consumption and carbon emission from the building construction sector in China, a process-based LCA approach. Energy Policy 134:110949
63. Horbach J, Rammer C (2020) Circular economy innovations, growth and employment at the firm level: Empirical evidence from Germany. J Ind Ecol 24(3):615–625
64. Shooshtarian S, Hosseini MR, Kocaturk T, Arnel T, Garofano NT (2023) Circular economy in the Australian AEC industry: investigation of barriers and enablers. Build Res Inf 51(1):56–68
65. Kirchherr J, Reike D, Hekkert M (2017) Conceptualizing the circular economy: an analysis of 114 definitions. Resour Conserv Recycl 127:221–232
66. Paiho S, Mäki E, Wessberg N, Paavola M, Tuominen P, Antikainen M, Heikkilä J, Rozado CA, Jung N (2020) Towards circular cities—conceptualizing core aspects. Sustain Cities Soc 59:102143
67. Lima L, Trindade E, Alencar L, Alencar M, Silva L (2021) Sustainability in the construction industry: a systematic review of the literature. Journal Clean Prod 289:125730
68. Anastasiades K, Goffin J, Rinke M, Buyle M, Audenaert A, Blom J (2021) Standardisation: an essential enabler for the circular reuse of construction components? A trajectory for a cleaner European construction industry. J Clean Prod 298:126864
69. Potting J, Hekkert MP, Worrell E, Hanemaaijer A (2017) Circular economy: measuring innovation in the product chain
70. Norouzi M, Chàfer M, Cabeza LF, Jiménez L, Boer D (2021) Circular economy in the building and construction sector: a scientific evolution analysis. J Build Eng 44:102704
71. Yaro NSA, Sutanto MH, Baloo L, Habib NZ, Usman A, Yousafzai AK, Ahmad A, Birniwa AH, Jagaba AH, Noor A (2023) A comprehensive overview of the utilization of recycled waste materials and technologies in asphalt pavements: towards environmental and sustainable low-carbon roads. Processes 11(7):2095
72. González A, Sendra C, Herena A, Rosquillas M, Vaz D (2021) Methodology to assess the circularity in building construction and refurbishment activities. Resour Conserv Recycl Adv 12:200051

Chapter 8
Emerging Circular Materials

8.1 Introduction

A consequence of the construction industry's reliance on the linear "take-make-dispose" cycle is the generation of approximately 40% of global material consumption and waste [1]. In response to the construction industry's pressing need to reduce its carbon footprint and environmental impact, the principles of circular economy (CE) present a promising pathway towards sustainability [2]. The principles of circular economy are applicable throughout the entire life cycle of construction projects [79]. The paradigm repurposes waste and recycled materials into resources, transforming economies from linear to circular models [3]. Circular economy concept intends ensure of humanity's welfare and habitability, focusing on responsible and efficient utilization of all resources. In advancing towards a circular economy model, strategic selection and engineering of materials are critical components [4]. Circular materials, the tangible manifestation of circular economy model, are materials designed, manufactured, and utilized based on the principles of "reuse", reduce", and "recycle" [5]. This chapter explores the various emerging circular materials and delves into discussion on the barriers to the uptake of these materials. The exploration is done based on three main categories,bio-based materials, recycled materials, and waste-based materials.

8.2 Overview of Emerging Circular Materials

Bio-based materials are sustainable and versatile alternative building materials [6]. The existence of bio-based building materials has been around for a long time but the creation of the modern policy framework for bio-sourced construction materials was formalized in 2012. France is noted for its regulatory push and intensification of the development of the bio-based construction material [7]. A material which

M. N. Addy et al., *Decarbonization of Building Energy in Developing Countries*, Green Energy and Technology, https://doi.org/10.1007/978-3-032-09113-0_8

incorporates plant or animal biomass is categorized as bio-based [8]. Examples of plant bio-based materials include hemp, flax, mycelium, ligning-based fibres, wood waste, and sunflower, while animal bio-based materials consist of sheep wool, camel wool, and goat wool [9]. Bio-based building materials have been extensively explored over the past decade, with the increasing impact and specificity of published documents each year underscoring its importance [10, 11]. Chen et al. [12] posited that, bio-based materials can reduce over 320,000 tonnes of carbon dioxide emissions by 2050. Further, bio-based materials demonstrate the potential of 40% decrease in water absorption, 8.7% reduction in energy consumption, enhancing acoustic absorption by 6.7%, and improving mechanical properties.

Circular economy on the other hand is enhanced through the integration of recycled materials into construction projects [13]. The utilization of recycled materials strengthens the resilience of built environments to manage climate change impacts [14]. Key terms associated with recycling include downcycling, materials recovery, and recycled Content [13]. Downcycling refers a specific form of recycling where recycled material is converted into a product of lower quality and reduced functionality. Materials recovery encompasses a broader range of activities beyond just recycling. Recycled content refers to the proportion of a product made from materials that have been recovered or diverted from the waste stream.

Examples of recycled materials as building materials include recycled plastics and recycled aggregates. Recycling plastics into pellets for construction materials such as pipelines, insulation, and roofing tiles has proven effective in reducing waste and its environmental impacts [14]. Recycled plastic aggregates can be successfully utilized in concrete bricks/pavement blocks non-structural panels [15]. Recycled aggregate (RA) are categorized into various types, including recycled concrete aggregate (RCA), fine recycled concrete aggregate (FRCA), recycled ceramic aggregate, recycled masonry aggregate (RMA), coarse mixed recycled aggregate (CMRA), and fine mixed recycled aggregate (FMRA) [16]. Repurposing waste materials for construction and soil stabilization has become an emerging field in the construction industry [17]. Efficient waste recycling can substantially impact on carbon reduction, approximately from 3.94 Mt CO_2e in 2000 to 58.65 Mt CO_2e in 2020 [18]. Incorporating waste-based materials as new building products is critical for sustainable construction, significantly reducing the consumption of resources, minimize energy usage, and decrease waste production [19].

Examples of waste-based building materials include recycled concrete aggregates, fly ash, slag, waste glass, plastics, rubber, construction and demolition (C&D) wastes, and reclaimed wood [14]. The performance of ash-based concretes can be enhanced through carbonation, promoting waste reutilization in the construction sector [20]. Green masonry bricks, converted from waste plastics and foundry sand, improves durability and strength indices of buildings and further mitigates pollution and address environmental littering issues [21]. By examining the concepts of bio-based materials, recycled materials and waste-based materials, this section of the chapter provides an overview of emerging circular materials. Subsequent sections of the chapter will highlight some examples of emerging circular materials identified.

8.3 Emerging Circular Materials: Bio-Based Materials, Recycled-Based Materials and Waste-Based Materials

The construction industry has been recognized as one of the primary contributors to environmental degradation and pollution, and also a significant consumer of non-renewable resources and a substantial generator of waste [10]. Sustainability has become a critical imperative for the construction industry in the drive towards reducing ecological footprint and carbon output [19]. Incorporating circular materials into the construction industry is vital in advancing environmentally friendly and resilient built environment [19]. This section of the chapter delves deeper into some emerging circular materials, discussing its composition and applicability in the built environment.

Bio-based materials are defined as "materials produced by plant or animal biomass that can be used as raw materials for building and decorative products, fixed furniture, and building materials in buildings" [7]. Bio-based materials are a sustainable opportunity in reducing the construction industry's environmental footprint [10], with its applicability in both new buildings and renovation [22]. Natural materials, fundamental to construction since antiquity, are now emerging in sustainable architecture towards the industry transitioning form linear to circular economy [19]. Table 8.1 provides an overview of some emerging circular materials.

There are plethora of studies highlighting the performance and strategies of integrating bio-based materials in the built environment. This is achieved by the different form and wide possibilities through which bio-based materials can be utilized in the complex construction industry [23]. In investigating sustainable building materials, Mouton et al. [24] demonstrated the carbon removal potentials of straw-based building elements and its applicability in external walls, internal floors, and flat floors. A combination of cross-laminated timber and straw in building envelope exhibited more energy efficiency during operational phase as compared to reinforced concrete. Yadav and Saini [25] highlighted the high insulation and thermal properties of plant-based biomaterials like bamboo, flax, hemp, etc., and their appropriateness for non-structural concrete mixtures. Results from a study by Mnasri et al. [26] identified the superior moisture regulator potential of wood-cement composite as compared to wood panels or fibrous materials. Yadav and Agarwal [23] also identified mycelium, a bio-based material cultivated from fungi, which can function as a material matrix within the composite materials.

Table 8.1 Emerging bio-based circular materials

S/N	Emerging bio-based circular material	Source
1	Straw-based building elements	[24]
2	Wood-cement composite	[26]
3	Mycelium	[23]
4	Hemp-infused concrete (Hempcrete)	[27–30]

Source Authors' own work

One of the emerging circular bio-based materials is hemp-infused concrete, hempcrete, offering an opportunity to carbon neutral and eco-friendly building practices [27]. Hempcrete is a bio-composite material with excellent environmental and thermal properties [28]. Hempcrete enclosures is efficient in solar heat gain and reducing cooling energy needs [29]. Hempcrete absorbs more CO_2 over its lifespan than is emitted during its production, transport, and construction [25]. The porous structure of hemp contributes to thermal insulation, soundproofing, and biological resistance of building materials [27]. Hempcrete can be effectively and efficiently incorporated into walls, floors, and roofs [25]. Hempcrete represents a shift towards environmentally conscious construction practices [30].

The reuse and recycling of by-products aligns with achieving the global ambitious "zero waste" goal [31]. Recycling is extracting new materials and objects from collecting and processing waste materials [13]. The growth of diverse recycled-based sustainable materials have been well documented in research. Dos Reis et al. [32] investigated the potentials of recycled aggregates, highlighting seven applications,sand production, pavement/road construction, ready mix concrete, concrete blocks, cement, ceramics/bricks, and low-cost adsorbent for wastewater treatment. Crushed concrete, made from recycled aggregates, is a sustainable solution for the concrete industry [33]. Recycled materials and industrial by-products can be incorporated as cement-based materials as an environmentally and technically promising solution for application to structural elements [34]. Recycling steel has a significant of carbon savings [35]. Recycled steel fibre reinforced concrete is an initiative that encompasses sustainable, environmental and health-friendly principles [36]. Table 8.2 provides an overview of recycled-based circular material.

Estimates commonly indicate that construction and demolition (C&D) activities account for roughly 30–40% of the global solid-waste stream, and WGBC reports that up to 32% of landfill waste can originate from construction sites [16]. Initiatives therefore to repurpose waste is significant for processing and recovery in a circular economy [41]. Gholampour et al. [37] investigated and proposed waste-based concrete (WBC) as a viable eco-friendly and sustainable alternative to conventional concrete. Waste-Based Green Cement (WGC) is a supplementary cementitious material which has significant influence on compressive strength of concrete [38]. Asa et al. [39] highlighted embraced wood as a method to facilitate the structural reuse of timber, increasing the amount of waste wood that has to be recycled. Reclaimed timber has embodied energy and cost savings potentials [42]. The structural reuse

Table 8.2 Emerging recycled-based circular material

S/N	Emerging recycled-based circular material	Source
1	Recycled aggregates	[32]
2	Crushed concrete	[33]
3	Recycled plastics	[14]
4	Recycled steel and recycled steel fibre reinforced concrete	[35, 36]

Source Authors' own work

Table 8.3 Emerging waste-based circular materials

S/N	Emerging waste-based circular material	Source
1	Waste-based concrete	[37]
2	Waste-based green cement	[38]
3	Embraced wood	[39]
4	Reclaimed wood	[40]

Source Authors' own work

of reclaimed wood can reduce the climate impact of the construction industry [40]. Bio-composites, created from gypsum and waste materials, offer the building sector a more sustainable and environmentally friendly alternative which is well-suited for use as interior cladding [43]. Table 8.3 provides an overview of emerging waste-based circular material.

8.4 The Role of Emerging Circular Materials in Energy Decarbonization

Decarbonization is an urgent and necessary effort towards aligning the construction industry, a significant contributor to global greenhouse gas (GHG), with international climate goals such as the Paris Agreement and the European Green Deal [44]. Buildings worldwide require a multi-sectoral, comprehensive strategy for decarbonization [45]. This is a crucial step to develop national and transnational roadmaps to achieve global emission reduction targets [46]. Wang et al. [47] posited that decarbonization strategies centred around commercial buildings operations is highly important for various countries to achieve net-zero emission goals.

Decarbonization of the building sector can be achieved through sustainable material selection, recycling, and reuse, targeting a minimum of 70% recycled content [48]. Assessing the environmental impact of building materials and strategies for its decarbonization is therefore crucial [49], with the choice of materials key in achieving energy efficiency in the building industry [50]. Mercader-Moyano and Esquivias [51] affirmed that a decarbonized society can be achieved with the provision of efficient construction materials and solutions which employ the principles of circular economy and resource efficiency. This is because local sustainable materials require less energy to produce and transport [52], with great environmental benefits [53]. This section of the chapter discusses the synergy between circular materials and energy decarbonization.

Emerging circular materials in construction practices are poised to transform the construction sector [54]. These materials are sustainable alternatives that can partially or completely replace natural resources like sand, cement, and gravel [14], leading to improved resource efficiency translating to energy and cost savings. Opoku et al. [55] demonstrated that 70% sandcrete with 30% treated palm fibre (P30) exhibited the best thermal performance with 38% reduction in thermal conductivity compared to the traditional sandcrete blocks. The integration of bio-based materials and waste-based materials into concrete production has the potential for significant carbon footprints reduction [56]. The use of local bio-based materials in wall assemblies exhibited lower life cycle impacts per functional unit, compared to traditional construction [57]. Almusaed et al. [50] highlight the potential for significant reductions in GHG emissions and energy use in the construction industry through the incorporation of recycled materials. Hempcrete, bio-based composites and recycled steel has the potential to reduce emissions during production and also actively sequester carbon over time [54]. Zouré and Genovese [58] found that hemp wool was the best-performing bio-sourced insulation, reducing annual energy consumption by up to 43,852 kWh. Essaghouri et al. [28] highlighted the performance of hempcrete to achieve a relatively low carbon emission of 484.42 tCO_2 over a lifetime of 100 years. Wood waste, used in waste-based concrete, is a strategy for the production of low carbon and circular materials in the concrete industry [59]. Natural fibre-based insulation materials are sustainable alternatives which is prominent in energy efficiency and thermal comfort [60].

The literature examined above highlights the synergy between circular materials and energy decarbonization. Emerging circular materials achieves sustainability in three ways; reduced extraction of raw materials, efficient management of waste and reduced energy consumption. Recycling ensures that there are readily available resources, from existing materials, for the production of new materials. This approach manages the depletion of natural resources and reduces the amount of energy needed in the production and manufacturing processes. The same is illustrated by bio-based materials and waste-based materials. Also, emerging circular materials is an effective method to repurpose waste generated from the construction industry, responsible for the generation of approximately 40% of global material consumption and waste [1]. The three main aspects of emerging circular materials, as discussed in this chapter, embodies waste to produce materials that are best in thermal performance and energy efficiency. The repurposing of waste is key in the management of natural resource depletion and the reduction of energy needed for production and manufacturing. Emerging circular materials highlighted in this chapter have demonstrated better thermal performances as compared to conventional building materials. The use of circular materials has been most prevalent in building envelopes, which play a critical role in regulating cooling demand and, consequently, the energy required for cooling. Evidently, emerging circular materials play a key role in energy decarbonization and further, the decarbonization of the entire construction industry.

8.5 Barriers to the Uptake of Emerging Circular Materials

Existing studies have extensively discussed circular materials and significant role in the decarbonization of the construction industry. However, the uptake of circular materials has faced some barriers. This section of the chapter discusses these barriers, highlighting case studies presented in literature.

8.5.1 Economic Barrier

Economic barriers are one of the significant challenges to the adoption of circular materials. This is mainly because, for a construction project, materials are typically priced from a short-term economic perspective, rather than by performance, environmental, aesthetic or human health, and well-being factors [61]. Osei-Tutu et al. [62] expanded economic barriers to encompass high cost of reclaimed materials, low market value, low landfill cost, limited market supply and demand of reclaimed materials budget and upfront cost, and design cost. One of the respondents, from [63], characterized bio-based materials with "higher production conversion cost and additional cost associated with the need for changing construction method". Munaro and Tavares [64] noted that recycling processes need high investments and limited incentive towards waste management. In this context, the adoption of emerging circular materials often entails higher initial investment requirements, which constitutes a critical limiting factor. This is more acute in developing countries where economic constraints are severe and investment decisions are heavily influenced by cost minimization priorities.

8.5.2 Political/Stakeholders' Barrier

The political/stakeholders' barrier highlights the absence of extra time, resources, and effort necessary for the consideration of sustainability and circularity [65]. In investigating the barriers to the uptake of circular economy in the construction industry by Hailu [66], the top ten major identified barriers were focused on stakeholders' commitment and interest. The study identified stakeholders' positions and levels of engagement as a critical barrier to effective waste management, thereby constraining the recovery of materials that could be reintegrated into circular economy practices. Munaro and Tavares [64] highlighted the lack of regulatory instruments from the government to encourage integrated management. This assertion is supported by Metinal and Gumusburun Ayalp [67], concluding that it hinders the widespread adoption of circular practices.

8.5.3 Technological Barriers

Technological barriers have also been identified as significant constraints to the uptake of circular materials. These include limitations in knowledge regarding material composition and recoverability, constraints associated with construction methods, challenges in integrating circular practices at different project phases, and considerations related to building lifespan [67]. Furthermore, the absence of standardized design approaches and cost structures for reducing and managing construction waste substantially diminishes the availability of materials for circular reuse [64]. In addition, the lack of sustainably managed forest resources continues to hinder the wider adoption of timber-based construction systems [68].

8.5.4 Culture Barrier

The culture barrier was identified as one of the barriers to uptake of circular materials. These barriers are linked to resistance to change, cultural beliefs and seeking of profit first [65]. Gottlieb et al. [63] discussed how myths and prejudices about sustainable materials were acting as deterrents to the uptake of bio-based materials. Some of the interviewees from the study describes bio-based materials as an unprofessional approach to building. Ingrained behaviours and preferences for established processes remains a challenge [67]. The easy market penetration of reclaimed materials is impacted by user preference [62].

In some cases, these barriers are interconnected, serving a complex network of constraints to the uptake of emerging circular materials. In the adoption of bio-based materials, Dams et al. [69] identified a network of economic barriers, political barriers, technological and cultural barriers as the major challenges. The barriers encompassed ability to upscale production by manufactures, problems surrounding initial costs, gaining accreditation for materials, the vested interests present in the construction industry and levels of knowledge among clients and construction professionals. In the New Zealand construction industry, regulatory constraints, insufficient funding and infrastructure, entrenched industry norms and culture, and a lack of training and awareness are some the barriers to plastic waste recycling [70]. Linh Le et al. [71] concluded that, in the Vietnam construction industry, some of the obstacles to the uptake of circular bio-based building materials are “building code restrictions and lack of regulation”, “lack of government promotion and incentives”, and “lack of design guides and tools for projects using Circular bio-based building materials”. In Europe and USA, the key barriers are contractor’s availability, a lack of professional knowledge, and cost-to-value perceptions [72]. The lack of political will, funding, and national policies on waste management is significant barrier to achieving circular economy in sub-Saharan Africa countries [73] (Table 8.4).

8.6 Overcoming Barriers to the Uptake of Emerging Circular Materials

This section of the chapter discusses approaches to managing barriers to the uptake of emerging circular materials. These approaches represent the construction industry’s commitment to the broader goal of achieving sustainability.

Table 8.4 Barriers to the uptake of emerging circular materials

S/N	Barrier	Discussion area	Sources
1	Economic barrier	1. Short-term economic perspective 2. High cost of reclaimed materials 3. Low market value 4. Limited market supply 5. High upfront cost 6. Limited incentives	[61–64]
2	Political/stakeholder barrier	1. Absence of interest for circularity 2. Lack of regulatory instruments for waste management	[64–67]
3	Technological barriers	1. Lack of knowledge on recoverability of waste materials 2. Lack of standard designs and cost structures 3. Absence of sustainably managed forests	[68, 67, 64]
4	Culture barrier	1. Resistance to change and culture beliefs 2. Myths and prejudices about sustainable materials 3. Ingrained behaviours and preferences	[65, 63, 67, 62]

Source Authors' own work

The development of a material bank with inventory control systems, product tracking, and monitoring protocols is an essential driver to manage economic barriers [64]. There should be established incentives or mandates geared towards recycling and waste reduction to motivate the uptake of circular materials [67]. Osei-Tutu et al. [62] asserted the sensitization for the promotion of reclaimed materials as an avenue to economically drive the uptake of circular materials. Dams et al. [69] posited that bio-based materials can be upscaled through collaborative economic initiatives such as long-term financing, grant policies, incentives, and local economic regeneration. Example of such initiatives is the Plastic Innovation Fund project in New Zealand, which exhibits potential for successfully recycling waste plastic products [70].

Long-term holistic circular ambitions of a city or state is an important starting point towards managing barriers imposed by stakeholders [64]. Public policies are an effective and efficient way to promote a rapid transition to low carbon bio-based materials [68]. Local-level involvement creates examples through construction projects that showcase circular practices [67]. Cross-sector stakeholder collaboration enhances the capacity and optimum resource utilization for waste management and the strengthening of more circular practices [74]. Multi-stakeholder collaboration is an important antecedent to the advancement of knowledge and practices in the implementation of circular economy practices in developing countries [75].

Technological barriers can be addressed through development of building codes and research [62]. The introduction of mandatory construction and demolition waste management plans is a significant approach to ensuring steady source of fit-for-use materials for the production of circular materials [64]. There is a need to develop new design typology to facilitate circular economy oriented decision-making in a building context [2]. This new design typology should incorporate adaptability of circular materials to ensure its suitability in the ever-growing buildings industry and its needs [76]. Circular materials should be manufactured to meet the requirements of formal standardization agencies such as ISO [77]. The advancement of technology has identified fast-growing materials for bio-based materials due to their short crop rotation periods [68]. Metinal and Gumusburun [67] asserted that standardized digital technologies can improve reclamation of construction wastes. The digital technologies will facilitate tracking of materials through specific datasets to ensure reuse of materials at their end-of-life [78].

Targeted awareness campaign and training programs is the necessary framework to manage the barriers imposed by the user preference towards circular materials [67]. These initiatives have to be backed by significant research development and innovation [64]. The promotion of awareness, knowledge, and understanding is needed to ensure the uptake of reclaimed materials and reused materials in the construction industry [62]. The awareness campaign and commitment to sustainable construction practices should be focused on the younger generation to ensure its place in the future [70] (Table 8.5).

The approaches discussed above serve an opportunity to intensify the advancement of sustainability in the construction industry. The need for collaborative efforts to incentivize stakeholders, formulate progressive circular policies, promote technological solutions and well-research, and targeted campaigns have been fully explored. Overcoming the barriers to the uptake of emerging circular materials is a step further towards a more sustainable construction industry.

8.7 Summary

Circular Economy is an innovative paradigm towards a sustainable built environment [78]. One of aspect promoted by circular economy is circular materials, which involves the reclamation and reuse of construction and other related waste materials. The benefits of circular materials have been well researched and documented,however, its uptake is hindered by some significant barriers. This chapter highlighted the role of emerging circular materials in the decarbonization of the building industry, the barriers to its uptake and solutions to facilitate the incorporation of these emerging circular materials in the future of sustainable construction practices.

In this chapter, the exploration of circular materials is done through three main categories; bio-based materials, recycled-base materials, and waste-based materials. Some of the emerging bio-based circular materials identified include straw-based elements, wood–cement composites, and hemp-infused concrete. Recycled-based examples comprise crushed concrete, recycled plastics, and recycled steel, while

Table 8.5 Strategies for overcoming barriers to the uptake of emerging circular materials

S/N	Driver	Discussion areas	Source
1	Economical	1. Development of material bank 2. Incentives 3. Long-term financing 4. Grant policies 5. Local economic regeneration	[69, 67, 64, 62, 70]
2	Policy development and stakeholder involvement	1. Long-term holistic circular ambitions 2. Public policies 3. Local-level involvement 4. Cross-sector stakeholder collaboration 5. Multi-stakeholder collaboration	[68, 74, 67, 75, 64]
3	Technological	1. Development of building codes and research 2. Mandatory construction and demolition waste management plan 3. Development of design topologies 4. Standardization	[77, 2, 68, 76, 76–78]
4	Culture and social	1. Targeted awareness and campaign 2. Training programs 3. Research development and innovation 4. Promotion of knowledge	[67, 64, 62, 70]

Source Authors' own work

waste-based options include waste-derived concrete, green cement, and reclaimed wood. Further in the chapter, the role of emerging circular materials towards energy decarbonization is discussed. Emerging circular materials achieves sustainability in three ways; reduced extraction of raw materials, efficient management of waste, and reduced energy consumption.

The chapter concludes with discussion on barriers to the uptake of the emerging circular materials and solutions to manage these hindrances. Central of these barriers are economic barriers. Stakeholders, thus manufacturers and consumers within the construction industry have low motivation towards adopting circular materials due to its upfront cost. Transitioning to circular materials require the development of a new building material framework which is heavy on the budget. From the case studies examined, however, advancement in technology and the provision of necessary incentives can facilitate the rapid transition towards circular materials. Emerging circular materials reduce the need for raw materials, creates an avenue to repurpose construction generated waste and has a major impact towards energy decarbonization. There is the need for more stakeholders' consultations in order to ensure its incorporation into the building industry.

References

1. Khadim N et al (2025) From circularity to sustainability: advancing the whole building circularity indicator with life cycle assessment (WBCI-LCA). Build Environ 269:112413. https://doi.org/10.1016/j.buildenv.2024.112413
2. Eberhardt LCM, Birkved M, Birgisdottir H (2022) Building design and construction strategies for a circular economy. Arch Eng Design Manag 18(2):93–113. https://doi.org/10.1080/17452007.2020.1781588
3. Reuter MA et al (2019) Challenges of the circular economy: a material, metallurgical, and product design perspective. Annu Rev Mater Res 49:253–274. https://doi.org/10.1146/annurev-matsci-070218-010057
4. Dumée LF (2022) Circular materials and circular design—review on challenges towards sustainable manufacturing and recycling. Circ Eco Sustain 9–23. Springer Nature. https://doi.org/10.1007/s43615-021-00085-2
5. Santos P et al (2024) Circular material usage strategies and principles in buildings: a review. Buildings. https://doi.org/10.3390/buildings14010281
6. Le DL, Salomone R, Nguyen QT (2024) Sustainability assessment methods for circular bio-based building materials: a literature review. J Environ Manage. https://doi.org/10.1016/j.jenvman.2024.120137
7. Li A et al (2024) Promoting circular economy of the building industry by the use of straw bales: a review. Buildings Multidisciplinary Digital Publishing Institute (MDPI). https://doi.org/10.3390/buildings14051337
8. Bourbia S, Kazeoui H, Belarbi R (2023) A review on recent research on bio-based building materials and their applications. Mater Renew Sustain Energy. https://doi.org/10.1007/s40243-023-00234-7
9. Parlato MCM, Pezzuolo A (2024) From field to building: harnessing bio-based building materials for a circular bioeconomy. Agronomy Multidisciplinary Digital Publishing Institute (MDPI). https://doi.org/10.3390/agronomy14092152
10. Arias-Cárdenas B, Lacasta AM, Haurie L (2024) Bibliometric analysis of research on thermal, acoustic, and/or fire behaviour characteristics in bio-based building materials. Elsevier Ltd., Construction and building materials. https://doi.org/10.1016/j.conbuildmat.2024.136569
11. Ferreira-Filipe DA et al (2021) Are biobased plastics green alternatives?—A critical review. Int J Environ Res Public Health. https://doi.org/10.3390/ijerph18157729
12. Chen L et al (2024) Biomaterials technology and policies in the building sector: a review. Environ Chem Lett. https://doi.org/10.1007/s10311-023-01689-w
13. Goh KC et al (2024) Harvesting valuable elements from solar panels as alternative construction materials: a new approach of waste valorization and recycling in circular economy for building climate resilience. Sustain Mater Technol. Elsevier B.V. https://doi.org/10.1016/j.susmat.2024.e01030
14. Shukla BK et al (2024) Sustainable construction practices with recycled and waste materials for a circular economy. Asian J Civil Eng. Springer Nature. https://doi.org/10.1007/s42107-024-01111-y
15. Lamba P et al (2022) Recycling/reuse of plastic waste as construction material for sustainable development: a review. Environ Sci Pollut Res. https://doi.org/10.1007/s11356-021-16980-y
16. Joseph HS et al (2023) A comprehensive review on recycling of construction demolition waste in concrete. Sustainability (Switzerland). MDPI. https://doi.org/10.3390/su15064932
17. Alzhanova GZ et al (2022) Development of environmentally clean construction materials using industrial waste. Materials 15(16). https://doi.org/10.3390/ma15165726
18. Liu H et al (2022) Quantifying the carbon reduction potential of recycling construction waste based on life cycle assessment: a case of Jiangsu Province. Int J Environ Res Public Health 19(19). https://doi.org/10.3390/ijerph191912628
19. Firoozi AA et al (2024) Emerging trends in sustainable building materials: technological innovations, enhanced performance, and future directions. Res Eng. Elsevier B.V. https://doi.org/10.1016/j.rineng.2024.103521

20. Al Salaheen M et al (2023) Advancing waste-based construction materials through carbon dioxide curing: a comprehensive review. Res Eng. https://doi.org/10.1016/j.rineng.2023.101591
21. Aneke FI, Shabangu C (2021) Green-efficient masonry bricks produced from scrap plastic waste and foundry sand. Case Stud Construct Mater 14. https://doi.org/10.1016/j.cscm.2021.e00515
22. Galimshina A et al (2022) Bio-based materials as a robust solution for building renovation: a case study. Appl Energy 316. https://doi.org/10.1016/j.apenergy.2022.119102
23. Yadav M, Agarwal M (2021) Biobased building materials for sustainable future: an overview. Mater Today Proc. https://doi.org/10.1016/j.matpr.2021.01.165
24. Mouton L, Allacker K, Röck M (2023) Bio-based building material solutions for environmental benefits over conventional construction products—life cycle assessment of regenerative design strategies (1/2). Energy Build 282. https://doi.org/10.1016/j.enbuild.2022.112767
25. Yadav M, Saini A (2022) Opportunities and challenges of hempcrete as a building material for construction: an overview. Mater Today Proc. https://doi.org/10.1016/j.matpr.2022.05.576
26. Mnasri F et al (2020) Building incorporated bio-based materials: experimental and numerical study. J Build Eng 28. https://doi.org/10.1016/j.jobe.2019.101088
27. Ansari H, Tabish M, Zaheer MM (2025) A comprehensive review on the properties of hemp incorporated concrete: an approach to low carbon footprint construction. Next Sustain 5:100075. https://doi.org/10.1016/j.nxsust.2024.100075
28. Essaghouri L, Mao R, Li X (2023) Environmental benefits of using hempcrete walls in residential construction: an LCA-based comparative case study in Morocco. Environ Impact Assess Rev 100. https://doi.org/10.1016/j.eiar.2023.107085
29. Shang Y, Tariku F (2021) Hempcrete building performance in mild and cold climates: integrated analysis of carbon footprint, energy, and indoor thermal and moisture buffering. Build Environ 206. https://doi.org/10.1016/j.buildenv.2021.108377
30. Shanbhag SS, Dixit MK (2025) Understanding carbon-negative potential of hempcrete using a life cycle assessment approach. In: Lecture notes in civil engineering. Springer Science and Business Media Deutschland GmbH, pp 169–182. https://doi.org/10.1007/978-3-031-69626-8_15
31. Branca TA et al (2020) Reuse and recycling of by-products in the steel sector: Recent achievements paving the way to circular economy and industrial symbiosis in Europe. Metals. MDPI AG. https://doi.org/10.3390/met10030345
32. Dos Reis GS et al (2021) Current applications of recycled aggregates from construction and demolition: a review. Materials. https://doi.org/10.3390/ma14071700
33. Lu L (2024) Optimal replacement ratio of recycled concrete aggregate balancing mechanical performance with sustainability: a review. Buildings Multidisciplinary Digital Publishing Institute (MDPI). https://doi.org/10.3390/buildings14072204
34. Frazão C et al (2022) Technical and environmental potentialities of recycled steel fiber reinforced concrete for structural applications. J Build Eng 45. https://doi.org/10.1016/j.jobe.2021.103579
35. Kanyilmaz A et al (2023) Reuse of steel in the construction industry: challenges and opportunities. Int J Steel Struct 23(5). https://doi.org/10.1007/s13296-023-00778-4
36. Liew KM, Akbar A (2020) The recent progress of recycled steel fiber reinforced concrete. Constr Build Mater. https://doi.org/10.1016/j.conbuildmat.2019.117232
37. Gholampour A et al (2021) Concretes containing waste-based materials under active confinement. Construct Build Mater 270. https://doi.org/10.1016/j.conbuildmat.2020.121465
38. Perera MAGP, Ranjith PG (2025) Greener horizons: revolutionizing construction materials with waste-based innovations—an experimental study. J Build Eng 103. https://doi.org/10.1016/j.jobe.2025.112211
39. Asa P et al (2024) Embraced wood: circular construction method for composite long-span beams from unprocessed reclaimed timber, fibers and clay. Construct Build Mater 416. https://doi.org/10.1016/j.conbuildmat.2024.135096

40. Bergsagel D, Heisel F (2023) Structural design using reclaimed wood—a case study and proposed design procedure. J Clean Prod 420. https://doi.org/10.1016/j.jclepro.2023.138316
41. Spišáková M et al (2022) Waste management in a sustainable circular economy as a part of design of construction. Appl Sci (Switzerland) 12(9). https://doi.org/10.3390/app12094553
42. Pronk A, Brancart S, Sanders F (2022) Reusing timber formwork in building construction: testing, redesign, and socio-economic reflection. Urb Plan 7(2). https://doi.org/10.17645/up.v7i2.5117
43. Balti S et al (2023) Advancing the circular economy: reusing hybrid bio-waste-based gypsum for sustainable building insulation. Buildings 13(12). https://doi.org/10.3390/buildings13122939
44. Karakosta C, Papathanasiou J (2025) Decarbonizing the construction sector: strategies and pathways for greenhouse gas emissions reduction. Energies 18(5):1285. https://doi.org/10.3390/en18051285
45. Ma M et al (2024) Challenges and opportunities in the global net-zero building sector. Cell Rep Sustain 1(8):100154. https://doi.org/10.1016/j.crsus.2024.100154
46. Camarasa C et al (2022) A global comparison of building decarbonization scenarios by 2050 towards 1.5–2 °C targets. Nat Commun 13(1). https://doi.org/10.1038/s41467-022-29890-5
47. Wang T et al (2025) Toward net zero: assessing the decarbonization impact of global commercial building electrification. Appl Energy 383. https://doi.org/10.1016/j.apenergy.2025.125287
48. Palomar-Torres A et al (2025) Decarbonizing near-zero-energy buildings to zero-emission buildings: a holistic life cycle approach to minimize embodied and operational emissions through circular economy strategies. Appl Sci 15(5):2670. https://doi.org/10.3390/app15052670
49. Sbahieh S, Zaher Serdar M, Al-Ghamdi SG (2023) Decarbonization strategies of building materials used in the construction industry. Mater Today Proc. https://doi.org/10.1016/j.matpr.2023.08.346
50. Almusaed A et al (2024) Assessing the impact of recycled building materials on environmental sustainability and energy efficiency: a comprehensive framework for reducing greenhouse gas emissions. Buildings 14(6). https://doi.org/10.3390/buildings14061566
51. Mercader-Moyano P, Esquivias PM (2020) Decarbonization and circular economy in the sustainable development and renovation of buildings and neighbourhoods. Sustainability (Switzerland). https://doi.org/10.3390/SU12197914
52. Sonebi M et al (2022) Trends and opportunities of using local sustainable building materials in the Middle East and North Africa region. RILEM Tech Lett 7. https://doi.org/10.21809/rilemtechlett.2022.169
53. Liang Y et al (2023) Decarbonization potentials of the embodied energy use and operational process in buildings: a review from the life-cycle perspective. Heliyon. https://doi.org/10.1016/j.heliyon.2023.e20190
54. Mikulėnas M, Šeduikytė L (2025) Circularity and decarbonization synergies in the construction sector: implications for zero-carbon energy policy. Energies 18(5):1164. https://doi.org/10.3390/en18051164
55. Opoku R et al (2020) Minimizing heat transmission loads and improving energy efficiency of building envelopes in sub-Saharan Africa using bio-based composite materials. Sci Afr 8. https://doi.org/10.1016/j.sciaf.2020.e00358
56. Schmidt W, Kamashanju K (2024) Circular, bio-based, intelligent ecosystems for climate friendly construction materials supplies
57. Keena N et al (2022) A life-cycle approach to investigate the potential of novel biobased construction materials toward a circular built environment. Energies 15(19). https://doi.org/10.3390/en15197239
58. Zoure AN, Genovese PV (2023) Comparative study of the impact of bio-sourced and recycled insulation materials on energy efficiency in office buildings in Burkina Faso. Sustainability (Switzerland) 15(2). https://doi.org/10.3390/su15021466
59. Caldas LR et al (2021) Building materials in a circular economy: The case of wood waste as CO2-sink in bio concrete. Res Cons Recycl 166. https://doi.org/10.1016/j.resconrec.2020.105346

60. O'Brien L et al (2025) Review of bio-based wood fiber insulation for building envelopes: characteristics and performance assessment. Elsevier Ltd., Energy and buildings. https://doi.org/10.1016/j.enbuild.2024.115114
61. Dams B et al (2023) Upscaling bio-based construction: challenges and opportunities. Build Res Inf 51(7):764–782. https://doi.org/10.1080/09613218.2023.2204414
62. Osei-Tutu S et al (2023) Barriers impeding circular economy (CE) uptake in the construction industry. Smart Sustain Built Environ 892–918. Emerald Publishing. https://doi.org/10.1108/SASBE-03-2022-0049
63. Gottlieb SC et al (2023) Roadmap for the transition to biogenic building materials: a socio-technical analysis of barriers and drivers in the Danish construction industry. J Clean Prod 414. https://doi.org/10.1016/j.jclepro.2023.137554
64. Munaro MR, Tavares SF (2023) A review on barriers, drivers, and stakeholders towards the circular economy: the construction sector perspective. Clean Respons Consump 8. https://doi.org/10.1016/j.clrc.2023.100107
65. Charef R, Morel JC, Rakhshan K (2021) Barriers to implementing the circular economy in the construction industry: a critical review. Sustainability (Switzerland). MDPI. https://doi.org/10.3390/su132312989
66. Hailu HA (2025) Unraveling the barriers to effective waste management in the ethiopian construction industry: a multi-method qualitative approach. J Eng Sustain Develop 29(1):1–7. https://doi.org/10.31272/jeasd.2739
67. Metinal YB, Gumusburun Ayalp G (2025) Uncovering barriers to circular construction: a global scientometric review and future research agenda. Sustainability (Switzerland), 17(4). https://doi.org/10.3390/su17041381
68. Göswein V et al (2022) ETH library barriers and opportunities of fast-growing biobased material use in buildings rights/license: creative commons attribution 4.0 international. https://doi.org/10.3929/ethz-b-000587227
69. Dams B et al (2021) Upscaling non-residential bio-based circular construction in the United Kingdom. In: IOP conference series: earth and environmental science. IOP Publishing Ltd. https://doi.org/10.1088/1755-1315/855/1/012015
70. Thomson P, Low JK, Berry TA (2024) Addressing barriers and unveiling opportunities for plastic waste recycling in the New Zealand construction industry. Front Sustain. Frontiers Media SA. https://doi.org/10.3389/frsus.2024.1487972
71. Linh Le D et al (2025) Circular bioeconomy in the global south: the case of circular bio-based building material use in Vietnam. https://ssrn.com/abstract=4983219
72. Carcassi OB et al (2024) Upscaling natural materials in construction: earthen, fast-growing, and living materials. Sustainability 16(18):7926. https://doi.org/10.3390/su16187926
73. Debrah JK, Teye GK, Dinis MAP (2022) Barriers and challenges to waste management hindering the circular economy in Sub-Saharan Africa. Urban Science. MDPI. https://doi.org/10.3390/urbansci6030057
74. Luthra S et al (2022) Overcoming barriers to cross-sector collaboration in circular supply chain management: a multi-method approach. Transp Res Part E Log Transp Rev 157. https://doi.org/10.1016/j.tre.2021.102582
75. Mishra JL, Chiwenga KD, Ali K (2019) Collaboration as an enabler for circular economy: a case study of a developing country. Manag Decis 59(8):1784–1800. https://doi.org/10.1108/MD-10-2018-1111
76. Hamida MB et al (2022) Circular building adaptability and its determinants—a literature review. Int J Build Pathol Adapt 41(6):47–69. https://doi.org/10.1108/IJBPA-11-2021-0150
77. Dorrego-Viera JI, Urbinati A, Lazzarotti V (2025) Transition towards circular economy: exploiting open innovation for circular product development. J Innov Knowl 10(2). https://doi.org/10.1016/j.jik.2025.100668
78. Rahla KM, Mateus R, Bragança L (2021) Implementing circular economy strategies in buildings—from theory to practice. Appl Syst Innov 4(2). https://doi.org/10.3390/asi4020026
79. Gamage I, Senaratne S, Perera S, Jin X (2024) Implementing circular economy throughout the construction project life cycle: a review on potential practices and relationships. Buildings 14(3):653

Chapter 9
Indigenous Construction Methods and Appropriate Technology

9.1 Introduction

There is a rising need and concern for sustainable construction activities in recent time due to the environmentally unfriendly ways of construction [1, 2]. The construction industry has been noted in literature as one of the major contributors to climate change and global warming [3]. The operation of buildings depends largely on electrical energy, and in some cases, some other forms of energy such as gas [4]. The conventional ways of construction which are not sustainable places a huge demand for energy to ensure a smooth operation of building [5]. This is seen in the demand for thermal comfort and artificial lighting. The methods of construction and materials used in construction greatly affect the performance of buildings and its overall energy efficiency. Moreover, the use of some materials poses a serious threat and demand for energy especially in their product stage [6, 7]. Santamouris and Vasilakopoulou [4] posits that future decarbonization potentials of the building sector depends largely on manufacturing efficiency and inculcation of circular economy principles in materials production. This implies that, building energy decarbonization has to be tackled on a life cycle basis from the product stage to the end of life of buildings. The energy consumed in the extraction, processing, manufacturing, and transportation of construction materials which is largely referred to as embodied energy must be critically considered. A wholistic approach would account for both the operational energy used and the embodied energy. It is important to note that building energy decarbonization is strongly influenced by the overall energy efficiency of buildings [8, 9]. This is seen in the reduction of the demand for energy for the production, construction and operation of buildings which in effect will reduce the CO_2 emissions associated. Even if the means of energy production is not yet sustainable in its entirety, there will be a reduction in emissions when buildings are built to be sustainable [10]. Indigenous construction methods refer to the methods of construction which are local and use vernacular materials [11]. Some of these materials include bamboo, timber, laterite, grass/thatch, clay, mud among others [1].

M. N. Addy et al., *Decarbonization of Building Energy in Developing Countries*, Green Energy and Technology, https://doi.org/10.1007/978-3-032-09113-0_9

Indigenous construction methods and appropriate technology leverage on local and traditional resources to construct buildings that are eco-friendly, socially responsible, and economical. Indigenous construction, locally sourced or regional materials and resources are used to create a building in a style typical to that area [12]. Additionally, these methods and technologies take into account local climate and geography. This results in the construction of buildings that are better adapted to their surroundings and require less energy demands. Indigenous construction methods can therefore be referred to as local construction methods or vernacular architecture in a broad sense.

The idea of technology tailored to local needs has roots in Gandhi's 1920s–1930s advocacy for small, village-based production in India, but the modern concept and label "appropriate (intermediate) technology" were articulated by E. F. Schumacher from the 1960s. In this sense, appropriate technology refers to context-appropriate solutions that are affordable, resource-efficient, and locally maintainable. It gained popularity and an adoption in sustainable development in the 2010s. In literature, Buitenhuis et al. [13] and Pearce [14] have stated that appropriate technology finds its application in many fields. These include: commerce, agriculture, water, medicine, transport, energy, construction among others. In the construction industry, Hazeltine and Bull [15] have reported that the term "appropriate technology" has over the years been developed. It starts with the works of Schumacher [16] through his book "Small is beautiful" where he discusses what is commonly referred to as Intermediate Technology. Other well-known people cited in relation to this technology include Julius Nyerere of Tanzania, Mao Zedong of China, and Mahatma Gandhi of India.

According to Willoughby [17], a synthesis of Schumacher's ideas gives a definition of appropriate technology as "a technology tailored to fit the psychosocial and biophysical context prevailing in a particular location and period". Nwalusi and Okeke [18] argue that this definition for appropriate technology is generic. Although it communicates the idea that appropriate technology founders had in mind, it leaves it open for any technology to be characterized as: "appropriate" which even what appropriate technology's founders and proponents would consider not to be. One of the proponents of appropriate technology Auroville Building Centre defines appropriate technology as "building processes and tools that are appropriate to the climate, socio-economic conditions and natural resources of an area, and which contribute to sustainable development" Chetia and Prashad [19] went further to give the characteristics appropriate technology (AT) for ease of identification as a technology that is energy-efficient, decentralized, locally controlled, small-scale, capital saving, environmentally sound, people-cantered, and labour-intensive. According to Tekpe et al. [20], an appropriate technology is not expensive and simple to build, maintain, and operate. It uses renewable resources rather than fossil fuels and does not require high energy concentrations. It relies primarily on people's skills, not on automated machinery, encourages human scale operations, small businesses, and community cohesion; is protective of human health, and is ecologically sound [18]. Appropriate technology tends to promote and motivate a self-reliant and eco-friendly system where local people, especially those in under-developed regions can effectively manage their needs [21]. Appropriate technology focuses on the available resources to meet the specific need of people in a particular setting. It may not be a modern or advanced technology but it is able to answer to the needs of people.

9.2 Overview of Indigenous Construction Methods and Appropriate Technology

The concept of indigenous construction methods and appropriate technology has been introduced to encourage sustainable construction activities to promote environmental friendliness and reduce building energy carbon emissions [22, 23]. ICM promotes the preservation of cultural heritage and local craftmanship. Indigenous Construction methods stem from indigenous knowledge of construction. Using indigenous, traditional, or local knowledge can greatly preserve domestic diversity and develop sustainable adaptations and mitigation techniques [23]. In the context of indigenous construction, the case of bioclimatic architecture surges. Bioclimatic architecture refers to how the buildings systems relates with the local climate and the occupants of the building for their comfort without or very little use of mechanical systems. Bioclimatic architecture can be traced to the design concepts of vernacular architecture and indigenous construction technologies across the globe, including sub-Saharan Africa.

Indigenous construction methods focus extensively on the use of local and sustainable building materials that are readily available. For instance, vernacular materials like clay and mud (Adobe) have received widespread usage within the construction industry worldwide for thousands of years [24]. In many climatic conditions all over the world, indigenous construction methods such adobe construction have been used extensively [25]. Additionally, the use of earth and laterite as an indigenous material for construction have been noted by Costa et al. [26] as a significant indigenous material for modern construction types. Also, vernacular techniques such as adobe and rammed earth construction, and wattle and Daub are well reported in the literature [25, 27, 28]. Indigenous construction methods have also been applied to modern sustainable construction for passive design strategies in specific climates. For example, in hot deserts, warm climates where cooling demand is high, various indigenous techniques have been used to provide passive cooling to reduce energy consumption by mechanical cooling means [29]. Such techniques are identified to include roof ponds, domes, air vents, cooling towers among others [29]. The selection of the techniques and materials for such buildings is usually dependent on the desired benefits, as well as the local availability of construction materials and skilled labour.

In the construction industry, appropriate technology and indigenous construction work in tandem to foster sustainable construction practices by utilizing local and vernacular materials to meet the needs of occupants in the building. These materials and methods are coherent with the environmental, cultural, social, and economic facets of a community. It is the simplest level of technology or method that can be used in any developmental project. As already stated above, it is a technology appropriate enough which does not use high energy consumption and utilizes renewable and local resources [18, 20]. It can then be inferred that, the application of the appropriate technology adopted in construction activities will hugely promote energy efficiency of buildings, thus reducing carbon emissions associated with energy production. Teartisup [30] opines that when appropriate technology is applied to a construction site, the results become appropriate construction technology which is indigenous and local in nature. Furthermore, out of appropriate technology comes appropriate housing design, which focuses on the engineering and architectural perspective of a structure that supports sustainable construction and development [20]. From the overview, the following characteristics of indigenous construction methods and appropriate technology include, use of passive design solutions adapted to local climatic conditions; use of local building materials; use affordable building techniques and technologies; and influenced by local culture and tradition.

9.3 Examples of Indigenous Construction Methods and Appropriate Technology in Application

9.3.1 Adobe Construction Method

Adobe construction has been an existing indigenous method of construction for many years [24]. While Agyekum et al. [1] cite the often-repeated estimate that roughly 30% of the world's population lives in adobe or earthen buildings, more recent analyses place the global share closer to 8–10% (up to 20–25% in developing countries). This type of construction is particularly prevalent in most developing countries, especially the African region because it comes with economic benefits for construction [26]. Adobe construction has very good building performance in terms of indoor air quality and indoor temperature [24]. This property makes it possible for building occupants to limit the use of mechanical means for thermal comfort. Houses that are constructed with adobe brick technology are still being built in either multi-storey buildings or smaller single-storey homes. Adobe homes have been used in both North America, South America and in Africa for thousands of years [20]. In a study conducted by Akomah et al. [31], abode construction and other earth-based construction was found to be an appropriate technology for sustainable construction. Figure 9.1 shows how adobe construction has evolved to fit into modern architecture.

Fig. 9.1 Typical adobe construction [80, 81]. *Source* https://www.thespruce.com/what-is-an-adobe-style-house-7151962

9.3.2 Wattle and Daub Technology

Wattle and daub method of construction is a wall construction method highly associated with developing countries and the African region. Essienyi [32] reported that, this method of construction is typical in West Africa, and other countries such as Sudan [33]. The construction technology involves the weaving of branches (wattle) as a support for mud plaster (daub). Wattle and daub construction is mainly an appropriate technology for dry and hot climates because of the perishability of the materials used [34]. It is mostly appropriate in the equatorial regions where there is little variation in precipitation and humidity levels. Wattle and daub have very good thermal mass. It absorbs heat in the day and dissipates it at night, keeping a cool indoor environment. A modern wattle and daub construction is the Casa Munita Gonzales building in Fig. 9.2.

Fig. 9.2 Typical wattle and daub construction method [79]. *Source* https://eartharchitecture.org/?cat=80

Fig. 9.3 Typical timber construction technology [77, 78]. *Source* https://theconstructor.org/building/timber-construction-popularity/21342/

9.3.3 Timber Construction Technology

Timber construction is another construction method that has been adopted for many years. It is one of the construction methods that has easily been adopted into modern construction methods among the other indigenous construction technologies. Timber construction method is a unique indigenous construction method and appropriate technology for forest regions. This method of construction requires a precise cutting of timber to create members with required sizes to be used for construction [1]. Timber construction can be for both structural and non-structural works. Structural timber construction is constructed as beams, and columns to carry loads while timber elements such as ceilings and floor finishes are not for load bearing. One of the major challenges of this construction is the propensity for forest degradation [35] and its weakness to fire and termite attacks. However, when treated very well, timber can resist fire for a good time and resist termite attacks. According to Fischer et al. [36], wood and timber buildings are energy-efficient due to their low embodied energy and reduced energy requirements for maintaining indoor climate. Additionally, Bucklin et al. [37] reports that timber framed buildings can provide energy efficiency, reducing greenhouse gas emissions in construction and operation of buildings (Fig. 9.3).

9.3.4 Pile Dwellings or Wetland Stilt Construction

Another appropriate technology for construction is indigenous pile dwellings, also known as floating buildings. This type of construction is usually required for wetlands or swampy areas [1]. This type of construction method is influenced by the climatic conditions within a specific area, as well as the ground conditions [38]. This construction requires a floating structural system than can accommodate vertical rise in water levels and its movement in all directions. Mostly indigenous or vernacular materials such as timber and thatch are used in the construction of these stilt houses. A classic example is the Nzulezu settlement in Ghana. Another example is seen at the

Fig. 9.4 Pile dwelling at Nzulezu, Ghana [76]. *Source* https://ghana-beach-cottage.com/points-of-interest/nzulezu-stilt-village-western-region/

Musi riverside of Palembang [39]. The indigenous materials used in this construction method have very low embodied energy and embodied carbon emissions. Furthermore, the cooler climatic condition of riverside area shown by Milošević et al. [40] provides the basis to posit that, stilt house require less demand for cooling and increases energy efficiency. The provision of fenestration allows enough air movement into the spaces for ventilation and passive cooling (Fig. 9.4).

9.3.5 Rammed Earth Construction Technology

Rammed earth construction is a mixture of indigenous raw materials such as soil, sand, and gravel. It is one of the popular earth-based construction. Rammed earth construction involves the compaction of several layers of earthen materials in formworks. It is similar to adobe construction; however, rammed earth construction is denser than adobe and are significantly more uniformed [20]. Rammed earth has a high non-combustible property. It is thermally massive and durable. It is a passive method that provides a stable, controlled temperature and humidity that reduces air-conditioning requirements. Rammed earth uses local and less expensive labour making it cost beneficial. Because of the uniformity, the compressed earth uses little mortar [20]. Similar to other indigenous construction, rammed earth has low embodied energy. It has a low life cycle effect which makes it valuable for sustainable construction practices. This technology if mostly used at regions where supply for wood is low. In literature, rammed earth has been proven to have less embodied energy. Cheikhi et al. [41] reports that rammed earth buildings exhibits very good energy performance and indoor thermal conditions. By adding insulations, energy efficiency is enhanced further especially in hot and warm climates (Fig. 9.5).

The above discussed indigenous construction methods demonstrate that these technologies were well-suited, or "appropriate", to their specific historical and

Fig. 9.5 Rammed earth construction technology [74, 75]. *Source* https://terraviva.pt/the-portuguese-way/

geographical contexts. However, the development of new appropriate technologies that respond to the evolving demands of the contemporary construction industry appears to be limited. It is therefore essential to draw lessons from past practices while embracing innovation, in order to develop contextually relevant technologies that address current needs and contribute to a sustainable future.

9.4 Barriers to Indigenous Construction Methods and Appropriate Technology

According to Genovese and Zoure [12], barriers to the adoption of construction technologies are more notable in developing countries. Understanding region specific barriers will aid to formulate developmental pathways adapted to the context of a particular area. The challenges of indigenous construction techniques can be identified as the challenges facing general green building adoption and its market [42]. This is because, applying indigenous or vernacular construction methods and materials are seen to be green building options for improved energy, environment, and thermal performances. A thorough review of the general obstacles of the green building market is featured in the adoption of ICM.

9.4.1 Lack of Skilled Personnel and Technical Know-How

One of the major barriers of indigenous construction methods and appropriate technology identified in literature is the lack of skilled personnel and technical know-how [12, 43]. Indigenous construction requires specialized skills and craftmanship which local artisans and craftsmen possessed. But with the passing of time and adoption of other building technologies, these craftsmen of local materials have gradually phased out with very few people possessing knowledge in that technology [18]. The loss of these indigenous skills has made it difficult to maintain and restore existing structures, let alone to apply it in new and modern construction. In a study by Durrant [44], lack of expertise in the production and use of indigenous method such as hemp is a major barrier facing its adoption. Similarly in other studies, lack of skilled personnel and expertise has been identified a major barrier [45, 46].

9.4.2 *Inadequate Strength, Quality Requirements, and Lack of Standards*

Material strength is an important quality for any construction. Indigenous constructions use methods and materials that may not have adequate strength for many types of structures [47]. Despite the environmental friendliness they possess, ICM adoption is disadvantaged due to the strength requirements of modern architecture [48]. In modern construction that seeks to incorporate ICM, composite and engineered products made from local materials have been introduced [27, 49–52]. However, in the absence of robust quality standards, building codes and a clear regulatory framework to guide their application, the widespread adoption of these innovations is likely to remain limited [53]. The government has a responsibility to establish standards and codes that will guide and support the adoption of ICMs. Standards are basic frameworks for promoting quality production, however, there is hardly any available standard for indigenous building materials used in indigenous construction technologies. Due to economic crisis in developing countries, high priority has been placed on economic success above ecological success. This has led to a lack of focus on pertinent issues such as the lack of standards for sustainable construction technologies [54]. Acheampong et al. [43], found that a significant problem surrounding the construction of buildings with vernacular materials is the lack of standards for both guidance, quality control and assurance. Implementation of sustainable technologies involves a life cycle assessment from the extraction of raw materials to construction, operation and end of life. Accordingly, standards needed for the adoption of ICM should address all these phases of a building's life span.

9.4.3 *Low Demand*

Clients and customer demands plays a major role in the adoption of sustainable construction technologies. In the construction industry, clients are the major decision-makers. If a client is not demanding for the use of indigenous building methods, it will be difficult or impossible for contractors to implement appropriate technologies. The lack of demand as a barrier to ICM has been reported by Addy et al. [42] in Ghana. This challenge can be attributed to the lack of awareness on the benefits of indigenous construction methods [55]. Moreover, clients who may opt for indigenous construction methods may find it very expensive due to the limited skilled expertise. Every client wants to breakeven in time, cost, and quality. With very little tolerance to high prices, clients are often deterred from patronizing ICM.

9.4.4 Lack of Awareness and Education on Benefits

It is certain about humans that, without a good understanding of the importance of any system or way of life, adoption of these systems is hindered. It is important to note that, the use and adoption of any vernacular or bio-based construction method should emphasize on all round benefits. These are environmental, economic, and social benefits [45]. In most instances, social importance is the least, however, the adoption of any technology does not only depend on the efforts of policymakers and professionals but also the client's acceptance [56]. Agliata et al. [57] posits that, there is a general lack of knowledge on indigenous building methods. Therefore, it is important for knowledge sharing, awareness raising, and education on the benefits of adopting ICM. Aside the low level of awareness on benefits, the severity of negative effects of non-sustainable construction activities has not been fully understand. Awareness of the imperative to address climate change through building energy decarbonization remains limited though increasing among professionals and stakeholders. Many construction practitioners and clients lacking a clear understanding of its significance or the measures required for its implementation.

9.4.5 High Initial Cost and Low Level of Investment by Investors

The rate at which technology is adopted is also affected by the funding and investment it receives. Moreover, such technologies often entail higher initial costs due to factors such as limited local expertise, the need for specialized equipment, and deviations from conventional construction practices. Many technologies in the building industry are not known and therefore not adopted because they lack sufficient funding for research, exposure, education, and upgrade. Danja et al. [58] stated that, lack of funding and research is one major barrier hindering the adoption of vernacular and indigenous construction methods in developing countries. It is also important to note that, investment and funding is directly linked to the government. In many developing countries, and especially African countries, the government is often the largest single client in the construction sector, primarily due to its role in financing and procuring major public infrastructure projects. It is important for allocations of the national wealth to be made to drive research and innovation into indigenous methods and materials for construction and even other industries. Additionally, it is even more difficult for the small commercial producers of local building materials. Nwalusi and Okeke [18] argued that, local producers find it difficult to receive loans due to high interest rates from financial institutions. Overall, high initial cost coupled with inadequate funding combined with limited access to affordable financing for local producers, remains a critical barrier to research, innovation, and the wider adoption of ICM in developing countries.

9.4.6 *Low Level of Relevance and Loss of Indigenous Knowledge*

It has been identified that, neglect from construction professionals, government, and stakeholders is another barrier facing the adoption of ICM. There is a lack of appreciation of cultural heritage and poor acceptability of indigenous building materials and methods [58]. This is compounded by a taste for foreign construction methods and materials. Ibrahim [59] reported that in most developing countries, indigenous architecture has been largely forgotten, sometimes ignored, and mostly replaced by conventional construction methods that may not be sustainable. Modern construction has embraced the high-technological material leading to craftsmen of local building material gradually phased out of construction business. The loss of indigenous knowledge plagues its adoption in modern construction works [18]. With the passage of time, many foreign carbon intensive construction materials have been adopted and little regard is given to local sustainable materials.

9.5 Overcoming Barriers and Developmental Pathways

The goal of this discussion on indigenous construction methods is to explore its potential for building energy decarbonization. As has already been established, ICM and appropriate technology is a green building technology that promotes sustainable construction actives. Following the discussion, certain barriers that hinder the adoption of ICMs have been identified. This makes it possible for strategies to be formulated to overcome these barriers. Apart from acknowledging the barriers facing developing countries in adopting ICMs, various developmental strategies should be put in place to increase the pace of adoption of ICMs. This subsection discusses the pathways that can be taken by construction stakeholders to overcome the specific barriers above.

One of the major barriers that has been identified is the lack of skilled personnel and expertise in indigenous construction technologies [55]. The knowledge and expertise that professionals possess is an important factor in the successful adoption of ICMs. Due to the sustainable demand in construction activities, there is a high demand for workers who possess skills in green building technologies [43]. Having experts with good understanding in ICM in organizations will propel its adoption.

Therefore, providing extensive training and education on these indigenous building systems is the best way to overcome the barrier of the lack of skilled personnel [60]. If a project team has a very good green building ability, the adoption of green boiling technologies such as ICM will be increased. These trained can serve as channels to educate clients on the benefits of ICM as well. Moreover, the loss of indigenous knowledge will be restored into the industry [61].

The training of skilled personnel can further overcome the barrier where there is a problem with quality and strength requirements of ICMs. Professionals who have thorough understanding in the mechanical properties of indigenous construction materials are better informed on the solutions they can employ [62]. The problem of quality and strength is therefore mitigated through the use of composite and engineered materials. Many indigenous materials in modern construction do not possess certain physical strength needed if used solely. However, with the fusing or combination of materials, strength can be achieved [50]. In literature, the physical and mechanical characterization of vernacular materials such as earth-based materials, stone, and timber is vast [63–65]. Therefore, the use of experimental testing helps to simulate the structural behaviour of these material for building elements such as walls, roofs, and floors [66]. This experimental testing will inform the solutions needed for structural integrity. It has been noted by Paul et al. [67] that the addition of natural fibres to earthen materials can greatly increase strength properties and durability. In some other regions of the world, vernacular materials have been engineered with improved strength enough to stand against seismic events, flooding, and extreme weather condition [68, 69]. Consequently, advanced research and the provision of funding in this direction will aid in closing the knowledge gap.

The adoption of ICMs can be achieved through the availability of information on cost and benefits [43]. One of the major barriers hindering the adoption of ICM discussed above is the lack of knowledge on the benefits of ICMs [43]. Due to the high initial cost and investment associated with ICMs, stakeholders who possess adequate knowledge and information on life cycle benefits in terms of cost and environmental sustainability are likely to adopt ICMs and other green building technologies [60]. This developmental pathway is highly associated with research and knowledge sharing in the construction industry on appropriate technologies. Conducting robust and credible studies on ICMs will accurately show, verify, and quantify the cost benefits of ICMs on the long. This will attract the interest of clients, professionals, and investors [70].

The adoption of ICM also depends on support from organizations and the government [61]. As noted earlier, in most developing countries, the government is a major consumer of construction. There is also the need for cooperation of the major stakeholders of the construction industry for the implementation of appropriate technologies. The adoption behaviour of clients and professionals can be guided through the support of organizations and stakeholders by providing enough research funding and investments [60]. The results of such investment will expose many innovative ways to use and construct building that will suit modern architecture using indigenous materials. There is the need for the establishment of research centres to strengthen general sustainable construction research. The problem of lack of ICMs database and standards will also be solved [71]. Research conducted will provide adequate information on appropriate technologies and ICMs for the development of databases and standards needed to guide its adoption.

Apart from the investment in research, incentives can be given to clients who adopt ICMs. This can help in mitigating the barrier of low client demand. Incentives from the government and other privates' entities interested in sustainable building developments has a high propensity to influence and drive construction stakeholders to embrace ICMs [60]. Incentive policies are considered as one of the important strategies for the advance of general sustainable construction practice [60]. Qian et al. [72] also confirms that incentive schemes are crucial for the promotion of sustainable building developments. Financial incentives can provide a good economic support for the adoption of ICMs since the initial cost of adopting green building technologies is often high compared to traditional construction methods. Financial incentives can be in tax exemption and grants. In other countries, tax incentives where green building adopters are offered tax deduction and exemption have been one of the most important pathways taken by the government to promote sustainable construction activities [73]. Other financial incentives such as reduced interests on loans will help local craftsmen overcome the barrier of lack of capital in the business that involves the use indigenous construction materials.

One other important pathway towards ICM adoption is the establishment of mandatory government policies and regulations [61]. Government policies and regulations is an important strategy and driver for adoption of any sustainable technology across all industries. Despite the hinderances of high cost, low demand, and interest in sustainable construction activities, people will not adopt ICMs unless there are legislations surrounding it [60]. Hence, the importance of imposing mandatory rules and regulations for sustainable developments. These policies made mandatory will motivate and force the construction industry professionals to take relevant actions in order to avoid fines or penalties or even get rewards.

This strategy will mitigate the low level of importance given to ICMs [58]. When policies and regulations are formed and made mandatory on the use of indigenous building materials and technologies, its importance level will be increased. More specifically, investment in research and the collating information to form a database could also be useful in green labelling and rating of ICMs. Building constructed with indigenous building materials have been shown to be sustainable to improve energy efficiency [1]. A green building rating will give owners some form of recognition which can motivate further adoption from other clients and developers [60]. Green building rating systems are important to GBTs adoption promotion because they provide useful information and helpful guidance concerning green building to the industry professionals and the public [60]. Although there are global green rating systems such as LEED, BREEAM, Green star, and others, it is essential for region specific and authoritative green systems to be established. This is necessary due to the variations climate, technology, policies, and regulations and resources. This argument is hinged on the concept of appropriate technology defined already in this chapter. Furthermore, green rating systems are valuable in producing more accredited and skilled green building professionals [61].

9.6 Summary

This chapter focused on indigenous construction methods and appropriate technologies that can be adopted towards building energy decarbonization. Indigenous construction methods are sustainable construction technologies that are being adopted in many developed countries. This is due to the strategies put in place to foster its adoption. Appropriate technology and indigenous construction methods go hand in hand. From the discussion, some of the indigenous construction methods identified in Africa were, adobe construction, rammed earth construction, timber construction, wattle and daub construction, and pile dwelling. These indigenous methods were also found to be appropriate technologies for certain places. Despite the potential of these indigenous construction methods, it was found that the adoption rate in developing countries, especially African countries is very low. The chapter then discussed some of the pertinent barriers hindering the adoption of ICMs. Lack of skilled personnel, lack of standards and codes, low demand, low quality and strength demand, and low investments were some of the major barriers discussed in this chapter. With these barriers identified, further discussion was done on strategies and developmental pathways that can be used to mitigate the barriers. Education and training of construction stakeholders, research, and funding into ICMs, financial incentives for adopters, and establishment of standards and codes, were some of the developmental pathways discussed. By proffering these pathways, the development of innovative and create construction methods may be enhanced towards meeting the energy demands of construction.

References

1. Agyekum K, Kissi E, Danku JC (2020) Professionals' views of vernacular building materials and techniques for green building delivery in Ghana. Sci Afr 8. https://doi.org/10.1016/J.SCIAF.2020.E00424
2. Yang X, Lyu H, Li J, Fu F, Zhou H (2022) Comprehensive design for a vernacular wood house with energy-saving process. Sustain Energy Technol Assess 53:102571. https://doi.org/10.1016/J.SETA.2022.102571
3. Kusi E, Boateng I, Danso H (2024) Energy consumption and carbon emission of conventional and green buildings using building information modelling (BIM), Int J Build Pathol Adapt
4. Santamouris M, Vasilakopoulou K (2021) Present and future energy consumption of buildings: challenges and opportunities towards decarbonisation. E-Prime Adv Electr Eng Electron Energy 1:100002. https://doi.org/10.1016/J.PRIME.2021.100002
5. Moran P, O'Connell J, Goggins J (2020) Sustainable energy efficiency retrofits as residenial buildings move towards nearly zero energy building (NZEB) standards. Energy Build 211:109816. https://doi.org/10.1016/J.ENBUILD.2020.109816
6. Cabeza LF, Boquera L, Chàfer M, Vérez D (2021) Embodied energy and embodied carbon of structural building materials: worldwide progress and barriers through literature map analysis. Energy Build 231:110612. https://doi.org/10.1016/J.ENBUILD.2020.110612
7. Dascalaki EG, Argiropoulou P, Balaras CA, Droutsa KG, Kontoyiannidis S (2021) Analysis of the embodied energy of construction materials in the life cycle assessment of Hellenic

residential buildings. Energy Build 232:110651. https://doi.org/10.1016/J.ENBUILD.2020.110651
8. Schwarz M, Nakhle C, Knoeri C (2020) Innovative designs of building energy codes for building decarbonization and their implementation challenges. J Clean Prod 248:119260. https://doi.org/10.1016/J.JCLEPRO.2019.119260
9. Shove E (2018) What is wrong with energy efficiency? Build Res Inf 46(7):779–789. https://doi.org/10.1080/09613218.2017.1361746
10. Obergassel W, Xia-Bauer C, Thomas S (2023) Strengthening global climate governance and international cooperation for energy-efficient buildings. Energ Effi 16(8):1–32. https://doi.org/10.1007/S12053-023-10177-7/TABLES/4
11. O'Rourke T (2020) Uses of the vernacular in the design of indigenous housing. Fabrications 30(1):68–91. https://doi.org/10.1080/10331867.2020.1721104
12. Genovese PV, Zoure AN (2023) Architecture trends and challenges in sub-Saharan Africa's construction industry: a theoretical guideline of a bioclimatic architecture evolution based on the multi-scale approach and circular economy. Renew Sustain Energy Rev 184:113593. https://doi.org/10.1016/J.RSER.2023.113593
13. Buitenhuis J, Zelenika I, Pearce JM (2010) Open design-based strategies to enhance appropriate technology development. In: Open: NCIIA 14th annual conference
14. Pearce JM (2012) The case for open source appropriate technology. Environ Dev Sustain 14:425–431
15. Hazeltine B, Bull C (1998) Appropriate technology; tools, choices, and implications. Academic Press, Inc.
16. Schumacher EF (1973) Small is beautiful: economics as if people mattered: 25 years later with commentaries. Hartley & Marks Publishers. Inc., Point Roberts, WA
17. Willoughby KW (2019) Technology choice: a critique of the appropriate technology movement. Routledge
18. Nwalusi DM, Okeke FO (2021) Adoption of appropriate technology for building construction in the tropics; a case of Nigeria. IOP Conf Ser Earth Environ Sci 730(1). https://doi.org/10.1088/1755-1315/730/1/012013
19. Chetia S, Prashad D (2010) New architecture and urbanism: development of Indian traditions. Cambridge Scholars Publishing
20. Tekpe E, Ansah SK, Akomah BB (2022) Appropriate technology and design: a solution for sustainable and affordable housing delivery in major cities of Ghana. J Eng Res Rep 23(8):18–29. https://doi.org/10.9734/jerr/2022/v23i8739
21. Patnaik J, Bhowmick B (2018) Appropriate technology: revisiting the movement in developing countries for sustainability. Int Sch Sci Res Innov 12(3):301–305
22. Ghadge VJ, Konnur B, Landage AB (2022) Sustainable development through appropriate technologies. J Inter Des Reg Plan 7(1):24–27. https://doi.org/10.46610/JOIDRP.2022.V07I01.002
23. Kodai Kaje B, Katini K, Krelo Peter Kajeo K, Andrew Thomas K (2023) Indigenous house construction of the Mao Nagas: a sustainable perspective. Mater Today Proc. https://doi.org/10.1016/J.MATPR.2023.06.038
24. Ramakrishnan S, Loganayagan S, Kowshika G, Ramprakash C, Aruneshwaran M (2021) Adobe blocks reinforced with natural fibres: a review. Mater Today Proc 45:6493–6499. https://doi.org/10.1016/J.MATPR.2020.11.377
25. Parisi F, Tarque N, Varum H, Vargas-Neumann J (2021) Adobe constructions in the world: a first overview. In: Structural characterization and seismic retrofitting of adobe constructions: experimental and numerical developments, pp 1–14
26. Costa C, Cerqueira Â, Rocha F, Velosa A (2018) The sustainability of adobe construction: past to future. Int J Arch Herit
27. Kosarimovahhed M, Toufigh V (2020) Sustainable usage of waste materials as stabilizer in rammed earth structures. J Clean Prod 277:123279
28. Wyżgoł M, Deptuła A (2020) Living outside the wall: The quarter of wattle-and-daub houses in Old Dongola of the Funj Period. J Afr Archaeol 18(1):109–132

29. Alrashed F, Asif M, Burek S (2017) The role of vernacular construction techniques and materials for developing zero-energy homes in various desert climates. Buildings 7(1):17
30. Teartisup S (2007) Technology for residence in chao phraya river estuary: a case study of Sakla Village. J Environ Des Plan 2:97–108
31. Akomah BB, Ansah SK, Tekpe E (2022) Appropriate technology and design: A solution for sustainable and affordable housing delivery in major cities of Ghana. J Eng Res Rep 25(8):18–29
32. Essienyi E (2012) Traditional building methods in Southern Ghana. Afford Hous
33. Ibrahim AO (2021) The mud traditional architecture of the Sudan and Saudi Arabia: the difference in employment techniques. Eximia 3(1):100–116. https://www.eximiajournal.pluscommunication.eu/index.php/eximia/article/view/83. Accessed 24 Jun 2024
34. Amicone S, Croce E, Castellano L, Vezzoli G (2020) Building Forcello: Etruscan wattle-and-daub technique in the Po Plain (Bagnolo San Vito, Mantua, northern Italy). Archaeometry 62(3):521–537. https://doi.org/10.1111/ARCM.12535
35. Salingaros A, María J, Pardo F (2023) Challenges and current research trends for vernacular architecture in a global world: a literature review. Buildings 13(1):162. https://doi.org/10.3390/BUILDINGS13010162
36. Fischer H, Aichholzer M, Korjenic A (2023) Ecological potential of building components in multi-storey residential construction: a comparative case study between an existing concrete and a timber building in Austria. Sustainability 15(8):6349
37. Bucklin O, Di Bari R, Amtsberg F, Menges A (2022) Environmental impact of a mono-material timber building envelope with enhanced energy performance. Sustainability 15(1):556
38. Puspitasari P, Kadri T, Indartoyo I, Kusumawati L (2018) Microclimate and architectural tectonic: vernacular floating house resilience in Seberang Ulu 1, Palembang. In: IOP conference series: earth and environmental science. IOP Publishing, p 12118
39. Wicaksono B, Siswanto A, Kusdiwanggo S, Anwar WFF (2019) The identification of the existence of stilt houses at Musi riverside settlement in Palembang. In: IOP conference series: materials science and engineering. IOP Publishing, p 12001
40. Milošević D, Trbić G, Savić S, Popov T, Ivanišević M, Marković M, Ostojić M, Dunjić J, Fekete R, Garić B (2022) Biometeorological conditions during hot summer days in diverse urban environments of Banja Luka (Bosnia and Herzegovina). Geographica Pannonica 26(1)
41. Cheikhi W, Baba K, Nounah A, Cherradi C (2020) Effect of adding insulation on the energy performances of rammed earth buildings in hot and arid climates. In: E3S Web of Conferences. EDP Sciences, p 1011
42. Addy M, Adinyira E, Danku JC, Dadzoe F (2020) Impediments to the development of the green building market in sub-Saharan Africa: the case of Ghana. Smart Sustain Built Environ 10(2):193–207. https://doi.org/10.1108/SASBE-12-2019-0170/FULL/PDF
43. Acheampong A, Hackman A, Ayarkwa J, Agyekum J (2014) Factors inhibiting the use of indigenous building materials (IBM) in the Ghanaian construction industry, Africa development and resources research institute (ADRRI). J ADRRI J (www.adrri.org) 8(8): 2343–6662
44. Durrant G (2020) Why hemp should be used as a sustainable building material. Innovation News Network. https://www.innovationnewsnetwork.com/why-hemp-shouldbeused-as-a-sustainable-building-material/6022
45. Agyekum K, Adinyira E, Oppon JA (2021) Factors limiting the adoption of hemp as an alternative sustainable material for green building delivery in Ghana. Int J Build Pathol Adapt 40(2):202–218
46. Debrah C, Owusu-Manu D-G, Kissi E, Oduro-Ofori E, Edwards DJ (2020) Barriers to green cities development in developing countries: evidence from Ghana. Smart Sustain Built Environ 11(3):438–453
47. Elert K, Baños EG, Velasco AI, Bel-Anzué P (2021) Traditional roofing with sandstone slabs: implications for the safeguarding of vernacular architecture. J Build Eng 33:101857
48. Chowdhooree I, Das KK (2022) Indigenous knowledge of mud architecture: experiences of surviving against multiple natural hazards. Int J Disaster Resil Built Environ 13(4):451–469. https://doi.org/10.1108/IJDRBE-12-2020-0128/FULL/PDF

49. Amin MN, Ahmad W, Khan K, Ahmad A (2022) A comprehensive review of types, properties, treatment methods and application of plant fibers in construction and building materials. Materials 15(12):4362
50. Fernandes J, Peixoto M, Mateus R, Gervásio H (2019) Life cycle analysis of environmental impacts of earthen materials in the Portuguese context: Rammed earth and compressed earth blocks. J Clean Prod 241:118286
51. Jannat N, Hussien A, Abdullah B, Cotgrave A (2020) Application of agro and non-agro waste materials for unfired earth blocks construction: a review. Constr Build Mater 254:119346
52. Sun X, He M, Li Z (2020) Novel engineered wood and bamboo composites for structural applications: state-of-art of manufacturing technology and mechanical performance evaluation. Constr Build Mater 249:118751
53. Chan APC, Darko A, Olanipekun AO, Ameyaw EE (2018) Critical barriers to green building technologies adoption in developing countries: the case of Ghana. J Clean Prod 172:1067–1079
54. Banihashemi S, Hosseini MR, Golizadeh H, Sankaran S (2017) Critical success factors (CSFs) for integration of sustainability into construction project management practices in developing countries. Int J Proj Manage 35(6):1103–1119
55. Darko A, Chan APC (2017) Review of barriers to green building adoption. Sustain Dev 25(3):167–179. https://doi.org/10.1002/SD.1651
56. Sijtsema SJ, Onwezen MC, Reinders MJ, Dagevos H, Partanen A, Meeusen M (2016) Consumer perception of bio-based products—an exploratory study in 5 European countries. NJAS Wageningen J Life Sci 77:61–69
57. Agliata R, Marino A, Mollo L, Pariso P (2020) Historic building energy audit and retrofit simulation with hemp-lime plaster—a case study. Sustainability 12(11):4620
58. Danja II, Dalibi SG, Safarov A (2019) Hindrances to vernacular architecture of Northern Nigeria
59. Ibrahim HGA (2016) Regeneration of sustainability in contemporary architecture: approach based on native function and activities to strengthen identity. Procedia Soc Behav Sci 216:800–809
60. Chan APC, Darko A, Ameyaw EE (2017) Strategies for promoting green building technologies adoption in the construction industry—an international study. Sustainability 9(6):969
61. Darko A, Chan APC (2018) Strategies to promote green building technologies adoption in developing countries: the case of Ghana. Build Environ 130:74–84. https://doi.org/10.1016/J.BUILDENV.2017.12.022
62. Debrah C, Darko A, Oduro-ofori E, Acquah C, Asamoah E, Debrah C, Darko A, Oduro-E (2023) Drivers for green cities development in developing countries: Ghanaian perspective drivers for green cities development in developing countries: Ghanaian perspective. Int J Constr Manag 23(6):1086–1096. https://doi.org/10.1080/15623599.2021.1955321
63. Abhilash HN, Hamard E, Beckett CTS, Morel J-C, Varum H, Silveira D, Ioannou I, Illampas R (2021) Mechanical behaviour of earth building materials. In: Testing and characterisation of earth-based building materials and elements: state-of-the-art report of the RILEM TC 274-TCE. Springer, pp 127–180
64. Liu L, Yao Y, Zhang L, Wang X (2022) Study on the mechanical properties of modified rammed earth and the correlation of influencing factors. J Clean Prod 374:134042
65. Parlato, M., Porto, S.M.C. and Cascone, G. (2021) Raw earth-based building materials: An investigation on mechanical properties of Floridia soil-based adobes, Journal of Agricultural Engineering, 52(2).
66. Galassi S, Ruggieri N, Dipasquale L, Tempesta G (2018) Assessment of the Moroccan vernacular timber roof: a proposal for an eco-friendly strengthening system. J Archit Conserv 24(3):224–248
67. Paul S, Islam MS, Elahi TE (2022) Comparative effectiveness of fibers in enhancing engineering properties of Earth as a building material: a review. Constr Build Mater 332:127366
68. Bothara J, Ingham J, Dizhur D (2022) Qualifying the earthquake resilience of vernacular masonry buildings along the Himalayan arc. J Build Eng 52:104339

69. D'Ayala D, Wang K, Yan Y, Smith H, Massam A, Filipova V, Pereira JJ (2020) Flood vulnerability and risk assessment of urban traditional buildings in a heritage district of Kuala Lumpur, Malaysia. Nat Haz Earth Syst Sci 20(8):2221–2241
70. Darko A, Chan APC, Yang Y, Shan M, He BJ, Gou Z (2018) Influences of barriers, drivers, and promotion strategies on green building technologies adoption in developing countries: the Ghanaian case. J Clean Prod 200:687–703. https://doi.org/10.1016/J.JCLEPRO.2018.07.318
71. Wang W, Zhang S, Su Y, Deng X (2019) An empirical analysis of the factors affecting the adoption and diffusion of GBTS in the construction market. Sustainability 11(6):1795
72. Qian, Q.K., Fan, K. and Chan, E.H.W. (2019) Regulatory incentives for green buildings: gross floor area concessions, in Building Governance and Climate Change. Routledge, pp. 173–190.
73. He L, Chen L (2021) The incentive effects of different government subsidy policies on green buildings. Renew Sustain Energy Rev 135:110123
74. Gomaa M, Jabi W, Soebarto V, Xie YM (2022) Digital manufacturing for earth construction: a critical review. J Clean Prod 338:130630
75. Terraviva (2023) Rammed earth—Terraviva. Available at: https://terraviva.pt/the-portuguese-way/. Accessed 2 Dec 2025
76. Strautmane E (2023) Nzulezu (Nzulezo): the stilt village of Ghana. GhanaTRVL. Available at: https://ghanatrvl.com/places-to-see/nature/nzulezu-nzulezo-the-stilt-village-of-ghana/. Accessed 2 Dec 2025
77. Sudbury S (2015) Sudbury's architecture school taking form. Sudbury Star. Available at: https://www.thesudburystar.com/2015/06/05/sudburys-architecture-school-taking-form. Accessed 2 Dec 2025
78. Neenu SK (2021) Why is timber construction popular in the 21st century? TheConstructor.org. Available at: https://theconstructor.org/building/timber-construction-popularity/21342/
79. Gonzalez CM (2013) Wattle and daub. Earth architecture. Available at: https://eartharchitecture.org/?tag=wattle-and-daub. Accessed 2 Dec 2025
80. Brinson L (2023) How adobe construction works. How Stuf Works. Available at: https://home.howstuffworks.com/home-improvement/construction/materials/adobe-construction.htm. Accessed 2 Dec 2025
81. Puisis E (2024) What is an adobe style house. Available at: https://www.thespruce.com/what-is-an-adobe-style-house-7151962. Accessed 2 Dec 2025

Chapter 10
Renewable Energy: Solar Energy

10.1 Introduction

Energy is a basic necessity for the human and socio-economic growth of a country [1]. The excessive use of fossil fuels for energy generation leads to many environmental problems affecting human life [2]. These problems range from the emission of greenhouse gases, deterioration of air quality, oil spills, acidic rain, and many other problems. In recent times, actions against climate change have gained popularity in both industry and academia The emissions of greenhouse gases from the various industrial sectors of the African economy are contributing to the deterioration of the environment [3]. For instance, the manufacturing industry and the construction industry depend heavily on energy generated [4, 5]. Particularly, electrical energy demand for manufacturing and operation of buildings has increased due to the increase in population and urbanization in Africa [6]. Energy generation in many African countries is by burning fossil fuels [7], This contributes large volumes of greenhouse gas emissions into the atmosphere. Efforts to reduce this consumption rate and contribution to climate change is an ongoing investigation. A notable strategy that is in adoption is the generation of electrical energy through renewable resources such as solar energy [8]. Solar energy offers regions an opportunity to improve their economic status and minimize their global carbon footprint and greenhouse emissions [9]. The abundance of solar energy and its accessibility hold strong potential in addressing the fight against climate change and energy poverty [10]. The sun can deliver more energy per hour than what the Earth uses in a year [11]. The sun can give 1.7×10^{22} Joules of energy in 1.5 days. This energy supplied is equivalent to all the energy that can be generated by 3 trillion barrels of total oil resources found on Earth. The total yearly energy used by the earth in a year is approximately 6×10^{20} Joules and this amount of energy is supplied by the sun in 1 h [12]. In essence, solar energy plays a very significant role in the future global energy needs and most especially, in developing countries [13].

M. N. Addy et al., *Decarbonization of Building Energy in Developing Countries*, Green Energy and Technology, https://doi.org/10.1007/978-3-032-09113-0_10

Africa is one of the highest receivers of solar energy at a level ranging from 2000KWh/M^2 to 2500Kh.m^2 yearly [14]. This enormous supply of solar energy positions the African continent to harness solar energy on a large scale to boost its economy, The countries located in the Sahel region, the southern region, and the Northern region show the highest solar potential. Neher et al. [15], providing a pathway and an opportunity for energy diversification and an increase in the supply of electrical energy (Fig. 10.1).

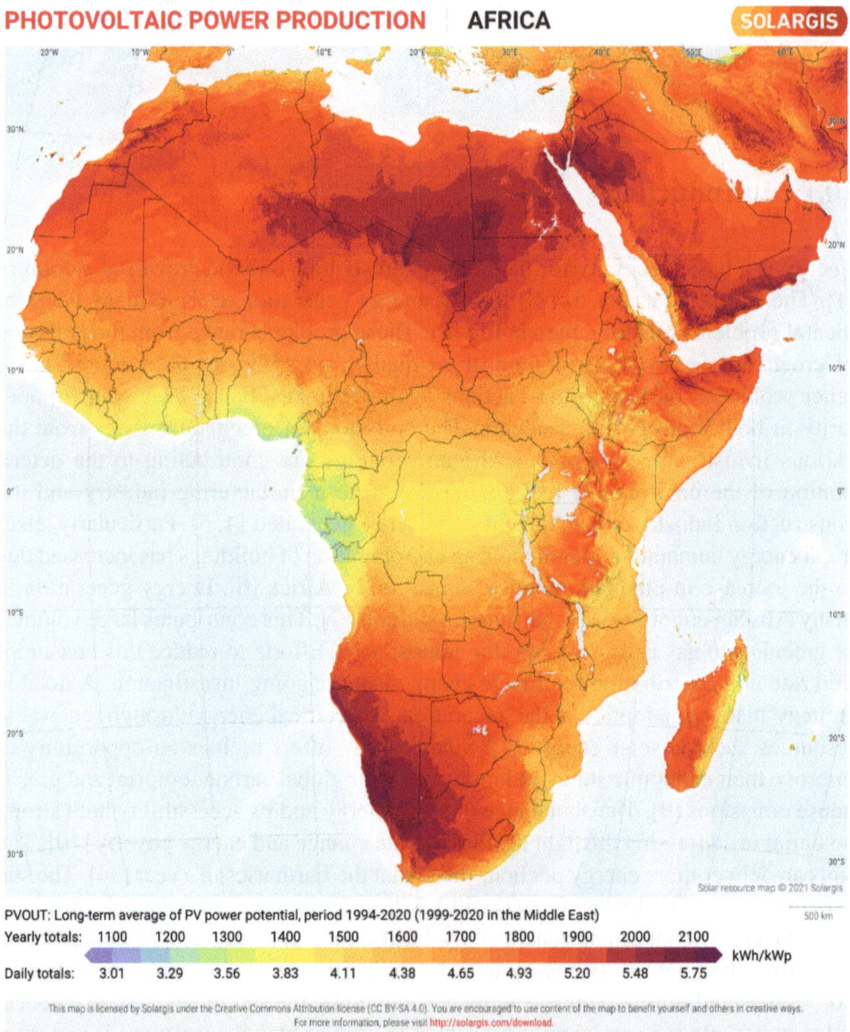

Fig. 10.1 Solar PV potential of Africa [160]. *Source* https://solargis.com/resources/free-maps-and-gis-data?locality=africa

In addition, it is estimated that solar photovoltaic potential in Africa exceeds 11 terawatts. Agbo et al. [16] and Urbina [17], hugely surpassing the continent's current and future electricity demand. Solar energy is therefore a reliable and consistent source and solution. Additionally, decentralized solar energy systems, such as off-grid and mini-grid systems, offer significant potential to electrify remote and underserved regions, contributing to social and economic development [18]. Despite these numerous advantages the limited access to financing, policy and regulatory issues, weak energy infrastructure support for solar energy, and low technical expertise militates against the usage of this renewable energy. This chapter explores the barriers to using solar-based renewable energy and situates it within the context of the African building energy market. It should be noted that in this book, the exploration of solar-based renewable energy is viewed from the demand side perspective and not the supply side. The chapter provides pathways to drive the uptake of solar-based renewables within the continent.

10.2 Applications of Solar-Based Renewables and Solar Energy Potential in Africa

Solar energy is received from solar radiation captured through devices such as photovoltaic panels to produce electricity [19]. Solar radiation in Africa surpasses the daily power needs for household lighting, water pumping, community and urban powering, and street lighting. Sub-Saharan Africa has a very high solar radiation intensity of 300 W/m^2 and positive trends of up to 5 W/m^2 per decade [15]. This can meet domestic needs. In addition, a country like Kenya possesses a significant solar energy potential of 4–6 $KWh/m^2/day$ which is equal to 250 million tons of oil daily [20]. The rising cost of oil-based power generation with its environmental factors is driving many African countries to solar energy adoption. Africa is adopting various solar-based technologies for different application [21]. For example, due to the high temperatures solar-driven cooling systems are receiving some attention, particularly in rural areas that do not have access to grid electricity through the installation of photovoltaic-powered ice makers in the coastal Kenyan areas [22]. Additionally, electricity is being generated by solar thermal energy technologies to provide hot water in both urban and rural areas [23]. The common solar technologies used for power generation from 2013 to 2022 were concentrated solar power (CSP) and solar photovoltaic systems [22]. CSP adoption has been slowly growing up to now because it is very expensive and is not well known in Africa [24]. Currently, countries such as Algeria, Egypt, Morocco, and South Africa are the ones developing CSP technologies in Africa [25]. However, PV technology is growing because it is simpler and less expensive as compared to other forms of solar technology in Africa [26]. Despite the abundance of solar resources in Africa, its adoption remains low as compared to other parts of the world with even lesser solar resource [27].

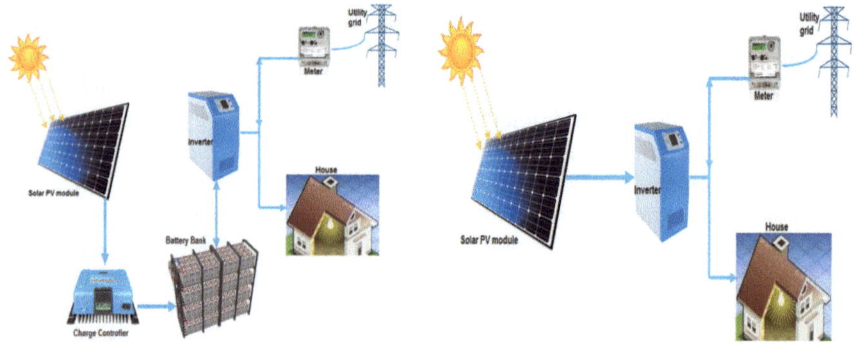

Fig. 10.2 Off-grid and grid connector PV system [35]

10.3 Typologies of Solar Energy Technologies

10.3.1 Photovoltaic Solar System

One of the major solar-based technologies is the photovoltaic system [28]. It is one of the most widely used solar technology for converting sunlight into electricity. PV technology fundamentally works using PV cells based on semiconduction, where photons that fall on the PV cells work on the electrons in the semiconductor, thus generating electricity [29]. PV technology is a reliable and clean way of generating electricity directly from the sun [30]. Its application ranging from small to large isolated and grid connections has become common in the world. Solar PV applications in buildings are mainly in two ways: Building integrated PV systems (BIPV) and building adapted PV systems (BAPV) [31]. With BIPV, the PV modules are integrated into the building elements such as walls and roofs [32] whereas BAPV allows for the PV modules to be mounted on the building elements [33]. The global installed capacity has increased from 483.1 GW in 2018 to 580.2 GW in 2019, revealing about 21% relative growth. The contribution of this installation is mainly from Asia, followed by China (175 GW), Japan (55.5 GW), and India (26.8 GW). Europe comes in the second rank in terms of PV installed capacity with considerable shares in Germany (45.9 GW), Italy (20.12 GW), and the UK (13.4 GW) [34]. Despite the enormous solar potential in Africa, the application of solar PV systems in the continent does not appear on a global scale. This further reveals how low application is in the region (Fig. 10.2).

10.3.2 Concentrated Solar Power (CSP)

Another type of solar-based technology for energy generation is the concentrated solar power (CSP) system. This technology does not use PV cells but by using optical devices such as lenses and mirrors to direct and concentrate sunlight into a small area [24]. The concentrated energy is transformed into heat which propels a steam turbine to generate electricity [36]. CSP is suitable for bulk power generation. This is particularly beneficial for balancing the intermittency of other renewable sources like wind and solar PV [37]. Various CSP technologies include:

1. **Parabolic trough systems** which consist of parallel rows of curved high-reflectance mirrors
2. **Linear Fresnel reflectors (LFRs)** are similar to the parabolic trough, but they use linear rows of mirrors to reflect the sun rays onto a flat fixed receiver,
3. **Solar towers or central receivers** use thousands of heliostats to concentrate the sun rays to one central receiver placed at a high level of the constructed tower
4. **Parabolic dish CSP** uses a dish to concentrate the direct normal irradiation to a central point [38]. CSP has been noticed to be a reliable and efficient system for high-insolation areas such as Northern Africa [25]. Furthermore, Africa has a very high potential to be an electricity seller in Europe due to the high solar irradiance [39]. One very important element of the CSP is thermal energy storage for storing excess heat from the solar field and using it when needed [36, 40]. According to recent reports from the IEA, CSP could generate 11.3% of global electricity with suitable government support [41]. Across the world, CSP application is seen being dominant in Spain (2.3 GW), the United States (1.5 GW), and China (596 MW) [40] (Fig. 10.3).

10.3.3 Solar Water Heating System

Solar water heating systems (SWHs) use solar thermal collectors to collect sunlight and convert it to heat which is then used to heat water for domestic, commercial, and industrial purposes [43]. This is one of the earliest solar-based technologies and has received numerous research and innovation [44]. The working principle of SWH systems is relatively straightforward. The solar collectors are mostly installed on roofs and open fields to capture sunlight and convert it into heat [45]. The heat energy is transferred to the water that is circulating through the system, either directly (unpressurized systems) [46] or through a heat exchanger (closed-loop system) [47]. Modern innovations of SWH systems integrate both flat-plate collectors and evacuated tube collectors [48]. Evacuated tube collectors have been noted to be more efficient [49]. SWH systems are suitable for residential and commercial applications. In developing countries and Africa, SWH offers a cost-effective remedy to cater for the daily hot water needs, thus, reducing dependence on centralized energy infrastructure [50]. Due to the difference in climate conditions, SWH designs are evolving. For example, in cold climates, storage tanks are insulated, and anti-freeze technologies are incorporated into the design to prevent heat loss and freezing [51]. In contrast, in hot and arid regions, the design is optimized to maximize heat absorption while maintaining minimum heat to prevent overheating [52] (Fig. 10.4).

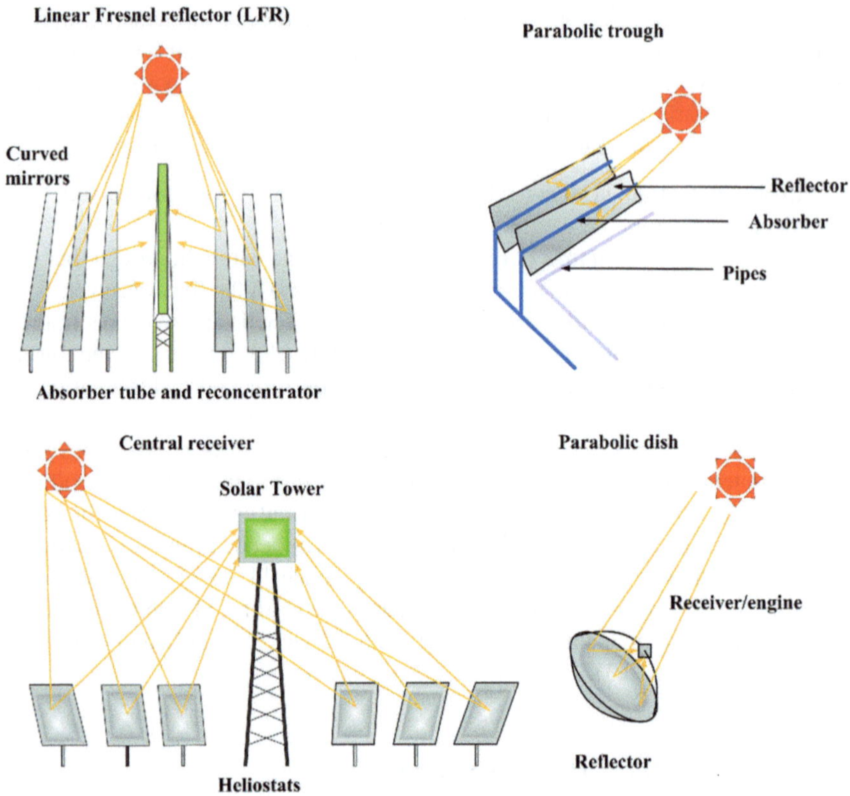

Fig. 10.3 Diagrams of CSD technologies [42]

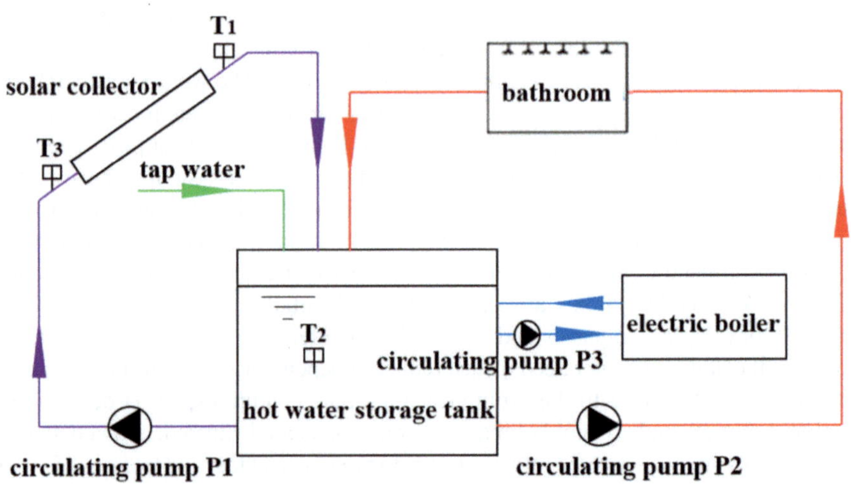

Fig. 10.4 Schematic diagram of solar-assisted electric boiler water heating system [44]

10.3.4 Solar Drying Systems

Solar energy is also applied in drying techniques across all industries [53]. Particularly in the agricultural sector for the drying and preservation of agricultural, marine, and herbal products [54]. The drying process advantages include sustaining the product flavor, nutrients, and quality, improving its appearance and life for human consumption [55]. The working process involves heat transfer by radiation from the direct sun, from the atmosphere to the wet material or product by convection and by conduction from the product surface to the interior parts or from the drying trays inside the drying chamber that is in contact with the product [56]. Solar dryers can be grouped according to two main aspects the movement of air inside the dryer, and product exposure to solar radiation. There is also a third classification which is a combination of the two, the hybrid system that is shown below. Apart from the agricultural sector, drying is also required in the timber industry. The two major drying methods are air drying and kiln drying. Solar drying systems can also be used in the drying process of timber to reduce moister before further wood processing [57]. In the industrial sector, drying is an important process in the food, chemical, textile, cement, and pharmaceutical products production [58]. This is a very energy-intensive process. The potential of this technology is great for adoption and the reduction of power demand produced by fossil fuels. However, at this point, both domestic and industries rely on open solar and conventional fuel for the drying process [55]. Numerous solar dryers have been developed and studied worldwide. Several authors have performed comprehensive reviews on solar dryers' design, types, and performances [59–61] (Figs. 10.5, 10.6 and 10.7).

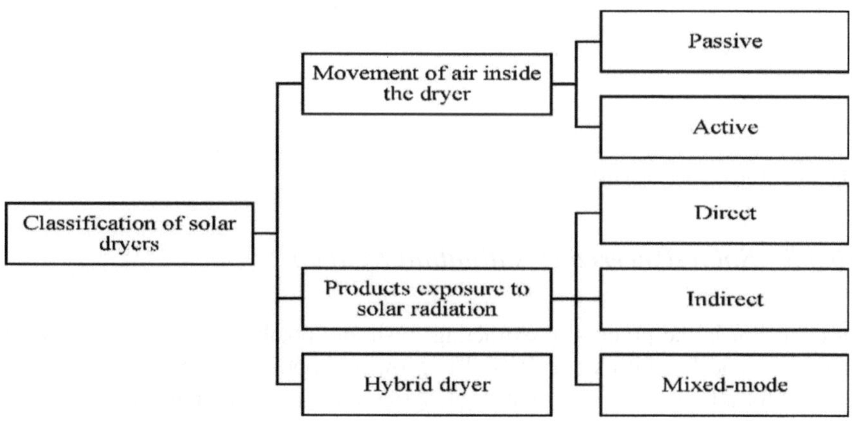

Fig. 10.5 Classifications of solar drying systems [55]

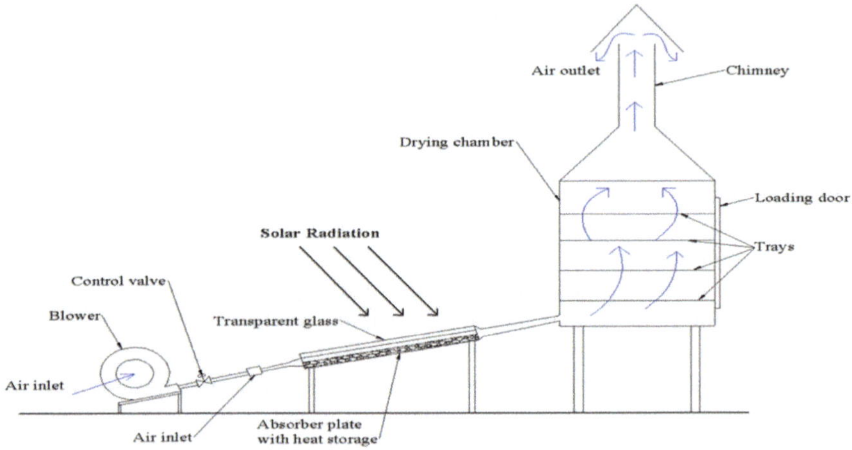

Fig. 10.6 Schematic diagram of an active solar dryer system [62]

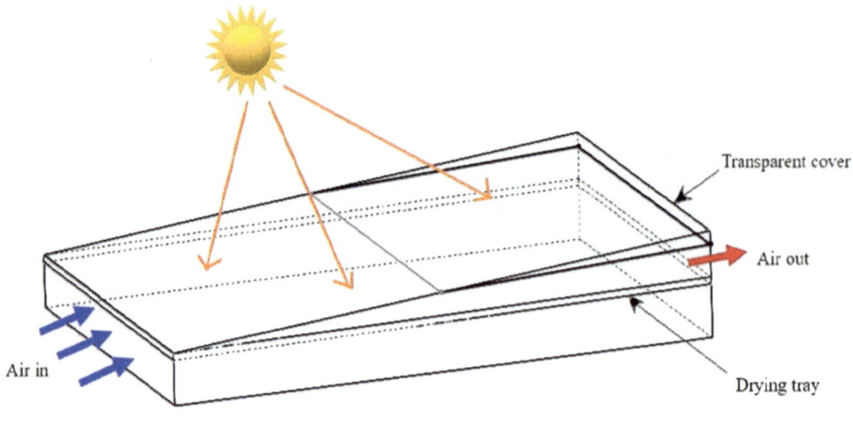

Fig. 10.7 Direct solar dryer system [63]

10.3.5 Solar-Powered Desalination Systems

Desalination is the process of extracting fresh and potable water from salt water (seawater) and brackish water [64]. The growing population increases freshwater demand for all vital activities: industrial, agricultural, and drinking, thereby, desalination capacity has grown significantly. Traditional desalination processes depend heavily on energy generated by fossil fuels, resulting in carbon emissions and global warming [65]. The development of new desalination technologies has received much attention and research recently [66, 67]. These systems are driven by heat from sources such as solar energy using environmentally friendly working fluids to generate potable water [68]. Solar power desalination is an emerging technology

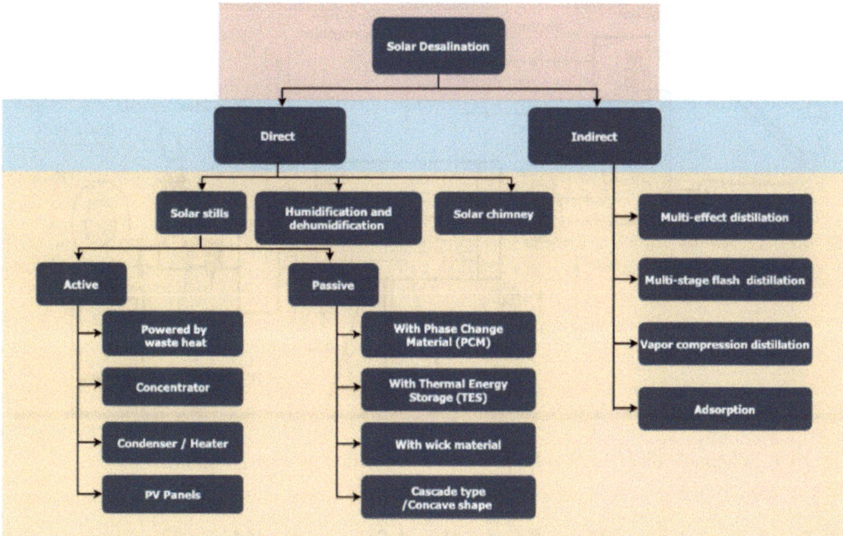

Fig. 10.8 Types of solar-powered desalination systems [68]

that addresses the dual challenges of water scarcity and energy access by using solar energy to drive the desalination of seawater or brackish water [67]. Given that many regions particularly Africa and the Middle East suffer from severe water shortages, this technology holds significant promise for sustainable water management [69]. The desalination process separates the salt from the salty water through solar radiation in evaporation and condensation [70]. The two major solar desalination classifications are direct and indirect processes. The direct process combines the desalination and solar radiation processes whereas in the indirect system, the process is separated into the solar collector part and the desalination part [71]. Solar desalination is a clean and useful way to solve the water shortage problems of the African continent, however, due to challenges, particularly in funding and technical expertise, the adoption of this technology is very low [72] (Figs. 10.8 and 10.9).

10.4 Solar Adoption in Africa

Most of the examples of solar energy projects undertaken in Africa are solar photovoltaic technology. Very few countries have adopted concentrated solar power systems or any other technology. Below is a list of solar projects in Africa out of the many projects undertaken so far (Table 10.1).

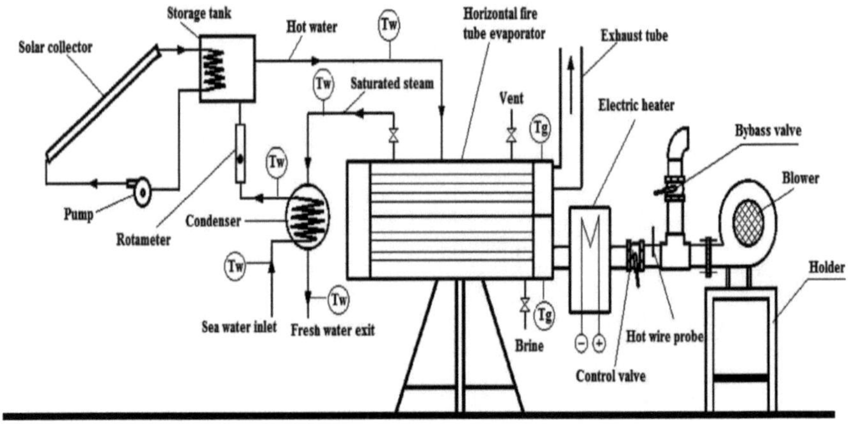

Fig. 10.9 Schematic diagram of solar-powered desalination process [73]

10.5 Importance/Benefits to the African Building Energy Market

Africa and developing countries have many opportunities to benefit from sustainable energy solutions such as solar energy. These benefits are seen in global alliances, technical and technological advancements, energy efficiency, resilience, and security, reduction in dependency on the national grid, economic growth, and many more [86]. Solar energy is a promising approach to addressing the energy crisis in Africa. Solar energy can provide communities with a dependable and constant supply of electricity, resulting in self-sufficiency in energy generation. Solar energy technology is a catalyst for socio-economic growth, allowing communities to establish businesses, provide power to educational facilities and healthcare [87]. Africa stands in a good position in the energy revolution due to the high amount of sunlight in the region. Embracing the opportunities in solar adoption will lead to the continent's energy resilience, security, and sustainable development goals [88].

10.5.1 Sustainable Energy Generation

Fossil fuels have a high negative impact on the environment when used for the generation of power. The carbon emissions associated with burning fossil fuels make it an unsustainable choice for energy production [89]. The use of renewable sources such as solar energy for power generation is a sustainable way that reduce the emissions of greenhouse gases into the atmosphere. It has a good potential for sustainable development and green growth in the African region [87]. Solar energy also aligns with global sustainable goals for clean energy and actions against climate change. By taking advantage of the sun's energy, Africa can contribute her quota to

Table 10.1 Solar projects in Africa

Country	Location	Solar project name	Generation capacity (Megawatt) (MW)	Type of solar technology
Morocco	Ouarzazate	Noor Ouarzazate Solar Complex	580	Concentrated Solar Power (CSP)
South Africa	Kenhardt	Kenhardt Hybrid Solar and Battery Facility	540	Photovoltaic (PV) + battery storage
Egypt	Aswan Governorate	Benban Solar Park	1,800	Photovoltaic (PV)
Ghana	Nzema	Nzema Solar Power Station	155	Photovoltaic (PV)
Burkina Faso	Ouagadougou	Zagtouli Solar Power Station	33	Photovoltaic (PV)
Kenya	Garissa	Garissa Solar Power Station	54.64	Photovoltaic (PV)
Namibia	Omburu	Omburu Solar Power Station	20	Photovoltaic (PV)
Rwanda	Rwamagana District	Agahozo-Shalom Youth Village Solar Plant	8.5	Photovoltaic (PV)
Senegal	Bokhol	Bokhol Solar Power Plant	20	Photovoltaic (PV)
Tanzania	Shinyanga	Shinyanga Solar Power Station	150	Photovoltaic (PV)
Uganda	Soroti	Soroti Solar Power Station	10	Photovoltaic (PV)
Zambia	Lusaka	Bangweulu Solar Power Station	54	Photovoltaic (PV)
Zimbabwe	Bwoni	Harava Solar Power Plant	20	Photovoltaic (PV)
Algeria	Hassi R'Mel	Hassi R'Mel Integrated Solar Combined Cycle Power Station	25	Parabolic trough CSP
Angola	Benguela Province	Biopio Solar Park	188	Photovoltaic (PV)
Benin	Illoulofin	Illoulofin Solar Power Station	50	Photovoltaic (PV)
Botswana	Letlhakane	Letlhakane Solar Power Station	100	Photovoltaic (PV)
Burundi	Mubuga	Mubuga Solar Power Plant	7.5	Photovoltaic (PV)

(continued)

Table 10.1 (continued)

Country	Location	Solar project name	Generation capacity (Megawatt) (MW)	Type of solar technology
Cameroon	Douala	Douala Solar Power Station	25	Photovoltaic (PV)
Cape Verde	Sal Island	Sal Solar Power Station	5	Photovoltaic (PV)
Chad	Djermaya	Djermaya Solar Power Plant	60	Photovoltaic (PV)
Côte d'Ivoire	Boundiali	Boundiali Solar Power Station	37.5	Photovoltaic (PV)
Democratic Republic of the Congo	Sombwe	Sombwe Hydroelectric and Solar Project	46	Photovoltaic (PV)
Djibouti	Grand Bara	Grand Bara Solar Power Plant	25	Photovoltaic (PV)
Eswatini	Lavumisa	Lavumisa Solar Power Station	13.5	Photovoltaic (PV)
Ethiopia	Metehara	Metehara Solar Power Station	100	Photovoltaic (PV)
Gabon	Ayémé	Ayémé Plaine Solar Power Station	120	Photovoltaic (PV)
Guinea	Khoumagueli	Khoumagueli Solar Power Station	40	Photovoltaic (PV)
Lesotho	Mafeteng	Mafeteng Solar Power Station	70	Photovoltaic (PV)
Liberia	Monrovia	Monrovia Solar Power Station	20	Photovoltaic (PV)

Source Author's compilation from web search [51, 74–85]

a sustainable future. Moreover, solar-powered energy production can replace fossil fuel dependency, thereby paving the way to a cleaner atmosphere and environment [90]. Solar energy enhances the holistic development of a country across all sectors of its economy such as agriculture, health, education, and manufacturing [91]. Due to the green potential of solar energy, industries, and businesses can take advantage of and to achieve their sustainable and energy-efficient goals. Industry can improve upon their competitiveness, reduce their operational cost, and reduce their impact on the environment [92].

10.5.2 Reduces Grid Dependency

Solar energy can significantly reduce the dependency on the national grid. This is achieved in various settings. For example, grid-connected photovoltaic systems with good energy storage systems allow surplus energy to be stored for later use, reducing the reliance on grid power during peak consumption hours [93]. Additionally, in regions with intermittent power supply such as Africa, low-cost hardware platforms and frameworks for optimization based upon low-cost sensing and real-time monitoring schemes can enhance efficient solar PV utilization and subsequently reduce grid dependency [94]. Implementing solar energy technology on the end-user's side can greatly reduce reliance on the national grid. There is the potential to achieve 97.5% energy savings in solar energy connected to the national grid as a backup solution [95]. The adoption of solar energy through the grid-connected system also reduces electricity costs and dependency on the grid for many communities by increasing the reliability of the supply of power [96]. For example, in remote areas with limited infrastructure development for grid connectivity, there is a guarantee of continuous electricity service which lessens reliance on the national grid [97].

10.5.3 Enhance Employment and Economic Growth

Solar energy implementation provides an improvement potential towards the generation of employment and enhancement of the economy [98]. It offers a wide range of job opportunities from the manufacturing of solar components such as solar panels and installation systems. Maintenance and technical support also provide avenues for employment. These employment opportunities ensure economic benefits across communities, fostering inclusivity among various geographical locations [99]. This goes on to empower communities with economic stability and development. Moreover, individuals in the jobs of installation and maintenance have the chance to acquire skills and expertise due to the rapid growth of the solar industry, empowering them in the future [100]. The use of solar energy is closely linked with economic resilience. It has the potential to reduce economic setbacks and shocks by serving as an economic diversification in the energy sector [101]. Because solar energy acts as a buffer against the downturns of other sectors of the economy, countries that adopt it have a better chance to stand against economic turmoil [102]. Adoption of solar energy generally improves living conditions, healthcare, and education, contributing to the growth of national wealth [103].

10.5.4 Energy Efficiency, Resilience, and Security

The adoption of solar-based technology and energy generation has the benefit of providing energy-efficient solutions and energy security to consumers [104, 105]. Energy security is the uninterrupted availability of power at an affordable price [106]. Solar energy can achieve a good level of energy generation and supply with reduced energy input, minimizing waste, and maximizing output [107]. With solar energy, there is the capacity of resilience where the system's capacity can withstand and recover from disruptions [108]. Solar energy's decentralized nature and reliance on renewable resources which are in abundance make it a key solution for meeting power demands and shortages in countries [109]. This is due to the diversification of the energy mix and integration with the national grid supply [110]. Because solar energy is decentralized, adoption is made easy for homes, offices, and industries [111]. This provides the consumer with an independent supply of energy which has the advantage of power supply during grid intermittent supply. As a result, companies and businesses will not experience profit losses and general economic growth will be achieved [112]. In addition, off-grid solar energy adoption and solar mini-grids offer access to electricity that is reliable and secured in remote and underdeveloped areas, thereby improving energy equity and reliance on centralized power systems [113].

10.5.5 Global Alliances

The shift to sustainable energy production in developing countries is greatly aided by international cooperation [114]. Energy projects that use sustainable means can receive benefits and financial and technical aid from partnerships with international organizations, financial agencies, and multilateral development banks [115]. This fosters global alliances among countries. These alliances are great ways to facilitate technology and knowledge transfer, provide access to technical skills, and help bridge financial gaps. For example, there are provisions for funds and capacity building opportunities from global organizations and programs such as the Climate Technology Centre and Network (CTCN) and the Green Climate for developing countries embarking on sustainable energy programs and projects [116]. Participating in these international programs and platforms can promote information and knowledge sharing, the diffusion of best practices, and the replication of effective sustainable energy models.

10.6 Barriers to the Use of Solar-Based Renewables

In many different countries, the barriers to solar energy and renewable energy adoption have been studied to propose a unique solution [109]. However, due to the difference in technological, environmental, social, political, and economic contexts, it is expedient to understand the barriers to solar adoption in the African context [22]. Developing countries face several difficulties when putting sustainable energy practices into practice. The transition from fossil fuel to renewable energy production in Africa is hindered by several factors. These include limited technological and technical skills, socio-cultural challenges, political and regulatory barriers, and financial limitations [91]. This section expounds on these barriers to understand the level of hindrance to coming out with strategic solutions and pathways toward solar energy adoption.

10.6.1 Financial and Economic Challenges

Africa is noted as one of the continents with financial and economic challenges. These include high initial investments, Lack of funding, high upfront costs, limited budget, and many others [117]. Infrastructure developments in many African countries are confronted with the challenge of limited funds and monetary resources. The adoption of solar energy in the region is equally plagued with such challenges [100]. Funding limitations and budgetary priorities pose a stringent and tight choice when it comes to financing energy projects. The upfront cost of conventional energy sources remains lower than the cost of investing in renewable energy sources [12]. This is because the equipment needed for renewable energy projects is not readily available in Africa [14]. The equipment is imported from other countries, which increases the upfront cost of the project. Although energy is a fundamental need of every country, most African countries face difficulties in implementing it due to economic challenges [92]. For economies with very limited financial resources, the huge upfront costs of these renewable energy projects can be very intimidating [118]. Furthermore, because of the uncertainties, risks and dangers in developing countries, securing funding from private investment from both local and foreign investors for energy projects can be difficult [91]. Very few financial institutions are willing to fund and invest in renewable energy projects in the continent. Moreover, the lack of inadequacy of credit facilities and the high interest rates they come with make it difficult to procure solar energy technology. Because of these challenges, the average cost of renewable electricity generation in Africa is expensive. For example, in 2016–2017, the cost of solar photovoltaic power amounted to USD 0.17/kWh, which is almost twice the price in Asia at USD 0.09/kWh [100].

10.6.2 Policy and Regulatory Barriers

When there are clear policies and regulations such as codes and standards for the adoption and implementation of renewable energy technologies, there is the creation of a stable environment that propels the development of renewable technologies [119]. However, Africa's regulatory and legal frameworks for renewable energy are weak and inconsistent, making adoption very difficult [120]. Africa lacks adequate policies to foster an efficient market for renewable energy technologies. Even if there are existing policies, there is a lack of implementation and enforcement due to a lack of rule of law, poor management of the energy sector, corruption, bureaucracy, and bad governance [92]. Further, the lack of clear policies on private investment and delays in the authorization of private sector projects hinder private sector participation in renewable energy projects in African countries [121]. African countries have unique national renewable energy policies, while regional policies are not fully formed due to an inappropriate implementation approach [122]. As a result, even though many renewable energy policies have been developed in most of these countries, their implementation has been challenging primarily due to their immaturity. The uncertainty regarding grid expansion and the process for integrating existing mini-grids when local grid expansion occurs can deter investors in mini-grid systems [123]. Additionally, a lack of political devotion and desire to develop and prioritize renewable energy transitions in developing countries due to policy delays in the implementation of energy measures can be caused by vested interests, short-term political considerations and opposition from the fossil fuel industry [124]. Consequently, the absence of a transparent and clear regulatory framework deters long-term investment in sustainable projects.

10.6.3 Socio-Cultural Challenge

The deployment of renewable energy in Africa may be hindered by socio-cultural particularities. To start, the cultural background of certain ethnic groups may have the potential to negatively hinder the development of renewable energy technologies. For example, in many rural areas where the population is primarily farmers [99], cooking activities are typically scheduled from early evening to late night when there is no solar radiation to make a solar cooker, especially in Kenya and Nigeria, which is why the development of solar cookers is said to have failed. Another barrier to the adoption of renewable energy technology in Africa is a lack of knowledge and awareness of the potential utility of renewable energy technologies [125]. People with little education are not likely to understand the environmental effects of non-renewable energy or the benefits of renewable energy [126]. This has reduced the region's rate of advancement and adoption of renewable energy technology and infrastructure. Land acquisition is also a notable problem against the adoption of solar energy technology. Nuru et al. [127] observed in their study that, families opposed the construction of a power plant on their land. Additionally, changing consumer behaviour and patterns of energy consumption is necessary to meet sustainable energy goals. It can be very challenging to promote renewable energy practices in places where energy use is deeply rooted in cultural norms and daily routines [99].

10.6.4 Technological, Technical and Human Skill Challenge

Renewable energy development in Africa is hindered by technical and technological challenges such as a lack of infrastructure to support solar energy technologies and a lack of skill and capacity [100]. Developing countries and regions such as Africa frequently face technological challenges when trying to put renewable energy infrastructure in place. Most African countries do not have the technology and technical capacity to construct, operate and maintain renewable projects such as solar energy [128]. This mostly concerns modern energy management systems, storage options, and grid integration [117]. For example, there is inadequate renewable energy connectivity to the national grid which leads to heavy loss in energy transmissions, this has a negative effect on the investment made [98]. Because most of the existing power grids of countries are not designed to integrate with solar energy or any other renewable energy, the existing grids need to be modified [129]. Grid integration is one of the main difficulties affecting renewable energy developments in Africa [130]. Ullah, [131] argues that financial constraints, corruption, and intellectual property rights are barriers that hinder technological transfer from developed countries into developing countries. Additionally, research and development capabilities and investment are insufficient to make solar energy adoption competitive on the market with fossil fuel. African governments and firms are reluctant to invest in research and development as renewable energy is still at a developmental stage in the region. As it stands now, there needs to be a rapid transfer and skill build-up for Africa to switch to a cleaner way of energy production [132].

10.7 Pathways to Solar-Based Renewables

The implementation of solar energy is crucial to the global transition to sustainable and low-carbon energy systems. However, there are barriers and challenges facing the adoption of solar energy particularly in the African continent [87]. Therefore, to achieve global sustainable development goals and transition to sustainable energy production, a multifaceted approach is required. This involves various pathways that will address technical and technological barriers, policies and regulatory frameworks, community awareness, capacity building, research and development, and many more [133]. These pathways serve as a strategic approach to overcome the present challenges facing solar adoption in Africa. Lessons could be learnt from other parts of the world that have effectively applied these pathways and have seen a change and increase in solar energy adoption. By following these pathways, stakeholders can accelerate the adoption of solar energy in Africa, making it more accessible, efficient, reliable, and economically viable [22]. These pathways play distinct roles, yet they are interconnected in facilitating solar energy development.

10.7.1 Cost Reduction Pathway

Government, financial organizations, and private agencies need to collaborate to establish a financial system where solar energy is easily adopted and made accessible to everyone. Subsidies, grants, and tax benefits are systems that can reduce the cost implication for solar energy to encourage individuals and businesses to adopt [133]. Additionally, microfinance institutions can help improve solar energy adoption by establishing financial systems to the local context and different income levels, providing avenues for investments [134]. Public private partnerships can also establish financial mechanisms that can reduce the upfront cost of solar investments [135]. It may also be essential to make available soft loans with low interest rates for solar energy projects [136]. With these, communities in the African region can transcend financial barriers and forge a future where solar energy empowers individuals, uplifts communities, and paves the way for sustainable progress. Micro-credit linked to micro-enterprises has proven very successful in enhancing renewable energy and reducing poverty in some African countries. Some countries provide these incentives towards fossil fuel power generation [137]. This puts solar energy at a disadvantaged point in the competitive market against fossil fuel [111]. If these incentives are to be given for solar energy, it is going to level the market's competition and make solar energy more attractive. For example, Denmark has established ambitious targets for renewable energy expansion, intending to be fossil fuel-free by 2050 [138]. The country's energy agreement in 2012 brought in a wide range of measures such as incentives and grid expansion to support renewable energy development. The country now produces over 40% of electricity from renewable sources [139]. This shows the power of cost-reducing measures and policies towards overcoming the adoption challenges of solar energy.

10.7.2 Capacity Building

The adoption of renewable energy in Africa requires capacity building and skill development because one of the major barriers facing solar energy implementation in Africa is the lack of technical skills [140]. A wide range of skills are needed to be built from designing of components, installation, and maintenance. Advancing a long-term capacity and skill build-up requires commitment and funding of education and training by the government. Additionally, capacity needs to be built in the area of research and development [141]. Research and development capacity will enhance new forms of technology in solar energy towards efficient generation, storage, and distribution of power. The Government, international organizations, and agencies can fund research and provide financial support for research and development activities in renewable energy [117]. These initiatives may include grants, scholarships, and funding for research centres and universities. These grants will support companies to develop new solar energy-based technologies, and also reduce the cost of testing and demonstration of projects. For example, the International Renewable Energy Agency (IRENA) provides capacity building support to its member countries, helping them develop the necessary skills and knowledge to navigate regulatory challenges in the renewable energy sector [142]. Additionally, research into anti-reflective coatings, self-cleaning surfaces, and temperature-regulating materials can enhance panel durability and performance in diverse environmental conditions [143].

10.7.3 Formulation and Enforcement of Favourable Government Policies

Favourable government policies are essential to improving the implementation of solar energy and development. Solar energy lifetime and its crediting period exceeds any government duration in power in any country [144]. Therefore, making sure that, favourable and stable laws are established and enforced is very essential. Investors may want to invest in solar projects in countries, but they need a high level of certainty that certain laws put in place that solar energy development will remain stable and enforced [145]. Additionally, there is a need for the development of essential processes in the political, and socio-economic structures to manage corruption and establish a high standard of transparency in the administration of public and foreign affairs [146]. Enforcing laws is very important in mitigating the barriers of policy and regulatory structures. One of the examples of successfully promoting renewable energy adoption and expansion through favourable government policies and legal frameworks is the case in Germany. In 2000, the country's energy policy guaranteed a fixed, above-market prices for renewable energy producers [147]. This provided a good environment for long-term investment security. This has led to a significant increase of about 40% of electricity produced from renewable sources. The African region can learn from this example [144].

10.7.4 Consistent Awareness Creation

Large scale adoption of solar energy technology can also be successfully undertaken with a good understanding and support of the public. It is therefore important that awareness of the benefits and opportunities of renewable energy in tackling climate change be increased to boost the desire and interest of the public [148]. The public is a great stakeholder and market for solar energy initiatives and development. Awareness is also found to be an important ingredient needed to help in specific solar project identification and will also assist in tackling the problem associated with the insecurity of solar plant infrastructures [149]. Public awareness can be achieved through various means of promotion and communication such as workshops, media, and community forums [150]. These meetings and promotions will also need financial support and dedication from the government and non-governmental institutions.

Stakeholder collaboration: Solar energy implementation requires dedicated efforts from all stakeholders. There needs to be a collaboration between policymakers, industry experts, legal experts and even other governments to foster solar energy innovation and development [151]. The knowledge, skills, and strategies required to boost solar energy adoption and mitigate challenges are possessed by all these stakeholders. A combined effort and knowledge will lead to drawing specific and unique solutions for the problems facing the solar industry in Africa [152]. Stakeholder engagement forums and discussions will bring together all the stakeholders to deliberate on challenges, share best practices, and enhance knowledge and technology transfer to develop innovative solutions for renewable energy development and adoption in Africa [153]. This approach has worked effectively in other countries. For example, in Denmark, the government has put up a collaborative initiative for energy planning, involving stakeholders from across the energy sector [144]. This approach has helped ensure that regulatory frameworks support the growth of renewable energy while addressing the challenges.

10.7.5 Technological Advancement and Innovation

Technological advancement is a crucial pathway towards solar energy implementation. It holds the solution to the problems in generation capacity and grid integration for a better and more reliable supply of energy [154]. Technological advancement enhances the performance, reliability, and economic viability of solar energy generation. One major technological advancement in solar energy is the development of an efficient PV technology which has significant energy conversion potential for maximum energy generation. Some of these technologies are bifacial solar panels that can capture sunlight from both surfaces [155]. Tandem solar cells, perovskite solar cells solar technologies have proven to be highly efficient, maximizing energy output than conventional solar cells [156]. Additionally, technological advancement in the integration of solar power with smart grid technologies and digitalization has the potential to improve the management and distribution of solar energy [157]. Advanced monitoring systems enhanced with artificial intelligence and Internet of Things (IoT) devices, enable real-time performance tracking, predictive maintenance, and optimized energy dispatch, reducing operational costs, and minimizing system downtimes [158]. Smart inverters and demand response systems can enhance grid stability by automatically adjusting solar energy output to match real-time energy usage, preventing voltage fluctuations and ensuring efficient energy use [159].

10.8 Summary

This chapter focused on renewable energy as a means to building energy decarbonization through demand side management. Although there are many renewable energy sources, this chapter levelled down to focus mainly on solar energy in Africa. Africa is one of the continents with very high solar resources due to the high solar radiation. The potential for solar energy adoption is great in the region. Numerous solar-based energy technologies can be adopted. Some typologies discussed in this chapter include solar photovoltaic (PV) technology. Other technologies include Concentrated solar power (CSP), solar drying technology, solar water heating system and solar power desalination technology. These technologies offer energy services to domestic, commercial and industrial needs which will otherwise be supplied by fossil fuel-based electricity. Solar energy technologies are sustainable ways of generating electrical energy for daily use. The chapter also explored the benefits of the adoption of solar-based energy. However, the region has not been able to maximize the full potential of this resource. This is due to certain challenges and barriers facing its adoption. These challenges include high initial investments and upfront costs, lack of technical skill and technological advancements, low level of awareness and understanding on the benefits of solar energy, lack of research and development, lack of favourable policies and standards, and many more. It is realized that almost all the challenges are closely linked to policies that are not available to promote adoption. The governments of many African countries have not dedicated intentional effort to solving these problems, as such, there are no funds allocated for training, research, subsidies, grants, awareness creation, and investment. If solar energy adoption is to be improved in Africa, it is necessary to follow pathways and strategies that address the specific challenges being faced. It is important to build knowledge and skill capacity, formulate cost reduction policies, increase awareness, invest in modern technologies and innovations, increase solar power generation capacity, and collaborate with stakeholders. Africa has the potential to foster economic growth. Solar resource is a revolutionary resource that can elevate the region to a high standard to improve the living standards of her citizens.

References

1. Vogel J, Steinberger JK, O'Neill DW, Lamb WF, Krishnakumar J (2021) Socio-economic conditions for satisfying human needs at low energy use: an international analysis of social provisioning. Glob Environ Chang 69:102287. https://doi.org/10.1016/J.GLOENVCHA.2021.102287
2. Yi S, Raza Abbasi K, Hussain K, Albaker A, Alvarado R (2023) Environmental concerns in the United States: can renewable energy, fossil fuel energy, and natural resources depletion help? Gondwana Res 117:41–55. https://doi.org/10.1016/J.GR.2022.12.021
3. Oladunni OJ, Mpofu K, Olanrewaju OA (2022) Greenhouse gas emissions and its driving forces in the transport sector of South Africa. Energy Rep 8:2052–2061. https://doi.org/10.1016/J.EGYR.2022.01.123
4. Johnsson F, Karlsson I, Rootzén J, Ahlbäck A, Gustavsson M (2020) The framing of a sustainable development goals assessment in decarbonizing the construction industry—Avoiding

'Greenwashing.' Renew Sustain Energy Rev 131:110029. https://doi.org/10.1016/J.RSER.2020.110029
5. Wang EZ, Lee CC, Li Y (2022) Assessing the impact of industrial robots on manufacturing energy intensity in 38 countries. Energy Econ 105:105748. https://doi.org/10.1016/J.ENECO.2021.105748
6. Gu D, Andreev K, Dupre ME (2021) Major trends in population growth around the world. China CDC Week 3(28):604. https://doi.org/10.46234/CCDCW2021.160
7. Mutezo G, Mulopo J (2021) A review of Africa's transition from fossil fuels to renewable energy using circular economy principles. Renew Sustain Energy Rev 137:110609. https://doi.org/10.1016/J.RSER.2020.110609
8. Farghali M, Osman AI, Israa, Mohamed MA, Chen Z, Chen L, Ihara I, Yap P-S, David, Rooney W (2023) Strategies to save energy in the context of the energy crisis: a review, vol 21, pp 2003–2039. https://doi.org/10.1007/s10311-023-01591-5
9. Mutombo NMA, Numbi BP (2019) Assessment of renewable energy potential in Kwazulu-Natal province, South Africa. Energy Rep 5:874–881. https://doi.org/10.1016/J.EGYR.2019.07.003
10. Belaïd F, Al-Sarihi A, Al-Mestneer R (2023) Balancing climate mitigation and energy security goals amid converging global energy crises: the role of green investments. Renew Energy 205:534–542. https://doi.org/10.1016/J.RENENE.2023.01.083
11. Dupont E, Koppelaar R, Jeanmart H (2020) Global available solar energy under physical and energy return on investment constraints. Appl Energy 257:113968. https://doi.org/10.1016/J.APENERGY.2019.113968
12. Hayat MB, Ali D, Monyake KC, Alagha L, Ahmed N (2019) Solar energy—a look into power generation, challenges, and a solar-powered future. Int J Energy Res 43(3):1049–1067. https://doi.org/10.1002/ER.4252
13. Hassan Q, Viktor P, Al-Musawi TJ, Mahmood Ali B, Algburi S, Alzoubi HM, Khudhair Al-Jiboory A, Zuhair Sameen A, Salman HM, Jaszczur M (2024) The renewable energy role in the global energy transformations. Renew Energy Focus 48:100545. https://doi.org/10.1016/J.REF.2024.100545
14. Gamil A, Li P, Ali B, Hamid MA (2022) Concentrating solar thermal power generation in Sudan: potential and challenges. Renew Sustain Energy Rev 161:112366. https://doi.org/10.1016/J.RSER.2022.112366
15. Neher I, Crewell S, Meilinger S, Pfeifroth U, Trentmann J (2020) Photovoltaic power potential in West Africa using long-term satellite data. Atmos Chem Phys 20(21):12871–12888. https://doi.org/10.5194/ACP-20-12871-2020
16. Agbo EP, Edet CO, Magu TO, Njok AO, Ekpo CM, Louis H (2021) Solar energy: a panacea for the electricity generation crisis in Nigeria. Heliyon 7(5):e07016. https://doi.org/10.1016/J.HELIYON.2021.E07016/ASSET/34EBF48D-A1CC-4DD9-8996-B23485C95880/MAIN.ASSETS/GR20.JPG
17. Urbina A (2022) Scenarios for solar electricity at the terawatt scale. Green Energy Technol 3–17. https://doi.org/10.1007/978-3-030-91771-5_1
18. González-García A, Ciller P, Lee S, Palacios R, de Cuadra García F, Pérez-Arriaga JI (2022) A rising role for decentralized solar minigrids in integrated rural electrification planning? Large-scale, least-cost, and customer-wise design of grid and off-grid supply systems in Uganda. Energies 15(13):4517. https://doi.org/10.3390/EN15134517
19. Herrando M, Wang K, Huang G, Otanicar T, Mousa OB, Agathokleous RA, Ding Y, Kalogirou S, Ekins-Daukes N, Taylor RA (2023) A review of solar hybrid photovoltaic-thermal (PV-T) collectors and systems. Prog Energy Combust Sci 97:101072
20. Ayora E, Munji M, Kaberere K, Thomas B (2023) Performance analysis of 600 kWp grid-tied rooftop solar photovoltaic systems at strathmore university in Kenya. Res Eng 19:101302. https://doi.org/10.1016/J.RINENG.2023.101302
21. Dahunsi SO, Fagbiele OO, Yusuf EO (2020) Bioenergy technologies adoption in Africa: a review of past and current status. J Clean Prod 264:121683. https://doi.org/10.1016/J.JCLEPRO.2020.121683

22. Agoundedemba M, Kim CK, Kim HG (2023) Energy status in Africa: challenges, progress and sustainable pathways. Energies 16(23):7708. https://doi.org/10.3390/EN16237708
23. Artur C, Neves D, Cuamba BC, Leão AJ (2020) Domestic hot water technology transition for solar thermal systems: an assessment for the urban areas of Maputo city, Mozambique. J Clean Prod 260:121043. https://doi.org/10.1016/J.JCLEPRO.2020.121043
24. Adak D, Bhattacharyya R, Barshilia HC (2022) A state-of-the-art review on the multi-functional self-cleaning nanostructured coatings for PV panels, CSP mirrors and related solar devices. Renew Sustain Energy Rev 159:112145. https://doi.org/10.1016/J.RSER.2022.112145
25. Azouzoute A, Alami Merrouni A, Touili S (2020) Overview of the integration of CSP as an alternative energy source in the MENA region. Energy Strat Rev 29:100493. https://doi.org/10.1016/J.ESR.2020.100493
26. Boamah F (2020) Desirable or debatable? Putting Africa's decentralised solar energy futures in context. Energy Res Soc Sci 62:101390. https://doi.org/10.1016/J.ERSS.2019.101390
27. Acheampong AO (2023) Governance, credit access and clean cooking technologies in Sub-Saharan Africa: implications for energy transition. J Policy Model 45(2):445–468. https://doi.org/10.1016/J.JPOLMOD.2023.01.002
28. Dambhare MV, Butey B, Moharil SV (2021) Solar photovoltaic technology: a review of different types of solar cells and its future trends. J Phys Conf Ser 1913(1):012053. https://doi.org/10.1088/1742-6596/1913/1/012053
29. Pastuszak J, Węgierek P (2022) Photovoltaic cell generations and current research directions for their development. Materials 15(16):5542. https://doi.org/10.3390/MA15165542
30. Kumar Behura A, Kumar A, Kumar Rajak D, Pruncu CI, Lamberti L (2021) Towards better performances for a novel rooftop solar PV system. Sol Energy 216:518–529. https://doi.org/10.1016/J.SOLENER.2021.01.045
31. Reddy P, Surendra Gupta MVN, Nundy S, Karthick A, Ghosh A (2020) Status of BIPV and BAPV system for less energy-hungry building in India—a review. Appl Sci 10(7):2337. https://doi.org/10.3390/APP10072337
32. Liu Z, Zhang Y, Yuan X, Liu Y, Xu J, Zhang S, He B (2021) A comprehensive study of feasibility and applicability of building integrated photovoltaic (BIPV) systems in regions with high solar irradiance. J Clean Prod 307:127240. https://doi.org/10.1016/J.JCLEPRO.2021.127240
33. Ding, L., Zhu, Y., Zheng, L., Dai, Q. and Zhang, Z. (2023) What is the path of photovoltaic building (BIPV or BAPV) promotion? --The perspective of evolutionary games, Applied Energy, 340, p. 121033. https://doi.org/10.1016/J.APENERGY.2023.121033.
34. Allouhi A, Rehman S, Buker MS, Said Z (2022) Up-to-date literature review on Solar PV systems: technology progress, market status and R&D. J Clean Prod 362:132339. https://doi.org/10.1016/J.JCLEPRO.2022.132339
35. Awasthi A, Shukla AK, Murali Manohar SR, Dondariya C, Shukla KN, Porwal D, Richhariya G (2020) Review on sun tracking technology in solar PV system. Energy Rep 6:392–405. https://doi.org/10.1016/J.EGYR.2020.02.004
36. Agyekum EB, Velkin VI (2020) Optimization and techno-economic assessment of concentrated solar power (CSP) in South-Western Africa: a case study on Ghana. Sustain Energy Technol Assess 40:100763. https://doi.org/10.1016/J.SETA.2020.100763
37. Khan MI, Gutiérrez-Alvarez R, Asfand F, Bicer Y, Sgouridis S, Al-Ghamdi SG, Jouhara H, Asif M, Kurniawan TA, Abid M, Pesyridis A, Farooq M (2024) The economics of concentrating solar power (CSP): assessing cost competitiveness and deployment potential. Renew Sustain Energy Rev 200:114551. https://doi.org/10.1016/J.RSER.2024.114551
38. Imran Khan M, Asfand F, Al-Ghamdi SG (2023) Progress in research and technological advancements of commercial concentrated solar thermal power plants. Sol Energy 249:183–226. https://doi.org/10.1016/J.SOLENER.2022.10.041
39. Zubair M, Awan AB (2021) Economic viability of solar energy export from the Middle East and North Africa to Europe and South Asia. Environ Dev Sustain 23(12):17986–18007. https://doi.org/10.1007/S10668-021-01424-X/METRICS

40. Alami AH, Olabi AG, Mdallal A, Rezk A, Radwan A, Rahman SMA, Shah SK, Abdelkareem MA (2023) Concentrating solar power (CSP) technologies: status and analysis. Int J Thermofluids 18:100340. https://doi.org/10.1016/J.IJFT.2023.100340
41. Bouckaert S, Pales AF, McGlade C, Remme U, Wanner B, Varro L, D'Ambrosio D, Spencer T (2021) Net zero by 2050: a roadmap for the global energy sector
42. Yan H, Wang A, Chong D, Liu M, Liu J, Yan J (2021) Review on performance analysis of the power block in concentrated solar power plants. Energ Technol 9(1):2000621. https://doi.org/10.1002/ENTE.202000621
43. Vengadesan E, Senthil R (2020) A review on recent development of thermal performance enhancement methods of flat plate solar water heater. Sol Energy 206:935–961. https://doi.org/10.1016/J.SOLENER.2020.06.059
44. Huang D, Ma H, Ma S, Chen M, Xu C, Zhang Y, Yan C, Calise F, Ma HD, Ma H, Chen S, Xu M, Zhang C, Yongyichuan, Cong Y (2022) Study on the energy efficiency improvement and operation optimization of a solar water heating system. Appl Sci 12(14):7263. https://doi.org/10.3390/APP12147263
45. Faisal Ahmed S, Khalid M, Vaka M, Walvekar R, Numan A, Khaliq Rasheed A, Mujawar Mubarak N (2021) Recent progress in solar water heaters and solar collectors: a comprehensive review. Therm Sci Eng Prog 25:100981. https://doi.org/10.1016/J.TSEP.2021.100981
46. Gambade J, Noël H, Glouannec P, Magueresse A (2023) Numerical model of intermittent solar hot water production. Renew Energy 218:119368. https://doi.org/10.1016/J.RENENE.2023.119368
47. Abu Mallouh M, AbdelMeguid H, Salah M (2022) A comprehensive comparison and control for different solar water heating system configurations. Eng Sci Technol Int J 35:101210. https://doi.org/10.1016/J.JESTCH.2022.101210
48. Greco A, Gundabattini E, Gnanaraj DS, Masselli C (2020) A comparative study on the performances of flat plate and evacuated tube collectors deployable in domestic solar water heating systems in different climate areas. Climate 8(6):78. https://doi.org/10.3390/CLI8060078
49. Olczak P, Matuszewska D, Zabagło J (2020) The comparison of solar energy gaining effectiveness between flat plate collectors and evacuated tube collectors with heat pipe: case study. Energies 13(7):1829. https://doi.org/10.3390/EN13071829
50. Fadzlin WA, Hasanuzzaman M, Rahim NA, Amin N, Said Z (2022) Global challenges of current building-integrated solar water heating technologies and its prospects: a comprehensive review. Energies 15(14):5125. https://doi.org/10.3390/EN15145125
51. Li T, Wang Z, Yu J, Mao Q (2023) Experimental and simulation on night heat dissipation characteristics and freezing process of evacuated tube solar water heaters. Int J Therm Sci 192:108455. https://doi.org/10.1016/J.IJTHERMALSCI.2023.108455
52. Mostafaeipour A, Zarezade M, Khalifeh Soltani SR, Hosseini Dehshiri SJ, Hosseini Dehshiri SS, Ao Xuan H, Arockia Dhanraj J, Techato K, Chowdhury S, Issakhov A (2022) A conceptual new model for use of solar water heaters in hot and dry regions. Sustain Energy Technol Assess 49:101710. https://doi.org/10.1016/J.SETA.2021.101710
53. Lingayat A, Balijepalli R, Chandramohan VP (2021) Applications of solar energy based drying technologies in various industries—a review. Sol Energy 229:52–68. https://doi.org/10.1016/J.SOLENER.2021.05.058
54. EL-Mesery HS, EL-Seesy AI, Hu Z, Li Y (2022) Recent developments in solar drying technology of food and agricultural products: a review. Renew Sustain Energy Rev. 157:112070. https://doi.org/10.1016/J.RSER.2021.112070
55. Kamarulzaman A, Hasanuzzaman M, Rahim NA (2021) Global advancement of solar drying technologies and its future prospects: a review. Sol Energy 221:559–582. https://doi.org/10.1016/J.SOLENER.2021.04.056
56. Yao Y, Pang YX, Manickam S, Lester E, Wu T, Pang CH (2022) A review study on recent advances in solar drying: mechanisms, challenges and perspectives. Sol Energy Mater Sol Cells 248:111979. https://doi.org/10.1016/J.SOLMAT.2022.111979
57. Lamrani B, Bekkioui N, Simo-Tagne M, Ndukwu MC (2023) Recent progress in solar wood drying: an updated review. Drying Technol 41(5):605–627. https://doi.org/10.1080/07373937.2022.2112048

58. Lingayat A, Zachariah R, Modi A (2022) Current status and prospect of integrating solar air heating systems for drying in various sectors and industries. Sustain Energy Technol Assess 52:102274. https://doi.org/10.1016/J.SETA.2022.102274
59. Kumar P, Singh D (2020) Advanced technologies and performance investigations of solar dryers: a review. Renew Energy Focus 35:148–158. https://doi.org/10.1016/J.REF.2020.10.003
60. Ortiz-Rodríguez NM, Condorí M, Durán G, García-Valladares O (2022) Solar drying technologies: a review and future research directions with a focus on agroindustrial applications in medium and large scale. Appl Therm Eng 215:118993. https://doi.org/10.1016/J.APPLTHERMALENG.2022.118993
61. Senger MKS, Dewangan AK, Dutt N, Yadav AK (2024) Research advances in solar dryer technologies integrated with solar air heater: a state-of-the art review on design variations, performance and feasibility assessments. J Therm Anal Calorim 149(22):12577–12607. https://doi.org/10.1007/S10973-024-13714-3
62. Vijayan S, Arjunan TV, Kumar A (2020) Exergo-environmental analysis of an indirect forced convection solar dryer for drying bitter gourd slices. Renew Energy 146:2210–2223. https://doi.org/10.1016/J.RENENE.2019.08.066
63. Kumar M, Sansaniwal SK, Khatak P (2016) Progress in solar dryers for drying various commodities. Renew Sustain Energy Rev 55:346–360. https://doi.org/10.1016/J.RSER.2015.10.158
64. Tian J, Zhao X, Gao S, Wang X, Zhang R (2021) Progress in research and application of nanofiltration (NF) technology for brackish water treatment. Membranes 11(9):662. https://doi.org/10.3390/MEMBRANES11090662
65. Elgendi M, Alhefeiti A, Almarshoodi H, Aldhaheri A, Alshehhi W, Selim M (2021) Embodiment design and detailed design of a pyramid solar still with an automatic feedwater system, In: 2021 6th international conference on renewable energy: generation and applications, ICREGA 2021, pp 44–47. https://doi.org/10.1109/ICREGA50506.2021.9388218
66. Ali ES, Alsaman AS, Tawfik MHM, Askalany AA, El-Maghlany WM, Zohir AE, Farid AM, Ghazy M (2024) Solar-powered hybrid adsorption desalination/humidification-dehumidification system. Therm Sci Eng Prog 51:102598. https://doi.org/10.1016/J.TSEP.2024.102598
67. Alsaman AS, Ghazy M, Ali ES, Askalany AA, Farid AM, Tawfik MHM (2024) Solar-powered adsorption desalination utilizing composite silica gel with a humidification-dehumidification desalination system. Desalination 582:117663. https://doi.org/10.1016/J.DESAL.2024.117663
68. AlMallahi MN, Asaad SM, Inayat A, Harby K, Elgendi M (2023) Analysis of solar-powered adsorption desalination systems: current research trends, developments, and future perspectives. Int J Thermofluids 20:100457. https://doi.org/10.1016/J.IJFT.2023.100457
69. Al-Addous M, Bdour M, Rabaiah S, Boubakri A, Schweimanns N, Barbana N, Wellmann J (2024) Innovations in solar-powered desalination: a comprehensive review of sustainable solutions for water scarcity in the Middle East and North Africa (MENA) Region. Water 16(13):1877. https://doi.org/10.3390/W16131877
70. Elgendi M, Selim MYE, Aldhaheri A, Alshehhi W, Almarshoodi H, Alhefeiti A (2022) Design procedures for a passive pyramid solar still with an automatic feed water system. Alex Eng J 61(8):6419–6431. https://doi.org/10.1016/J.AEJ.2021.12.002
71. Elgendi M, Selim MYE (2022) Determination of the effect of water depth on the yield of a solar still using an automatic feedwater system. Environ Sci Pollut Res 29(10):14595–14603. https://doi.org/10.1007/S11356-021-15707-3/METRICS
72. Vushe A (2021) Proposed research, science, technology, and innovation to address current and future challenges of climate change and water resource management in africa, climate change and water resources in Africa: perspectives and solutions towards an imminent water crisis, pp 489–518. https://doi.org/10.1007/978-3-030-61225-2_21
73. Elminshawy NAS, Siddiqui FR, Sultan GI (2015) Development of a desalination system driven by solar energy and low grade waste heat. Energy Convers Manage 103:28–35. https://doi.org/10.1016/J.ENCONMAN.2015.06.035

74. Addy M, Adinyira E, Danku JC, Dadzoe F (2021) Impediments to the development of the green building market in sub-Saharan Africa: the case of Ghana. Smart Sustain Built Environ 10(2):193–207
75. Akcay EC (2023) Barriers to undertaking green building projects in developing countries: a Turkish perspective. Buildings 13(4):841. https://doi.org/10.3390/BUILDINGS13040841
76. Anshebo MA, Mengesha WJ, Sokido DL (2023) Selection of the most appropriate sustainable buildings assessment categories and criteria for developing countries: case of ethiopia. J Urban Plan Dev 149(1): 04022057. https://doi.org/10.1061/JUPDDM.UPENG-4189
77. Debrah C, Owusu-Manu D-G, Kissi E, Oduro-Ofori E, Edwards DJ (2020) Barriers to green cities development in developing countries: evidence from Ghana. Smart Sustain Built Environ 11(3):438–453
78. Ferrari S, Zoghi M, Blázquez T, Dall'O G (2022) New level(s) framework: assessing the affinity between the main international green building rating systems and the European scheme. Renew Sustain Energy Rev 155:111924. https://doi.org/10.1016/J.RSER.2021.111924
79. Harisankar R, Rakesh P (2021) Simulation-based comparative study on energy efficiency criteria for LEED and BREEAM systems. J Phys Conf Ser 2070(1):012222. https://doi.org/10.1088/1742-6596/2070/1/012222
80. Hill Civil R, Bowen P, Opperman L (n.d.) Sustainable building assessment methods in South Africa: an agenda for research
81. Kylili A, Georgali PZ, Christou P, Fokaides P (2022) An integrated building information modeling (BIM)-based lifecycle-oriented framework for sustainable building design. Construct Innov. https://doi.org/10.1108/CI-02-2021-0011
82. Li X, Feng W, Liu X, Yang Y (2023) A comparative analysis of green building rating systems in China and the United States. Sustain Cities Soc 93:104520. https://doi.org/10.1016/J.SCS.2023.104520
83. Lobos Calquín D, Mata R, Vielma JC, Beaumont-Sepulveda JC, Correa C, Nuñez E, Forcael E, Blanco D, Pulgar P (2024) A simplified framework to integrate databases with building information modeling for building energy assessment in multi-climate zones. Sustainability 16(14):6123. https://doi.org/10.3390/SU16146123
84. Maqbool R, Thompson C, Ashfaq S (2023) LEED and BREEAM green building certification systems as possible game changers in attaining low-cost energy-efficient urban housing projects. J Urban Plan Develop 149(3):04023024. https://doi.org/10.1061/JUPDDM.UPENG-4292
85. Zhu B, Liu G (2023) The development model of sustainable campus based on green buildings: a systematic comparative study between Japan and China. Eng Constr Archit Manag 32(2):805–823. https://doi.org/10.1108/ECAM-03-2022-0236/FULL/PDF
86. Njoh AJ, Etta S, Ngyah-Etchutambe IB, Enomah LED, Tabrey HT, Essia U (2019) Opportunities and challenges to rural renewable energy projects in Africa: lessons from the Esaghem Village, Cameroon solar electrification project. Renew Energy 131:1013–1021. https://doi.org/10.1016/J.RENENE.2018.07.092
87. Adenle AA (2020) Assessment of solar energy technologies in Africa-opportunities and challenges in meeting the 2030 agenda and sustainable development goals. Energy Policy 137:111180. https://doi.org/10.1016/J.ENPOL.2019.111180
88. Abdullahi D, Renukappa S, Suresh S, Oloke D (2022) Barriers for implementing solar energy initiatives in Nigeria: an empirical study. Smart Sustain Built Environ 11(3):647–660. https://doi.org/10.1108/SASBE-06-2020-0094/FULL/PDF
89. Bishoge OK, Kombe GG, Mvile BN (2020) Renewable energy for sustainable development in sub-Saharan African countries: challenges and way forward. J Renew and Sustain Energy 12(5). https://doi.org/10.1063/5.0009297/1017497
90. Mungai EM, Ndiritu SW, Da Silva I (2022) Unlocking climate finance potential and policy barriers—a case of renewable energy and energy efficiency in Sub-Saharan Africa. Res Environ Sustain 7:100043. https://doi.org/10.1016/J.RESENV.2021.100043
91. Falcone PM (2023) Sustainable energy policies in developing countries: a review of challenges and opportunities. Energies 16(18):6682. https://doi.org/10.3390/EN16186682

92. Ukoba K, Yoro KO, Eterigho-Ikelegbe O, Ibegbulam C, Jen TC (2024) Adaptation of solar energy in the global south: prospects, challenges and opportunities. Heliyon 10(7):e28009. https://doi.org/10.1016/J.HELIYON.2024.E28009/ASSET/3CA175EB-A9E6-4C2E-8329-904F837CFAEA/MAIN.ASSETS/GR9.JPG
93. Schilive de Souza A, Urbanetz Junior J (2021) Benefit analysis of grid connected photovoltaic solar system with energy storage. Braz Arch Biol Technol 64:2021. https://doi.org/10.1590/1678-4324-75years-2021210146
94. Nasir M, Khan HA, Khan I, Hassan N, Zaffar NA, Mehmood A, Sauter T, Muyeen SM (2019) Grid load reduction through optimized PV power utilization in intermittent grids using a low-cost hardware platform. Energies 12(9):1764. https://doi.org/10.3390/EN12091764
95. Dushimimana JJ, Niyonsaba JDA, Kayibanda V, Bikorimana JMV, Urimubenshi F, Nduwamungu A (2022) PV system grid as backup connected management using automatic changeover switch. In: 2022 IEEE PES/IAS PowerAfrica. PowerAfrica. https://doi.org/10.1109/POWERAFRICA53997.2022.9905320
96. Nelson J, Sandwell P, Beath H, Winchester B, Markides CN, Mittal S (2024) Opportunities for decentralised solar power to improve reliability, reduce emissions and avoid stranded assets. https://doi.org/10.21203/RS.3.RS-4349203/V1
97. Kumar K (2022) Development of economic processes utilizing solar energy, Acta Academiae Beregsasiensis. Economics (2):32–41. https://doi.org/10.58423/2786-6742/2022-2-32-41
98. Adebanji B, Ojo A, Fasina T, Adeleye S, Abere J (2022) Integration of renewable energy with smart grid application into the Nigeria's power network: issues, challenges and opportunities. Euro J Eng Technol Res 7(3):18–24. https://doi.org/10.24018/EJENG.2022.7.3.2792
99. Elahi E, Khalid Z, Zhang Z (2022) Understanding farmers' intention and willingness to install renewable energy technology: a solution to reduce the environmental emissions of agriculture. Appl Energy 309:118459. https://doi.org/10.1016/J.APENERGY.2021.118459
100. Ouedraogo NS (2019) Opportunities, barriers and issues with renewable energy development in Africa: a comprehensible review. Curr Sustain Renew Energy Rep 6(2):52–60. https://doi.org/10.1007/S40518-019-00130-7/METRICS
101. El Anshasy AA, Khalid U (2023) From diversification resistance to sustainable diversification: lessons from the UAE's public policy shift. Manag Sustain 2(1):47–66. https://doi.org/10.1108/MSAR-06-2022-0025/FULL/PDF
102. Azad ASMS, Hayat A, Ahmed HJA (2024) Does the energy sector serve as a hedge and safe haven? Ann Operat Res 339(1–2):369–395. https://doi.org/10.1007/S10479-023-05707-6/METRICS
103. Soto EA, Hernandez-Guzman A, Vizcarrondo-Ortega A, McNealey A, Bosman LB (2022) Solar energy implementation for health-care facilities in developing and underdeveloped countries: overview, opportunities, and challenges. Energies 15(22):8602. https://doi.org/10.3390/EN15228602
104. Bouaguel W, Alsulimani T (2022) Understanding the factors influencing consumers' intention toward shifting to solar energy technology for residential use in Saudi Arabia using the technology acceptance model. Sustainability 14(18):11356. https://doi.org/10.3390/SU141811356
105. Lee J, Shepley MMC (2020) Benefits of solar photovoltaic systems for low-income families in social housing of Korea: renewable energy applications as solutions to energy poverty. J Build Eng 28:101016. https://doi.org/10.1016/J.JOBE.2019.101016
106. Alemzero DA, Sun H, Mohsin M, Iqbal N, Nadeem M, Vo XV (2021) Assessing energy security in Africa based on multi-dimensional approach of principal composite analysis. Environ Sci Pollut Res 28(2):2158–2171. https://doi.org/10.1007/S11356-020-10554-0/METRICS
107. Mohammadnia A, Rezania A, Ziapour BM, Sedaghati F, Rosendahl L (2020) Hybrid energy harvesting system to maximize power generation from solar energy. Energy Convers Manage 205:112352. https://doi.org/10.1016/J.ENCONMAN.2019.112352
108. Aldieri L, Gatto A, Vinci CP (2021) Evaluation of energy resilience and adaptation policies: an energy efficiency analysis. Energy Policy 157:112505. https://doi.org/10.1016/J.ENPOL.2021.112505

109. Mostafaeipour A, Alvandimanesh M, Najafi F, Issakhov A (2021) Identifying challenges and barriers for development of solar energy by using fuzzy best-worst method: a case study. Energy 226:120355. https://doi.org/10.1016/J.ENERGY.2021.120355
110. Gitelman L, Kozhevnikov M, Visotskaya Y (2023) Diversification as a method of ensuring the sustainability of energy supply within the energy transition. Resources 12(2):19. https://doi.org/10.3390/RESOURCES12020019
111. Shastry V, Rai V (2025) Beyond the business case: impacts and resilience post-adoption of decentralized renewable energy for rural livelihoods. Energy Sustain Dev 84:101600. https://doi.org/10.1016/J.ESD.2024.101600
112. Mohsin M, Taghizadeh-Hesary F, Iqbal N, Saydaliev HB (2022) The role of technological progress and renewable energy deployment in green economic growth. Renew Energy 190:777–787. https://doi.org/10.1016/J.RENENE.2022.03.076
113. Cartland R, Sendegeya AM, de Dieu Khan Hakizimana J (2022) Socio-economic analysis of solar photovoltaic-based mini-grids in rural communities: a Ugandan case study. J Energy South Afr 33(3):36–50 https://doi.org/10.17159/2413-3051/2022/V33I3A10441
114. Alam MM, Murad MW (2020) The impacts of economic growth, trade openness and technological progress on renewable energy use in organization for economic co-operation and development countries. Renew Energy 145:382–390. https://doi.org/10.1016/J.RENENE.2019.06.054
115. Basílio M, Bas M (2025) The support of multilateral development banks to renewable energy projects in developing countries. J Environ Plan Manag 68(3):585–611. https://doi.org/10.1080/09640568.2023.2263904
116. Lee WJ, Mwebaza R (2020) The role of the climate technology centre and network as a climate technology and innovation matchmaker for developing countries. Sustainability 12(19):7956. https://doi.org/10.3390/SU12197956
117. Amir M, Khan SZ (2022) Assessment of renewable energy: status, challenges, COVID-19 impacts, opportunities, and sustainable energy solutions in Africa. Energy Built Environ 3(3):348–362. https://doi.org/10.1016/J.ENBENV.2021.03.002
118. Sankaran S, Clegg S, Müller R, Drouin N (2022) Energy justice issues in renewable energy megaprojects: implications for a socioeconomic evaluation of megaprojects. Int J Manag Proj Bus 15(4):701–718. https://doi.org/10.1108/IJMPB-06-2021-0147/FULL/PDF
119. Hyginus V, Eze U, Favour Uzoma O, Tamball JS, Sarah NI, Robert O, Wisdom OO (2023) Assessing energy policies, legislation and socio-economic impacts in the quest for sustainable development. Int J Educ Sci Technol Eng (IJESTE) 6(2):68–79. https://doi.org/10.36079/LAMINTANG.IJESTE-0602.594
120. Nwaiwu F (2021) Digitalisation and sustainable energy transitions in Africa: assessing the impact of policy and regulatory environments on the energy sector in Nigeria and South Africa. Energy Sustain Soc 11(1):1–16. https://doi.org/10.1186/S13705-021-00325-1/TABLES/3
121. Zulu E, Mutwale J, Zulu SL, Musonda I, Kavishe N, Moobela C (2023) Challenges, drivers and incentives to private sector participation in public-private partnership projects in developing countries: evidence from Zambia. J Eng Des Technol 23(2):386–405. https://doi.org/10.1108/JEDT-03-2023-0092/FULL/PDF
122. Keshavadasu SR (2023) Regulatory and policy risks: Analyzing the uncertainties related to changes in government policies, regulations, and incentives affecting solar power project development and operations in Kenya. Energy Policy 182:113760. https://doi.org/10.1016/J.ENPOL.2023.113760
123. Abada I, Othmani M, Tatry L (2021) An innovative approach for the optimal sizing of mini-grids in rural areas integrating the demand, the supply, and the grid. Renew Sustain Energy Rev 146:111117. https://doi.org/10.1016/J.RSER.2021.111117
124. Gaulin N, Le Billon P (2020) Climate change and fossil fuel production cuts: assessing global supply-side constraints and policy implications. Climate Policy 20(8):888–901. https://doi.org/10.1080/14693062.2020.1725409
125. Oluoch S, Lal P, Susaeta A, Vedwan N (2020) Assessment of public awareness, acceptance and attitudes towards renewable energy in Kenya. Sci Afr 9:e00512. https://doi.org/10.1016/J.SCIAF.2020.E00512

126. Özbay F, Duyar I (2022) Exploring the role of education on environmental quality and renewable energy: do education levels really matter? Curr Res Environ Sustain 4:100185. https://doi.org/10.1016/J.CRSUST.2022.100185
127. Nuru JT, Rhoades JL, Gruber JS (2021) The socio-technical barriers and strategies for overcoming the barriers to deploying solar mini-grids in rural islands: evidence from Ghana. Technol Soc 65:101586. https://doi.org/10.1016/J.TECHSOC.2021.101586
128. Murshed M (2020) An empirical analysis of the non-linear impacts of ICT-trade openness on renewable energy transition, energy efficiency, clean cooking fuel access and environmental sustainability in South Asia. Environ Sci Pollut Res 27(29):36254–36281. https://doi.org/10.1007/S11356-020-09497-3/TABLES/13
129. Shafiullah M, Ahmed SD, Al-Sulaiman FA (2022) Grid integration challenges and solution strategies for solar PV systems: a review. IEEE Access 10:52233–52257. https://doi.org/10.1109/ACCESS.2022.3174555
130. Seetharaman, Moorthy K, Patwa N, Saravanan, Gupta Y (2019) Breaking barriers in deployment of renewable energy. Heliyon 5(1):e01166. https://doi.org/10.1016/J.HELIYON.2019.E01166/ATTACHMENT/89DD8F8B-A822-4083-BBA1-90185C33E8EA/MMC1.DOCX
131. Ullah B (2020) Financial constraints, corruption, and SME growth in transition economies. Q Rev Econ Finance 75:120–132. https://doi.org/10.1016/J.QREF.2019.05.009
132. Asante D, Ampah JD, Afrane S, Adjei-Darko P, Asante B, Fosu E, Dankwah DA, Amoh PO (2022) Prioritizing strategies to eliminate barriers to renewable energy adoption and development in Ghana: a CRITIC-fuzzy TOPSIS approach. Renew Energy 195:47–65. https://doi.org/10.1016/J.RENENE.2022.06.040
133. Oyewo AS, Aghahosseini A, Ram M, Lohrmann A, Breyer C (2019) Pathway towards achieving 100% renewable electricity by 2050 for South Africa. Sol Energy 191:549–565. https://doi.org/10.1016/J.SOLENER.2019.09.039
134. Manko K, Watkins TA (2022) Microfinance and SDG 7: financial impact channels for mitigating energy poverty. Dev Pract 32(8):1036–1048. https://doi.org/10.1080/09614524.2020.1863338
135. Fleta-Asín J, Muñoz F (2021) Renewable energy public–private partnerships in developing countries: determinants of private investment. Sustain Dev 29(4):653–670. https://doi.org/10.1002/SD.2165
136. Waidelich P, Steffen B (2024) Renewable energy financing by state investment banks: evidence from OECD countries. Energy Econ 132:107455. https://doi.org/10.1016/J.ENECO.2024.107455
137. Zhang J (2024) Energy access challenge and the role of fossil fuels in meeting electricity demand: promoting renewable energy capacity for sustainable development. Geosci Front 15(5):101873. https://doi.org/10.1016/J.GSF.2024.101873
138. Liu J, Love PED, Smith J, Matthews J, Sing C-P (2016) Praxis of performance measurement in public-private partnerships: moving beyond the iron triangle. J Manag Eng 32(4):04016004. https://doi.org/10.1061/(ASCE)ME.1943-5479.0000433
139. Lund H, Thellufsen JZ, Sorknæs P, Mathiesen BV, Chang M, Madsen PT, Kany MS, Skov IR (2022) Smart energy Denmark. A consistent and detailed strategy for a fully decarbonized society. Renew Sustain Energy Rev 168:112777. https://doi.org/10.1016/J.RSER.2022.112777
140. George A, Boxiong S, Arowo M, Ndolo P, Chepsaigutt-Chebet, Shimmon J (2019) Review of solar energy development in Kenya: opportunities and challenges. Renew Energy Focus 29:123–140. https://doi.org/10.1016/J.REF.2019.03.007
141. Obuseh E, Eyenubo J, Alele J, Okpare A, Oghogho I (2025) A systematic review of barriers to renewable energy integration and adoption. J Asian Energy Stud 9:26–45. https://doi.org/10.24112/JAES.090002
142. Asmelash E, Prakash G, Gorini R, Gielen D (2020) Role of IRENA for global transition to 100% renewable energy. Lect Notes Energy 74:51–71. https://doi.org/10.1007/978-3-030-40738-4_2
143. Singh BP, Goyal SK, Kumar P (2021) Solar PV cell materials and technologies: analyzing the recent developments. Mater Today Proc 43:2843–2849. https://doi.org/10.1016/J.MATPR.2021.01.003

144. Oduro P, Uzougbo NS, Ugwu MC (2024) Renewable energy expansion: Legal strategies for overcoming regulatory barriers and promoting innovation. Int J Appl Res Soc Sci 6(5):927–944
145. Inês C, Guilherme PL, Esther MG, Swantje G, Stephen H, Lars H (2020) Regulatory challenges and opportunities for collective renewable energy prosumers in the EU. Energy Policy 138:111212. https://doi.org/10.1016/J.ENPOL.2019.111212
146. Villanueva PAG (2020) Why civil society cannot battle it all alone: the roles of civil society environment, transparent laws and quality of public administration in political corruption mitigation. Int J Public Adm 43(6):552–561. https://doi.org/10.1080/01900692.2019.1638933
147. Xia-Bauer C, Vondung F, Thomas S, Moser R (2022) Business model innovations for renewable energy prosumer development in Germany. Sustainability 14(13):7545. https://doi.org/10.3390/SU14137545
148. Ali EB, Anufriev VP, Amfo B (2021) Green economy implementation in Ghana as a road map for a sustainable development drive: a review. Sci Afr 12:e00756. https://doi.org/10.1016/J.SCIAF.2021.E00756
149. Le TT, Sharma P, Bora BJ, Tran VD, Truong TH, Le HC, Nguyen PQP (2024) Fueling the future: a comprehensive review of hydrogen energy systems and their challenges. Int J Hydro Energy 54:791–816. https://doi.org/10.1016/J.IJHYDENE.2023.08.044
150. Zamiri M, Esmaeili A (2024) Methods and technologies for supporting knowledge sharing within learning communities: a systematic literature review. Adm Sci 14(1):17. https://doi.org/10.3390/ADMSCI14010017
151. Sanderink L, Nasiritousi N (2020) How institutional interactions can strengthen effectiveness: the case of multi-stakeholder partnerships for renewable energy. Energy Policy 141:111447. https://doi.org/10.1016/J.ENPOL.2020.111447
152. Martinez N, Komendantova N (2020) The effectiveness of the social impact assessment (SIA) in energy transition management: stakeholders' insights from renewable energy projects in Mexico. Energy Policy 145:111744. https://doi.org/10.1016/J.ENPOL.2020.111744
153. Pandey N, de Coninck H, Sagar AD (2022) Beyond technology transfer: innovation cooperation to advance sustainable development in developing countries. Wiley Interdiscip Rev Energy Environ 11(2):e422. https://doi.org/10.1002/WENE.422
154. Oyekale J, Petrollese M, Tola V, Cau G (2020) Impacts of renewable energy resources on effectiveness of grid-integrated systems: succinct review of current challenges and potential solution strategies. Energies 13(18):4856. https://doi.org/10.3390/EN13184856
155. Hasan A, Dincer I (2020) A new performance assessment methodology of bifacial photovoltaic solar panels for offshore applications. Energy Convers Manage 220:112972. https://doi.org/10.1016/J.ENCONMAN.2020.112972
156. Jošt M, Kegelmann L, Korte L, Albrecht S (2020) Monolithic Perovskite Tandem solar cells: a review of the present status and advanced characterization methods toward 30% efficiency. Adv Energy Mater 10(26):1904102. https://doi.org/10.1002/AENM.201904102
157. Kataray T, Nitesh B, Yarram B, Sinha S, Cuce E, Shaik S, Vigneshwaran P, Roy A (2023) Integration of smart grid with renewable energy sources: opportunities and challenges—a comprehensive review. Sustain Energy Technol Assess 58:103363. https://doi.org/10.1016/J.SETA.2023.103363
158. Kayalvizhi N, Santhosh M, Thamodharan R, Dhileep M (2024) IoT-enabled real-time monitoring and predictive maintenance for solar systems: maximizing efficiency and minimizing downtime. In: International conference on smart systems for applications in electrical sciences, ICSSES 2024. https://doi.org/10.1109/ICSSES62373.2024.10561454
159. Albogamy FR, Paracha MYI, Hafeez G, Khan I, Murawwat S, Rukh G, Khan S, Khan MUA (2022) Real-time scheduling for optimal energy optimization in smart grid integrated with renewable energy sources. IEEE Access 10:35498–35520. https://doi.org/10.1109/ACCESS.2022.3161845
160. Solargis (2021) Solar resource maps & GIS data for 200+ countries. Solargis. Available at: https://solargis.com/resources/free-maps-and-gis-data?locality=africa. Accessed 2 Dec 2025

Chapter 11
Renewable Energy: Non-solar-Based

11.1 Introduction

Energy demand globally is rising as a result of the need for buildings for residential, commercial, and industrial purposes. The African continent presently is witnessing a very high increase in population, thus the increase in the need for energy for buildings and houses [9]. Apart from the traditional fossil fuel-based technology for electrical energy generation, efforts are being made by the continent to adopt more sustainable and reliable energy sources. A notable technology is renewable energy which includes solar, wind, hydro, and many more [108]. Solar energy is one of the major renewable energy solutions being adopted as seen in the previous chapter. However, there exists a good potential in other forms of non-solar-based renewable energy technologies such as wind, hydro, biogas, and geothermal. These technologies have seldom been applied in the African context [113]. Given the high demand in energy, it is profitable to investigate and leverage the prospects of non-solar-based renewable energy.

Renewable energy technologies have been adopted by advanced countries to turn their energy mix into a more sustainable one, calling for a worldwide decarbonization of energy generation [35]. The vast renewable sources in Africa places the continent at a position of driving a huge change towards building energy decarbonization. As international organizations are calling for reduction in carbon emissions and mitigation of climate change, the African continent is striving to contribute to the achievement of that goal. However, this cannot be achieved without at least seven times the growth rate of renewable energy sources by 2060 [61]. This means that a continent like Africa needs to leverage on all the renewable energy sources available. This will contribute to supply of clean energy for domestic, commercial, and industrial uses [113]. Solar power generation is the most common renewable energy technology actively being adopted [60]. Other renewables such as wind, geothermal, hydro, and biomass have been gaining attention and investments in other parts of the world [113]. The figure below shows a general increase in renewable energy installed capacity over the last two decades for wind, hydro, biomass, and geothermal energies.

M. N. Addy et al., *Decarbonization of Building Energy in Developing Countries*, Green Energy and Technology, https://doi.org/10.1007/978-3-032-09113-0_11

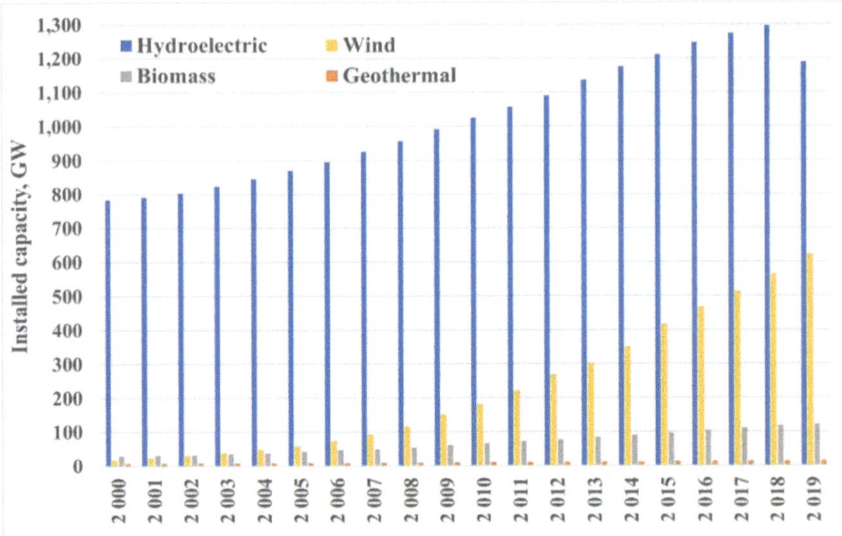

Fig. 11.1 Installed capacity of non-solar-based renewable energy [62]

Further, hydroelectric and tidal energy have previously seen considerable attention towards them being the first commercialized and mature renewable system compared with the rise of the other renewable systems [62] (Fig. 11.1).

The convergence of these non-solar renewable sources leads to a holistic energy efficiency and decarbonization pathway [84]. A successful integration therefore requires coherent and bespoke policies, technological innovation and expertise, and community involvement. The entire energy future of Africa cannot rely on a single resource [108]. Instead, a concerted effort must be made to enhance an adaptive energy system through the integrated use of wind, hydro, biomass, biogas, and geothermal energy so that this system is capable of meeting the increasing demands towards climate resilience [113]. In particular, decentralized energy solutions may restore electricity access to nearly 600 million Africans while simultaneously increasing their alternatives to diesel generators and fossil-fuel-powered grids [89]. In addition, the circular economy potential of bioenergy and biogas systems supports sustainability goals of turning waste into useful product and promoting local-level economic development [14]. Also, the reliable nature of geothermal energy can complement the intermittent features of wind power and small hydroelectric plants, thereby balancing the energy portfolio and stabilizing the grid [39]. However, to seize this opportunity, key barriers such as poor financing mechanism, weak regulatory frameworks, and lack of technical skills, inadequate research and development and community disengagement must be overcome [6]. Through international collaboration, good financing models, knowledge exchange, and development of technical capacity through training and education, community involvement can help overcome the challenges, ensuring that the African renewable energy transition is an inclusive one [138, 142].

11.2 Overview of Renewables: Wind, Earth, Hydro, Biomass and Biogas, Geothermal and Ocean/Tidal Energy

Hydroelectric power is a conversion from the energy of flowing water to the electrical energy through multi-step conversion processes. A dam is constructed to hold back water from river and forms an elevated reservoir with gravitational potential energy [112]. When the water in the reservoir is released, it flows down through penstocks (large pipes) and gains kinetic energy because of the potential head or height difference available [64]. This high-speed water then hits the turbine blades and rotates them, depending on the nature of head and flow, types of turbines used may be Francis, Kaplan, or Pelton [64]. The rotational mechanical energy of the turbine shaft is transferred to a generator, where, by virtue of electromagnetic induction according to Faraday's law, it is converted into electrical energy [132]. The generators have their rotor rotating about their stator, and generally currents are generated when magnetic fields cut conducting coils [33]. Then transformers step up the voltage for transmission via the grid to end-users. Hydropower offers great prospects throughout the African continent, especially sub-Saharan African (SSA) countries [143]. Approximately, only 11% of Africa's technical hydropower has been developed leaving the majority unused [73]. Furthermore, in Ethiopia, Malawi, Mozambique, and Zambia, about 80–90% of electricity is produced by hydropower, with Ethiopia producing the most at 3.8 GW and increasing with new projects underway [53, 68]. Historically, the greatest and most established green energy source utilized in Africa's electrical supply sector has been hydropower [20]. SSA's electricity output of 20% comes from hydropower, a figure that has been steadily increasing [144] (Fig. 11.2).

Wind energy is regarded as an environmentally friendly source of energy [69]. It is among the earliest RESs, and a lot of research has already been done on it [50, 99]. Both terrestrial and marine settings can accommodate the use of wind energy technology [18]. Wind energy technology uses wind turbine, it harnesses wind speed to produce electricity. When there is sufficient wind speed, the wind turbines blades rotate [110]. A rotor of an electric generator is connected to the turbine, and as the blades rotate the turbine, the rotor also rotates to generate electricity. Wind turbines are classified into two types according to the axis of rotation, whether vertical (vertical axis wind turbines, VAWTs) or horizontal (horizontal axis wind turbine, HAWT). The advantage of using a VAWT type is its ability to capture wind from all directions. However, HAWTs are mostly adopted in wind energy because of their higher power capacity [74]. In 2019 to 2020, around 651 GW of electricity are produced globally by wind turbines [108] with current estimates exceeding 1 TW. Wind is placed second only to hydropower (Fig. 11.3).

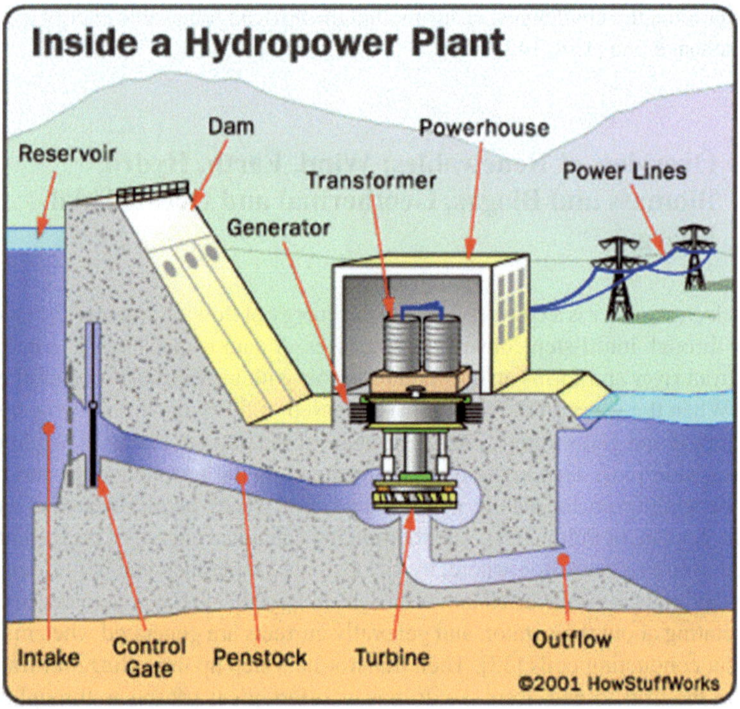

Fig. 11.2 Hydroelectric power generation [125]

Among other non-solar-based renewable energy technologies (RETs) is bioenergy (biogas and biomass). Biomass is processed from organic materials that are usually burned or disposed in landfills such as wood residue, ethanol, manure, switchgrass, wheat straw, human waste, and food waste [106]. As an alternative, biomass, such as millet, maize, miscanthus, etc., can be grown exclusively for energy production [105]. Hydrocarbon is the primary chemical component of biomass. The chemical compound stored in these biomass materials is converted into electrical energy [139, 140]. For example, traditional coal power plants can be used to burn biomass for the generation of electrical power since it has an unlimited supply and is considered as a renewable energy source [135]. Additionally, biogas contains methane and carbon dioxide which is found in sewage treatment plants, waste landfills, and animal manure processing schemes. Biogas may contain methane and carbon dioxide in itself, but when it is burned, it is converted to electricity, hence reducing the overall greenhouse emission [77] (Fig. 11.4).

Biopower comes from liquid biofuels or biomass [21]. There are two common forms of biofuels: bioethanol and biodiesel. Because bioethanol and biodiesel are biodegradable, readily available from renewable resources, and reduce combustion emissions, they are preferred as transportation fuels [127]. The adoption of bio-based renewable technology for energy generation is still not widely used in the African

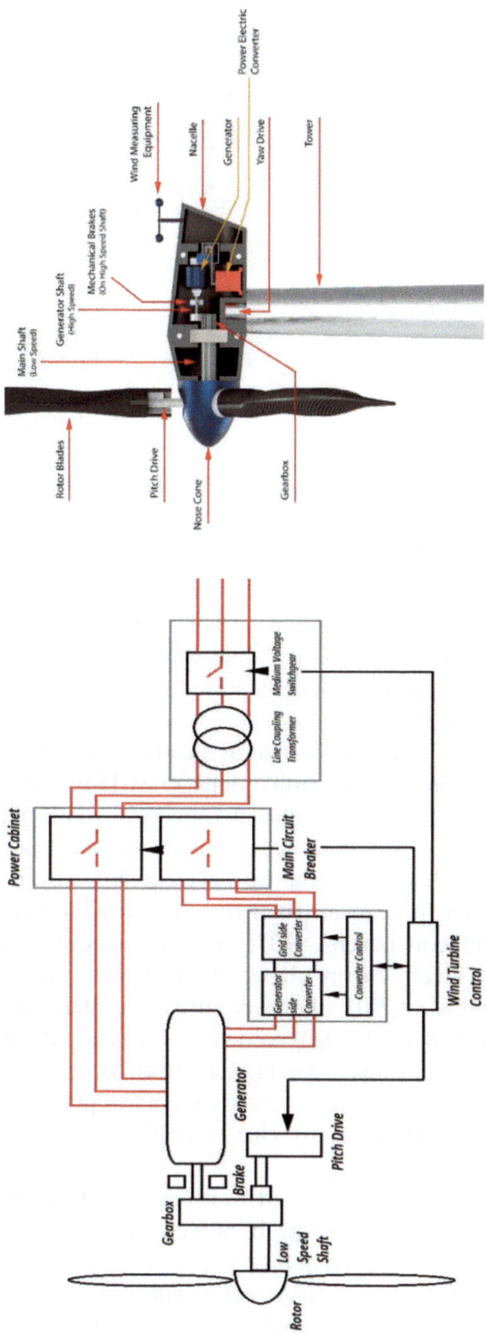

Fig. 11.3 Grid connected wind energy system and typical components [24]

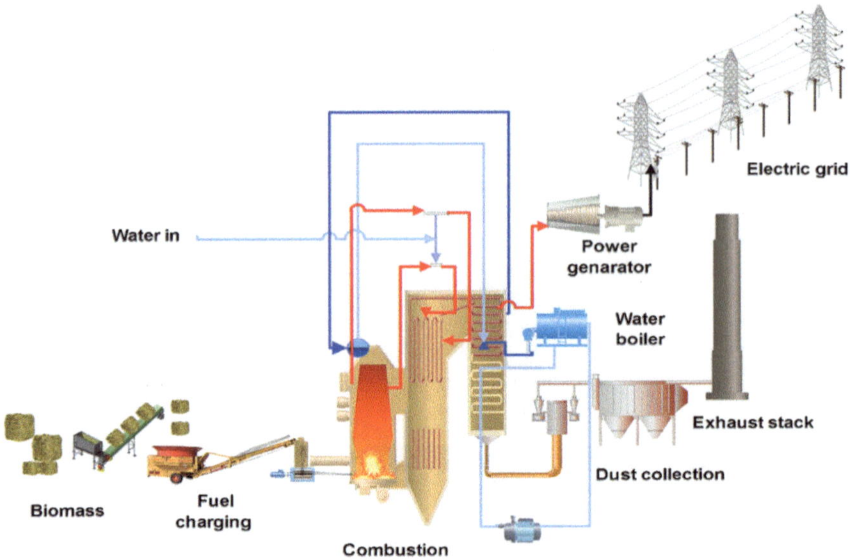

Fig. 11.4 Biomass electricity generation process [103]

context [57]. There are many raw materials available in the area that can be used to produce electricity, including municipal solid waste and agricultural and forest wastes [43]. Even though there are a lot of biopower resources in the SSA region, its harnessing remains limited [57].

Additionally, other non-solar RETs such as geothermal power plants use the thermal energy that is stored in the earth's outer crust [55]. Geothermal energy potentials are identified through a thorough and deep investigation of the geological, geothermal, and geophysical data of an area. This investigation is very expensive and complex, requiring high level of technical and technological expertise [39]. Hot springs or other geothermal fluids located between 300 and 8000 m below the surface of the earth capture geothermal energy [107]. In some cases, space heating can be accomplished directly with this energy [5]. Geothermal power plants typically have very low emissions, and as of 2019, they were used to capture about 95 TWh of electrical energy [108]. Heating of spaces accounts for the majority of the 8% rise in direct geothermal energy utilization in recent years. Even though geothermal energy is available all the time, it can only be extracted from a few places on earth. Kenya, Indonesia, Turkey, and a few other countries have recently adopted geothermal power technologies [108] (Fig. 11.5).

Additionally, one of the most amazing RESs is ocean- and tidal-based energy. It makes use of the water's rise and fall during the phases of flood and ebb tide [97]. The most notable technologies for capturing tidal energy are the tidal stream and tidal barrage [30]. In contrast to wind and solar energy, tidal energy is easily forecasted [136]. However, a major drawback is that only a small number of places show encouraging potential for tidal energy due to coastal proximity. Similar to wind turbines that

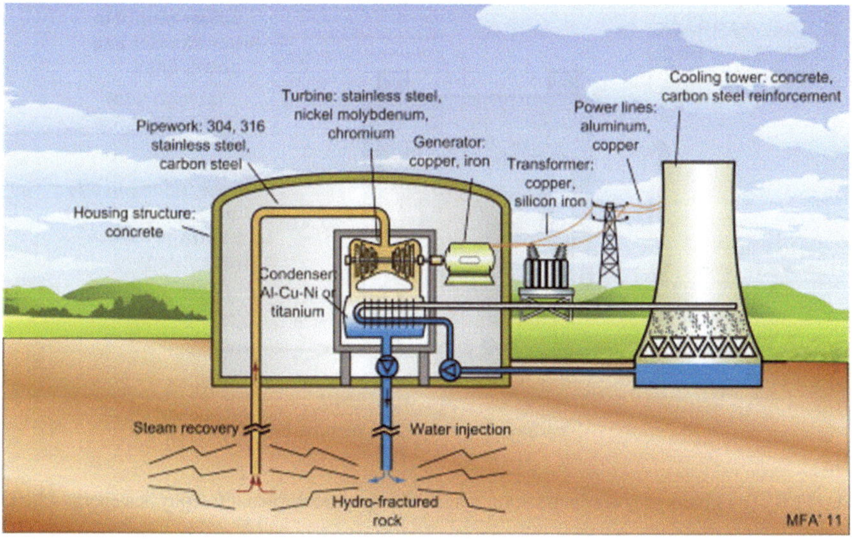

Fig. 11.5 Geothermal power generation [19]

use the kinetic energy of the wind, tidal stream turbines employ the kinetic energy of flowing water to produce electricity [58]. Unlike other renewable energy sources that take up natural space, tidal stream generators are either completely buried or integrated onto already-existing bridges [122]. Inlets and straits are common places to find high water velocity, making them suitable sites for tidal turbine installation [100, 111]. Both horizontal and vertical placements are possible for these turbines. Water is 784 times denser than air, and hence, the energy extracted from tidal streams is greater than that from wind [108]. With similar turbine systems, a tidal current of about 60 kph might produce as much energy as or more than a wind speed of about 145 kph. In a tidal barrage technology, small dams are built near coastlines and estuarine regions to retain water during flooding which is allowed to flow through a turbine. There are only 40 sites across the whole planet that are eligible for tidal barrage installation because at least a 5-m minimum water level or tidal range difference is needed for an economical operation [108]. The biggest tidal barrage power stations are the La Rance and Sihwa stations, having capacities of 240 MW and 254 MW, respectively [47]. Although having great potential, tidal barrage development is limited due to the large initial investment costs, lengthy construction times, and various environmental effects [104] (Fig. 11.6).

Ocean-based RESs have the least installed capacity globally but have the greatest potential. Ocean energy is largely predictable and always available [2]. 535 MW of ocean-based power plants were in operation as of 2019 [108]. Oceanic wave, ocean current, and ocean thermal are the three most impressive methods of using ocean energy. Ocean thermal energy uses the temperature difference between warm surface water and deep cold water to create electricity [56]. Oceanic wave energy has very large availability than many other renewable energy sources, which is one

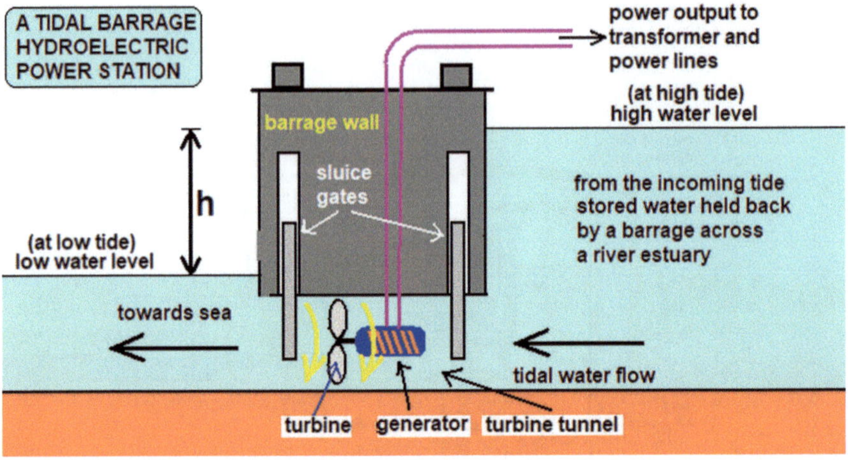

Fig. 11.6 Tidal barrage power generation diagram (https://docbrown.info/ephysics/energy4.htm)

of its main benefits [88]. One of the most promising renewable energy sources is oceanic wave energy, which has the capacity to provide 1–10 TW of electricity [42]. Electrostatic, electromagnetic, and piezoelectric technologies are mostly utilized to harness ocean wave energy [70]. The devices used for ocean-based energy generation are either totally submerged, partially submerged, or floating [42] (Fig. 11.7).

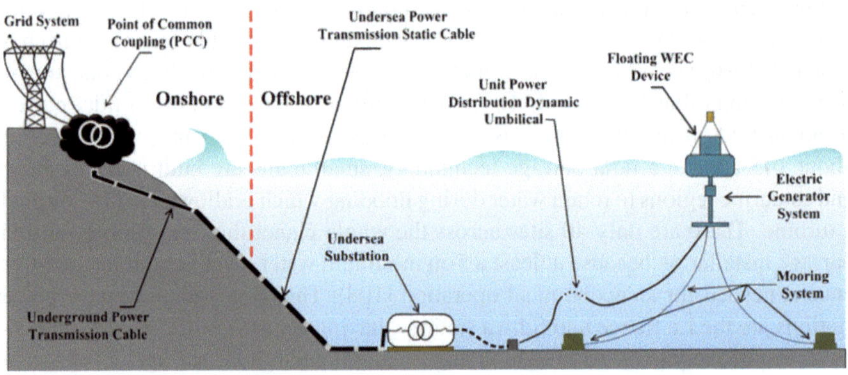

Fig. 11.7 Wave energy conversion system with underground substation [8]

11.3 Case Studies of Non-Solar-Based Renewables

11.3.1 Wind Power

Africa's wind energy sector is a promising and transformative endeavour to provide solutions to the energy challenges and shortages, as well as the high cost of electricity and dependence on fossil fuel-based electricity generation [108]. The eastern, northern, and southern parts of the continent possess the highest wind energy potential due to their consistent wind pattern and expansive terrain which are favourable conditions for wind power projects [36]. In these regions, the Kenyan Lake Turkana wind power project [54], Morocco's Tarfaya wind farm [41], South Africa's Renewable Energy Independent Power Producer Procurement Programme (REIPPPP) [27], and Ethiopia's Ashegoda and Adama wind farms [133] are some of the notable wind power projects. These projects exhibit the potential of how wind power can drive sustainable and clean energy production, reduce electricity cost, and enhance power supply reliability.

Kenya's Lake Turkana wind power project, situated in Loiyangalani in Marsabit County, being one of the largest wind energy projects in Africa with a capacity of 310 MW has been a transformative solution in the energy mix and crisis of the country, offering an avenue away from the costly fossil fuel-based generation [115]. As a result, the country has been able to reduce energy costs leading to a cost savings of about 150 million dollars moving from the usage of fossil fuel [32, 59]. According to the European Investment Bank (EIB), the project has contributed to the reduction of about 700,000 tonnes of CO_2 emissions annually [38]. Similarly, Morocco's 300 MW Tarfaya wind farm has played a tremendous role in the country's energy solutions, aiming at deriving 52% of its electricity from renewable resources by 2030 [98]. Morocco's reliance on imported fossil fuel which makes up to 95% of its energy generation and consumption has been reduced, leading to the potential of reducing 900,000 CO_2 tonnes/year [31, 52]. The Tarfaya wind farm operates in synergy with the country's solar power plants such as the Noor Ouarzazate to provide a balanced hybrid system to enhance grid stability [52]. This highlights the potential of hybrid systems which serve as a solution to the intermittency challenges of wind and solar power generation [17].

In the case of South Africa, facing the challenges of a debt of USD 24 billion, old infrastructure, and corruption, national power supply has been hindered leading to frequent load shedding as energy demands increased [90]. Dependency on coal and other fossil fuel in South Africa, accounting for 77% in electricity generation, has led to the degradation of the environment and has increased energy costs [126]. The Renewable Energy Independent Power Producer Procurement Programme (REIPPPP) in South Africa has supported investments in wind energy from the private sector, including projects such as the 138 MW Gouda Wind Farm and 140 MW Kangnas Wind Farm, which have considerably developed the renewable capacity in the country [119]. Wind power offers a viable solution, with the South African wind energy association estimating a 62% energy need of the country to be met

[123]. According to the REIPPPP, a 42% reduction in wind energy production costs is possible, bringing down the cost of electricity to 74 cents per kWh under competitive bidding procedures [141]. Moreover, these projects have incorporated strong local content requirements, benefiting job creation and industrial development in areas suffering from high unemployment [22]. Still, the South African experience exposes the challenges of integrating variable renewable energy in a grid dominated by the use of coal as [75] posits that, the REIPPPP has failed by supplementing fossil power generation instead of replacing it. This by implication requires investments in grid flexibility, battery storage, and demand-side management for stability. Ethiopia on the hand has also leveraged its wind resources to build a 120 MW Ashegoda wind farm and the 153 MW Adama wind farms [45]. This has diversified the country's hydropower-dependent grid which is often challenged by climatic conditions and drought [80]. By incorporating wind energy, the country has enhanced its energy security and supply [4]. The Adama project was built with Chinese financing and technology transfer, illustrating the need and opportunities of international partnerships to drive sustainable energy in Africa since one of the major challenges facing RET projects in Africa is lack of funding and technology [26, 28].

11.3.2 Hydropower

Hydropower has become the main beneficiary of the development of renewable energy across Africa. It offers many transforming opportunities but also presents complex challenges. With a capacity of 5,150 MW, the Grand Ethiopian Renaissance Dam (GERD) stands as the largest hydropower project in the continent, surpassing the High Aswan Dam (HAD) of Egypt, thus affirming Africa's quest for energy independence [40]. The population of Ethiopia as of 2023 stands at 128 million with a predicted increase of 75% to 225 million by 2050 according to the World Health Organization [131]. This will be accompanied by increase in energy demand for buildings. As of 2011–2015, the country saw a 32% rise in energy demand. The GERD was officially launched in 2011 to meet the rise in energy demand through more sustainable means away from fossil fuel power generation [25]. The GERD has gone through four reservoir filling phases and has now been completed with the fourth filling as of September 2023. The project adopted novel financing mechanisms through public–private partnerships to raise funds internally through bond sales, athletic events, and lottery draws to raise USD 460 million and Chinese loans [7, 118]. The project has however placed relations under significant tension concerning the rights over the waters of the Nile. The High Aswan Dam belonging to Egypt shares the same water source with the Ethiopian GERD [130]. Moreover, the dam, which could restrict water flow to downstream countries, poses an existential threat to Egypt because of the Nile's water resources and cultural importance (Haas-[51]). This called for an unusual level of diplomatic engagement among Ethiopia, Egypt, and Sudan, but efforts made on agreement have failed. It is recommended that negotiations founded on good faith and sincerity, unbiased mediation from external stakeholders, and good politics from the countries will help solve the current impasse [87].

Another notable hydroelectric power plant in west Africa is Ghana's Akosombo Dam constructed on the largest human-made reservoir in the world, Lake Volta. The dam has an installed capacity of currently 1,020 megawatts [93]. It generated over 90% of the nation's electricity at its inception [11]. However, reductions of about 40% of power have been experienced due to drought and sedimentation [82]. Currently with the increase in population and energy demand, the country is relying on thermal power plants to generate 60% to 65% per cent of electricity [48]. One major challenge the dam poses is the flooding challenge to dwellers at the downstream of the river when it is spilled during maximum water levels, [37]. Nevertheless, Ghana has made headways with energy policies to increase renewable energy in its energy mix [12]. The country has a potential to introduce floating solar PV hybrid systems on the Akosombo Dam to increase renewable energy generation. Such a project can generate a total annual electricity of 520,233 MWh and save 73,327 cubic metres of water from evaporating which can produce approximately 10 MWp hydroelectric power annually. Additionally, it can reduce electricity cost to US$ 0.10/kWh and reduce GHG emissions by 308,904.5 tCO 2/MWh annually [11].

The Inga complex of the Democratic Republic of Congo, with a theoretical capacities of Inga I-351 MW, Inga II-1424 MW, and Inga III-11 GW is another remarkable case of hydropower renewable energy adoption [116]. However, the project has also faced the frustrating realities of African hydropower, whereby political instability, environmental concerns, and financing challenges have hindered the development of even the Inga III expansion [49]. The Rusumo Falls 80 MW Hydroelectric Project stands in a different light. Developed under the Nile Basin Initiative, run-of-river technology was utilized to ensure minimal disruption of the environment in providing affordable power to the three nations (Burundi, Rwanda, and Tanzania) through carefully negotiated power-sharing agreements [94]. On the other side and perhaps, the greatest indicator of the operating efficacy of sustainable hydropower development in the continent today is the Zambia Kafue Gorge Lower Dam of 750 MW, erected in 2021 [81]. The project presents the most advanced environmental water flow management schemes [92] and benefited local communities [29]. Such measures now stand as a standard for all new projects across the continent. Nevertheless, with climate change-induced droughts reducing hydropower generation in Zambia, there is an urgent call for the hybrid systems of renewables [67].

11.3.3 Bio-Based Power

Bioenergy and biomass resources are essential avenues for Africa's energy transition. They provide decentralized means to make food, generate electricity, and run industrial processes, all while eliminating waste and creating jobs in rural areas [16]. In Kenya, Mumias Sugar Company's bagasse cogeneration plant represents an example of industrial biomass use, where sugarcane waste is used for cogeneration, producing 38 MW of electricity [65]. The power is primarily for internal use in the factory and exporting excess to the grid, hence increasing energy security in sugar-growing areas and reducing dependency on fossil fuels [66]. Just like every other power plant, the sugarcane power plant experienced a reduction in generation due to a decrease in sugarcane supply [63]. This raises the urgent need for a diversification of feedstock strategy in these plants [72, 117]. Koshe waste-to-energy, generating 50 MWh per day and eliminating over 1.4 million kilograms of waste every day, is the first waste-to-energy plant in Africa [137]. This facility goes a long way in alleviating Addis Ababa's dual urban crisis of waste management and power shortage. Due to financial constraints, the Koshe WTE operates at a reduced generation capacity of 25 MW [137]. Waste-to-energy (WTE) plants will thrive if effective waste segregation systems are implemented [124] and public awareness is raised on how to manage waste, shifting perceptions from viewing waste as nothing to valuing it [76]. It is important to note that, although WTE plants do produce green power, the process of generation, especially during incineration of waste, can release some harmful gases such as nitrous oxide (N_2O), carbon dioxide (CO2 in the atmosphere [86, 120]. It is important to initiate gas trapping and recirculation mechanisms to reduce the harmful emissions to the air [128]. Bioenergy is not merely another energy generation means but also integral part of the circular bioeconomy future of Africa, balancing climate action with energy access.

11.3.4 Geothermal Energy

According to Elbarbary et al. [39], geothermal power serves as a major resource along East African regions but has remained largely unexplored and underutilized. The resource offers clean energy options, energy security and supply, and economic and sustainable development [114]. Among the major users of geothermal power is Kenya, where the Olkaria Geothermal Complex near Naivasha is found [102]. The Olkaria field is the major production field with installed capacity standing at 920.5 MW (94.6% of total) in its five sections (Olkaria I-V). Olkaria produces nearly half of Kenya's power, reducing the cost and environmentally unfriendliness of thermal production. Its success is rooted in public–private partnership model, where government-owned KenGen collaborates with independent power producers like Ormat Technologies [109]. Also supplemented by World Bank and African Development Bank financing that buffered early exploration risk [3, 134]. In other instances, the feed-in tariff system and establishment of a Geothermal Development Company in Kenya have further accelerated the development of geothermal energy [23, 95]. Challenges for geothermal power generation involve expensive upfront drilling costs, lands issues, and community opposition to the use and allocation of resources [79].

Also in Ethiopia, the first activities in geothermal explorations relate to Corbetti and Aluto-Langano fields, but the development later slowed down due to financing issues and technical capacity constraints [46]. Nevertheless, the existence of a 7

MW Aluto-Langano pilot plant, together with the 10 MW pilot plant project at Tendaho Dubt, marks a bright future for growth [53]. There has been a recent agreement between the Ethiopian government and Reykjavik Geothermal for a 1,000 MW Corbetti project. This is a breakthrough that leverages Iceland's expertise in geothermal power drilling technology to make good use of the geothermal resource of the country [34]. Djibouti's Assal Rift project is another example of international cooperation, entering into partnership with Japan's JICA and the World Bank to provide a 50 MW generation capacity by 2030 [1]. Tanzania's Ngozi and Songwe projects also show the area's unexploited potential with estimates around 650 MW of exploitable capacity [96]. Challenges faced by Tanzania are a repetition of others, such as the lack of geological data and donor-financed dependencies [85]. On the other hand, the existence of Geothermal Risk Mitigation Facility (GRMF) for East Africa can contribute to de-risking the initial investments [71]. This development will meet the nation's total current power usage and terminate its reliance on external power.

11.3.5 Ocean and Tidal Energy

The African continent, with its long coastline spanning beyond 30,500 km, along with the dependence on strong marine currents, holds excellent promise in ocean and tidal energy generation [91]. Yet, in Africa, the very concept of implementing an ocean energy project is still nascent while much has already been done in the field of solar and wind energy [2]. Nigeria through continuous wavelet transform analysis, studied the tidal stream potential along the Calabar coast and discovered some sites with economically viable levelized costs comparable with those of conventional power [13]. Nevertheless, challenges to actual formulation plagues the project, namely financing challenges and maritime jurisdictional issues with Cameroon [15, 101]. In addition, the Agulhas Current of the east coast of South Africa ranks among some with the highest energy densities in the world. Such undersea highways bear maximum currents with power estimated to run into thousands of megawatts, enough to power millions of households [78]. However, the generally adverse marine conditions and the cost of infrastructure will continue to render any idea of a large-scale scheme with pilot projects such as the proposed 5 MW demonstration plant impossible [129]. Similarly, the Atlantic waves in Morocco, especially Tarfaya, offer great potential for wave energy. Pilot projects leveraging European technology transfers are being carried out. Though grid integration and environmental concerns over marine ecosystems remain unresolved, these pilot projects show the potential for ocean wave energy growth [10]. Also, Ghana's marine energy policy, influenced by ECOWAS renewable targets, has identified tidal and wave resources as future priorities [44] Ghana has a coastline measuring 550 km with the high potential of generating electricity from tidal wave. However, this resource potential is yet to be assessed due to already persisting challenges in technology, funding, and regulatory frameworks [83, 121].

11.4 Pathways to Non-Solar-Based Renewables

Renewable energy adoption is stalled and hindered due to many challenges in funding, technical expertise, technology inadequacy, and many more. It is important to identify pathways and solutions to these problems to increase the adoption and success of renewable energy projects in Africa. It is expedient to invest in advanced research and development into new ways and methods of building, operating, maintaining, and upgrading renewable power plants. Further, collaborations can foster innovative financing mechanisms, risk-sharing systems and technology sharing. From the cases explored in the various types of non-solar-based energy, these pathways are further explained below:

11.4.1 Research and Development

An important pathway to advance the adoption of renewables in Africa is research and development. It serves as the main discovery hub that researches into new ways and technologies to mitigate challenges and enhance generation capacities. The importance of R&D is demonstrated in the Lake Turkana Power project where through research in advanced meteorological modelling and grid integration, and there can be an energy saving of USD 150 million when fuel fossils are not used [32, 59]. Similarly, the Tarfaya wind farms have used hybrid systems to integrate wind and solar to enhance the intermittency challenges and improve grid stability and supply [17, 52]. Furthermore, challenges like the sedimentation in the hydrogeneration side necessitate research and development in sediment management and transboundary water impact assessment [130].

11.4.2 Collaboration

Renewable energy projects in Africa need collaborations and partnership between stakeholders to foster adoption and implementation. These collaborations are done to provide funding and technology sharing, risk sharing, and resource allocation across all stakeholders in the public. Private and international sectors. The establishment of the Ethiopia Adama wind farms shows the importance of collaboration, where Chinese financing and technology transfer enabled rapid development of the project [26, 28]. The need for local stakeholder collaboration is also seen in the challenged faced by the Lake Turkana wind project over land ownership. The same project however shows multilateral financial collaboration between European Investment Banks and private investments to mitigate high capital cost [38, 115]. One very important collaboration is the Nile Basin Initiative for the Rusumo Falls Project. Rwanda, Burundi and Tanzania jointly managed run-of-river technology to reduce environmental issues and share benefits according to negotiated agreements [94]. On the opposite side, there exist a tension between Egypt and Ethiopia over the GERD project without any agreements yet [51]. Additionally, south-south collaboration is highlighted in the Geothermal Risk Mitigation Facility (GRMF). Collaboration is seen between Ethiopia and Iceland for the Corbetti project in Ethiopia where Iceland's drilling technology was used in the Reykjavik partnership. This shows how collaboration can bridge skill gaps in the adoption of renewable energy [1, 34]. The effectiveness of collaboration thrives on sincere and fair agreements between parties. It is important to formulate favourable agreements that meet each party's needs while sharing in risk and potential losses.

11.4.3 Technical Expertise and Technology Advancement

Technical expertise and technological evolution are important avenues to further develop renewable energy potential of Africa into reliable and affordable generation assets. Thus, requiring training to develop local skills, rehabilitation of infrastructure, and bespoke technologies suitable to operational challenges. The wind energy development in Africa faces persistent skill gaps, necessitating the need for technical expertise. The Ethiopia Ashegoda wind farms face problems with grid integration systems due to inadequate infrastructure and limited engineering capacity in the local sector. This constrains demand for smart grid investments and training programmes [45, 133]. Although collaboration is important to alleviate certain challenges, the over reliance on foreign engineers for building, operation, and maintenance of projects such as the Lake Turkana and Ethiopia geothermal Corbetti project exposes the need for vocational training to equip and enhance local expertise. This will also help in reducing cost on technical expertise as compared to using foreign engineers [1, 54]. In addition, the need for technological advancement and technical expertise will help mitigate emission problems of waste-to-energy bioenergy generation during incineration. This is seen in the inadequate flue gas capture and recirculation systems lacking in Ethiopia's Koshe waste-to-energy plant [137]. Technical skills and technology remain a challenge in Africa's energy sector, international policies should be drawn by governments to introduce modern and digital skills in the education system. The government can also partner with engineering companies in their countries to provide student internships and field experience.

It is worth noting that unlike solar PV and other solar-related renewable energy system, which is widely deployable at the household or corporate scale in Africa, most non-solar renewable energy technologies are not. As a result, renewable energy such as hydropower, wind, or geothermal is far less suited to individual or private installation due to site-specific resource requirements, high capital costs, and technical complexity. Consequently, their contribution to demand-side energy management is limited to niche contexts, with most non-solar renewables being developed at utility scale as part of supply-side electrification strategies.

11.5 Summary

This chapter focuses non-solar-based renewable energy sources. Solar energy is one of the major renewable energy sources in Africa. However, the continent is rich in other forms of renewable energy sources such as wind, hydro, geothermal, bioenergy, and ocean-tidal energy sources. There is a high potential for Africa to solve its energy crisis and reduce demand-side energy by adopting these renewable sources. Countries like Egypt, Ethiopia, Kenya, Ghana, Morocco, and South Africa are making headways in adopting these sources. Renewable sources have been proven to be environmentally, thus the interest from all these African countries. However, the adoption and implementation of these energy sources faces challenges similar to the ones reviewed in the previous chapter. The review and exploration of cases and examples of these non-solar-based renewable technologies in Africa exposed challenges such as lack of funding, cross-country disagreement, opposition from local communities, lack of technical expertise and technology, and environmental challenges. From the case studies, best practices and pathways were realized to mitigate these challenges and foster adoption. Research and development, collaboration, technical, and technological advancement are important pathways that can mitigate the challenges highlighted. Governments need to enter into favourable agreements and partnerships that will benefit their country. There is the need for resource allocation and funds for advanced research and development to discover new technologies or upgrade existing one to suit present challenges. Technical skills should be a major priority to boost the expertise of local engineers and students as well. The overdependence on foreign countries for technology, skill, and technology has an adverse effect on the overall budget of renewable energy projects. Cost can be reduced if technology and skill are readily available locally. Africa is rich in all these resources, and effective usage will boost economic growth across all sectors and increase the standard of living of the citizens.

References

1. Abdi AM, Murayama T, Shigeo Nishikizawa D, Suwanteep K (2023) Gaps between Stakeholders‘ concerns and mitigation measures in ESIA reports of geothermal energy projects: a case study of geothermal energy projects in Djibouti, Ethiopia and Kenya. East African J Environ Nat Resour 6(1):491–513. Available at: https://doi.org/10.37284/EAJENR.6.1.1619
2. Adesanya A, Misra S, Maskeliunas R, Damasevicius R (2021) Prospects of ocean-based renewable energy for West Africa's sustainable energy future. Smart Sustain Built Environ 10(1):37–50. Available at: https://doi.org/10.1108/SASBE-05-2019-0066/FULL/PDF
3. AfDB (2018) The African development bank and the climate investment funds support their first private sector-led geothermal power plant in Kenya | Banque africaine de développement. Available at: https://www.afdb.org/fr/news-and-events/the-african-development-bank-and-the-climate-investment-funds-support-their-first-private-sector-led-geothermal-power-plant-in-kenya-18233. Accessed 10 June 2025
4. Agajie TF, Fopah-Lele A, Amoussou I, Khan B, Bajaj M, Zaitsev I, Tanyi E (2024) Enhancing Ethiopian power distribution with novel hybrid renewable energy systems for

sustainable reliability and cost efficiency. Scientific Reports 14(1):1–21. Available at: https://doi.org/10.1038/S41598-024-61413-8; SUBJMETA=166,4077,639,705;KWRD= ENERGY+SCIENCE+AND+TECHNOLOGY,ENGINEERING,MATHEMATICS+AND+ COMPUTING.

5. Aggarwal V, Meena CS, Kumar A, Alam T, Kumar A, Ghosh A, Ghosh A (2020) Potential and future prospects of geothermal energy in space conditioning of buildings: India and worldwide review. Sustainability 12(20):8428. Available at: https://doi.org/10.3390/SU12208428
6. Agoundedemba M, Kim CK, Kim HG (2023) Energy status in Africa: challenges, progress and sustainable pathways. Energies 16(23):7708. Available at: https://doi.org/10.3390/EN16237708
7. Akamo JO (2022) The GERD from an Ethiopian perspective: actors, Interests and Instruments. Available at: https://doi.org/10.1080/23311886.2021.1875598
8. Alcorn R, O Sullivan D (2013) Electrical design for ocean wave and tidal energy systems
9. Alemzero DA, Sun H, Mohsin M, Iqbal N, Nadeem M, Vo XV (2021) Assessing energy security in Africa based on multi-dimensional approach of principal composite analysis. Environ Sci Pollut Res 28(2):2158–2171. Available at: https://doi.org/10.1007/S11356-020-10554-0/METRICS
10. Alex-Oke T, Bamisile O, Cai D, Adun H, Ukwuoma CC, Tenebe SA, Huang Q (2025) Renewable energy market in Africa: opportunities, progress, challenges, and future prospects. Energy Strategy Rev 59:101700. Available at: https://doi.org/10.1016/J.ESR.2025.101700
11. Alhassan MO, Opoku R, Uba F, Obeng GY, Sekyere CKK, Nyanor P (2023) Techno-economic and environmental estimation assessment of floating solar PV power generation on Akosombo dam reservoir in Ghana. Energy Reports 10:2740–2755. Available at: https://doi.org/10.1016/J.EGYR.2023.09.073
12. Amoah Nyasapoh M, Gyamfi S, Debrah SK, Gabbar HA, Derkyi N (2024) Ghana's path to a sustainable energy future: assessing renewable (Solar and Wind) energy penetration in the generation mix. Available at: https://doi.org/10.2139/SSRN.4707112
13. Amoo LM (2018) Techno-economic assessment of energy production potential from tidal streams in Nigeria. Int J Energy Environ Eng 9(1):81–98. Available at: https://doi.org/10.1007/S40095-017-0256-2/TABLES/2
14. Antar M, Lyu D, Nazari M, Shah A, Zhou X, Smith DL (2021) Biomass for a sustainable bioeconomy: an overview of world biomass production and utilization. Renew Sustain Energy Rev 139:110691
15. Article R, Ateme ME (2021) Developing marine and coastal resources in Nigeria: prospects and challenges. Maritime Technol Res 3(4):335–347. Available at: https://doi.org/10.33175/MTR.2021.244473
16. Awaafo A, Awafo EA, Mahdavi M, Akolgo G, Jurado F, Vera D, Amankwah E (2024) Crops production and the contribution of agricultural biomass power generation to Africa's clean energy transition: analysis of trends from 1990 to 2021. Biomass Bioenergy 186:107244. Available at: https://doi.org/10.1016/J.BIOMBIOE.2024.107244
17. Bakht MP, Salam Z, Gul M, Anjum W, Kamaruddin MA, Khan N, Bukar AL (2022) The potential role of hybrid renewable energy system for grid intermittency problem: a techno-economic optimisation and comparative analysis. Sustainability 14(21):14045. Available at: https://doi.org/10.3390/SU142114045
18. Bento N, Fontes M (2019) Emergence of floating offshore wind energy: technology and industry. Renew Sustain Energy Rev 99:66–82. Available at: https://doi.org/10.1016/J.RSER.2018.09.035
19. Bhatia SC (2014) Geothermal power generation. Adv Renew Energy Syst, 334–388. Available at: https://doi.org/10.1016/B978-1-78242-269-3.50014-0
20. Bishoge OK, Kombe GG, Mvile BN (2020) Renewable energy for sustainable development in sub-Saharan African countries: challenges and way forward. J Renew Sustain Energy 12(5). Available at https://doi.org/10.1063/5.0009297/1017497
21. Bishoge OK, Zhang L, Mushi WG (2018) The potential renewable energy for sustainable development in Tanzania: a review. Clean Technol 1(1):70–88. Available at: https://doi.org/10.3390/CLEANTECHNOL1010006

22. Bohlmann HR, Horridge JM, Inglesi-Lotz R, Roos EL, Stander L (2019) Regional employment and economic growth effects of South Africa's transition to low-carbon energy supply mix. Energy Policy 128:830–837. Available at: https://doi.org/10.1016/J.ENPOL.2019.01.065
23. Cariaga C (2024) GDC planning geothermal industrial park at Menengai, Kenya. Available at: https://www.thinkgeoenergy.com/gdc-planning-geothermal-industrial-park-at-menengai-kenya/. Accessed 10 June 2025
24. Chaudhuri A, Datta R, Kumar MP, Davim JP, Pramanik S (2022) Energy conversion strategies for wind energy system: electrical, mechanical and material aspects. Materials 15(3):1232. Available at: https://doi.org/10.3390/MA15031232
25. Chen H, Swain A (2014) The grand Ethiopian renaissance dam: evaluating its sustainability standard and geopolitical. Significance 3(1):11–19
26. Chen Y (2018) Comparing North-South technology transfer and South-South technology transfer: the technology transfer impact of Ethiopian Wind Farms. Energy Policy 116:1–9. Available at: https://doi.org/10.1016/J.ENPOL.2017.12.051
27. Chetty K, Davids YD, Kanyane M, Madzivhandila T, Moosa T, Ndaba L (2023) Fostering a just energy transition: lessons from South Africa's renewable energy independent power producer procurement programme. South African J Int Affairs 30(2):225–244. Available at: https://doi.org/10.1080/10220461.2023.2229293
28. Chiyemura F (2020) Contextualizing African agency in Ethiopia-China engagement in wind energy infrastructure financing and development. Available at: http://www.open.ac.uk/ikd/sites/www.open.ac.uk.ikd/files/files/working-papers/Contextualizing%20African%20Agency%20IKD%2088.pdf. Accessed 9 June 2025
29. Chomba MJ, Hill T, Nkhata B, Nel A (2019) A social exchange analysis of adaptive governance in water allocation processes, the kafue flats, Zambia. Int J Commons 13(2):949–961. Available at: https://doi.org/10.5334/IJC.900
30. Chowdhury MS, Rahman KS, Selvanathan V, Nuthammachot N, Suklueng M, Mostafaeipour A, Habib A, Akhtaruzzaman M, Amin N, Techato K (2021) Current trends and prospects of tidal energy technology. Environ Dev Sustain 23(6):8179–8194. Available at: https://doi.org/10.1007/S10668-020-01013-4/METRICS
31. Climate Action (2017) Africa's largest wind farm to offset 900,000 tonnes of CO2 annually—Climate Action. Available at: https://www.climateaction.org/news/africas-largest-wind-farm-to-offset-900000-tonnes-of-co2-annually. Accessed 9 June 2025
32. Cormack Z, Kurewa A (2018) The changing value of land in Northern Kenya: the case of Lake Turkana Wind Power. Critical African Studies 10(1):89–107. Available at: https://doi.org/10.1080/21681392.2018.1470017
33. Cumbajín M, Sánchez P, Hidalgo A, Gordón C (2020) Suitable electric generator for Pico-hydro power plant in Ambato–Huachi–Pelileo water irrigation channel. Adv Intell Syst Comput 1137 AISC, pp 189–199. Available at: https://doi.org/10.1007/978-3-030-40690-5_19
34. Dambly MLT, Samrock F, Grayver A, Eysteinsson H, Saar MO (2024) Geophysical imaging of the active magmatic intrusion and geothermal reservoir formation beneath the Corbetti prospect, Main Ethiopian Rift. Geophys J Int 236(3):1764–1781. Available at: https://doi.org/10.1093/GJI/GGAD493
35. Domínguez R, Carrión M, Oggioni G (2020) Planning and operating a renewable-dominated European power system under uncertainty. Appl Energy 258:113989. Available at: https://doi.org/10.1016/J.APENERGY.2019.113989
36. Doorga JRS, Hall JW, Eyre N (2022) Geospatial multi-criteria analysis for identifying optimum wind and solar sites in Africa: towards effective power sector decarbonisation. Renew Sustain Energy Rev 158:112107. Available at: https://doi.org/10.1016/J.RSER.2022.112107
37. Dumevi CY, Owusu-Asenso CM, Amoah BD, Asiamah JJ, Vicar EK, Kretchy J-P, Dayie NTKD, Ayeh-Kumi PF (2024) Spillage of Akosombo and Kpong dams in Ghana: perspectives on public health impacts on affected populations and proposed mitigation strategies. Int

J TROPICAL DISEASE Health 45(7):55–67. Available at: https://doi.org/10.9734/IJTDH/2024/V45I71560
38. EIB (2019) EIB and lake turkana wind power. Available at: https://www.eib.org/en/videos/eib-and-lake-turkana-wind-power. Accessed: 9 June 2025
39. Elbarbary S, Abdel Zaher M, Saibi H, Fowler AR, Saibi K (2022) Geothermal renewable energy prospects of the African continent using GIS. Geothermal Energy 10(1):1–19. Available at: https://doi.org/10.1186/S40517-022-00219-1/FIGURES/7
40. Eldardiry H, Hossain F (2021) Evaluating the hydropower potential of the Grand Ethiopian Renaissance Dam. J Renew Sustain Energy 13(2). Available athttps://doi.org/10.1063/5.0028037/15707032/024501_1_ACCEPTED_MANUSCRIPT.PDF
41. Elmahmoudi F, Abra OEK, Raihani A, Serrar O, Bahatti L (2020) GIS based fuzzy analytic hierarchy process for wind energy sites selection in tarfaya Morocco. Proceedings—2020 IEEE international conference of Moroccan Geomatics, MORGEO 2020 [Preprint]. Available at: https://doi.org/10.1109/MORGEO49228.2020.9121921
42. Farrok O, Ahmed K, Tahlil AD, Farah MM, Kiran MR, Islam MR (2020) Electrical power generation from the oceanic wave for sustainable advancement in renewable energy technologies. Sustainability 12(6):2178. Available at: https://doi.org/10.3390/SU12062178
43. Ferronato N, Torretta V (2019) Waste mismanagement in developing countries: a review of global issues. Int J Environ Res Public Health (6):1060. Available at: https://doi.org/10.3390/IJERPH16061060
44. Foli BAK, Ansong JK, Addo KA, Wiafe G (2022) A WAVEWATCH III® model approach to investigating ocean wave source terms for West Africa: input-dissipation source terms. Remote Sens Earth Syst Sci 5(1–2):95–117. Available at: https://doi.org/10.1007/S41976-021-00065-Y/METRICS
45. Gebreslassie MG, Tesfay AH (2024) Unlocking wind power potential to improve energy security in Ethiopia. Sustain Energy Res11(1):1–16. Available at: https://doi.org/10.1186/S40807-024-00137-9
46. Genetu M (2025) Assessment of geothermal energy in Aluto-Langano and Tulu-Moye, central Ethiopia. Geol J 60(1):64–72. Available at: https://doi.org/10.1002/GJ.5073
47. Ghaedi A, Mirzadeh M (2020) The impact of tidal height variation on the reliability of barrage-type tidal power plants. Int Trans Electr Energy Syst 30(9):e12477. Available at: https://doi.org/10.1002/2050-7038.12477
48. Ghana Energy Commission (2024) 2024 National energy statistical bulletin. Available at: https://www.energycom.gov.gh/index.php/planning/energy-statistics?download=641:2024-energy-statistics
49. Gnassou L (2019) Addressing renewable energy conundrum in the DR Congo: focus on Grand Inga hydropower dam project. Energy Strategy Rev 26:100400. Available at: https://doi.org/10.1016/J.ESR.2019.100400
50. Gualtieri G (2019) A comprehensive review on wind resource extrapolation models applied in wind energy. Renew Sustain Energy Rev 102:215–233. Available at: https://doi.org/10.1016/J.RSER.2018.12.015
51. Haas-Goldberg S (2025) BIPR | dam completion will elevate Egypt-Ethiopia tensions. Available at: https://bipr.jhu.edu/BlogArticles/articleprofile/18-Dam-Completion-Will-Elevate-Egypt-Ethiopia-Tensions.cfm. Accessed 10 June 2025
52. El Hafdaoui H, Khallaayoun A, Al-Majeed S (2025) Renewable energies in Morocco: a comprehensive review and analysis of current status, policy framework, and prospective potential. Energy Conv Manag X 26: 100967. Available at: https://doi.org/10.1016/J.ECMX.2025.100967
53. Hailu AD, Kumsa DK (2021) Ethiopia renewable energy potentials and current state. Aims Energy 9(1)
54. Hashimshony Yaffe N, Segal-Klein H (2023) Renewable energy and the centralisation of power. The case study of Lake Turkana Wind Power, Kenya. Political Geogr 102:102819. Available at: https://doi.org/10.1016/J.POLGEO.2022.102819

55. Hassan H (2022) Fundamentals of geothermal energy extraction, utilization of thermal potential of abandoned wells: fundamentals. Appl Res, 11–33. Available at: https://doi.org/10.1016/B978-0-323-90616-6.00002-6
56. Herrera J, Sierra S, Ibeas A (2021) Ocean thermal energy conversion and other uses of deep sea water: a review. J Marine Sci Eng 9(4):356. Available at: https://doi.org/10.3390/JMSE9040356
57. Hlangwani E, Mpye KL, Matsuro L, Dlamini B (2023) The use of technological innovation in bio-based industries to foster growth in the bioeconomy: a South African perspective. Sustain Sci Practice Policy 19(1). Available at: https://doi.org/10.1080/15487733.2023.2200300;WGROUP:STRING:PUBLICATION
58. Ibrahim WI, Mohamed MR, Ismail RMTR, Leung PK, Xing WW, Shah AA (2021) Hydrokinetic energy harnessing technologies: a review. Energy Reports 7:2021–2042. Available at: https://doi.org/10.1016/J.EGYR.2021.04.003
59. Imbali F (2020) A different Wind of Change—harnessing Africa's largest wind project for climate action | Rapid Transition Alliance. Available at: https://rapidtransition.org/stories/a-different-wind-of-change-harnessing-africas-largest-wind-project-for-climate-action/. Accessed 9 June 2025
60. Iqbal A, Mahmoud MS, Sayed ET, Elsaid K, Abdelkareem MA, Alawadhi H, Olabi AG (2021) Evaluation of the nanofluid-assisted desalination through solar stills in the last decade. J Environ Manag 277:111415. Available at: https://doi.org/10.1016/J.JENVMAN.2020.111415
61. IRENA B (2017) Turning to renewables: climate-safe energy solutions. Int Renew Energy Agency Abu Dhabi
62. IRENA GEC (2020) Renewable capacity statistics 2020. Int Renew Energy Agency [Preprint]
63. Jeremiah M, Kabeyi B, Olanrewaju OA (2023) Bagasse electricity potential of conventional sugarcane factories. J Energy 2023(1):5749122. Available at: https://doi.org/10.1155/2023/5749122
64. Kabeyi MJB, Olanrewaju OA, Kabeyi MJB, Olanrewaju OA (2025) Hydroelectric power plants and their sustainability. Adv Hydropower Technol [Preprint]. Available athttps://doi.org/10.5772/INTECHOPEN.1008773
65. Kabeyi MJB, Olanrewaju OA (2021a) Performance analysis of a sugarcane bagasse cogeneration power plant in grid electricity generation. In: 11th annual international conference on industrial engineering and operations management Singapore, pp 7–11
66. Kabeyi MJB, Olanrewaju OA (2021b) Preliminary design of a bagasse based firm power plant for a sugar factory. 2021 Southern African universities power engineering conference/robotics and mechatronics/pattern recognition association of South Africa, SAUPEC/RobMech/PRASA 2021 [Preprint]. Available at: https://doi.org/10.1109/SAUPEC/ROBMECH/PRASA52254.2021.9377242
67. Kachunga A (2022) Application of weap model to assess future water demands and water balance of the middle Kafue river basin. Available at: http://dspace.unza.zm/handle/123456789/8107. Accessed 10 June 2025
68. Kaunda CS, Kimambo CZ, Nielsen TK (2012) Potential of small-scale hydropower for electricity generation in sub-Saharan Africa. Int Scholarly Res Notices 2012(1):132606. Available at: https://doi.org/10.5402/2012/132606
69. Khan FA, Pal N, Saeed SH (2021) Stand-alone hybrid system of solar photovoltaics/wind energy resources: an eco-friendly sustainable approach. Renew Energy Syst, 687–705. Available at: https://doi.org/10.1016/B978-0-12-820004-9.00030-9
70. Kiran MR, Farrok O, Abdullah-Al-Mamun M, Islam MR, Xu W (2020) Progress in piezoelectric material based oceanic wave energy conversion technology. IEEE Access 8:146428–146449. Available at: https://doi.org/10.1109/ACCESS.2020.3015821
71. Klagge B, Nweke-Eze C (2020) Financing large-scale renewable-energy projects in Kenya: investor types, international connections, and financialization. Geografiska Annaler, Series B: Human Geogr 102(1):61–83. Available at: https://doi.org/10.1080/04353684.2020.1729662;CSUBTYPE:STRING:SPECIAL;PAGE:STRING:ARTICLE/CHAPTER

72. Klein BC, de Mesquita Sampaio IL, Mantelatto PE, Filho RM, Bonomi A (2019) Beyond ethanol, sugar, and electricity: a critical review of product diversification in Brazilian sugarcane mills, Biofuels. Bioproducts and Biorefining 13(3):809–821. Available at: https://doi.org/10.1002/BBB.1969; JOURNAL:JOURNAL:19321031;PAGEGROUP:STRING:PUBLICATION
73. Korkovelos A, Mentis D, Siyal SH, Arderne C, Rogner H, Bazilian M, Howells M, Beck H, De Roo A (2018) A geospatial assessment of small-scale hydropower potential in sub-Saharan Africa. Energies 11(11):3100. Available at: https://doi.org/10.3390/EN11113100
74. Kouloumpis V, Sobolewski RA, Yan X (2020) Performance and life cycle assessment of a small scale vertical axis wind turbine. J Cleaner Prod 247:119520. Available at: https://doi.org/10.1016/J.JCLEPRO.2019.119520
75. Lawrence A. (2020) REIPPPP: renewables' Rise, or REIPPPP RIP?, South Africa's Energy Transition, pp 99–114. Available at: https://doi.org/10.1007/978-3-030-18903-7_5
76. Liu Y, Ge Y, Xia B, Cui C, Jiang X, Skitmore M (2019) Enhancing public acceptance towards waste-to-energy incineration projects: lessons learned from a case study in China. Sustain Cities Soc 48:101582. Available at: https://doi.org/10.1016/J.SCS.2019.101582
77. López F, Rodríguez L, Abdolmaleki F, Martínez G, Bugallo B, Fernández Morales J, Francisco López A, Lago Rodríguez T, Faraji Abdolmaleki S, Galera Martínez M, Bello Bugallo PM (2024) From biogas to biomethane: an in-depth review of upgrading technologies that enhance sustainability and reduce greenhouse gas emissions. Appl Sci 14(6):2342. Available at: https://doi.org/10.3390/APP14062342
78. Marais E, Chowdhury S, Chowdhury SP (2011) Theoretical resource assessment of marine current energy in the Agulhas Current along South Africa's East coast. IEEE Power Energy Soc Gen Meeting [Preprint]. Available at: https://doi.org/10.1109/PES.2011.6039502
79. Masikonte LN (2020) Potential environmental and socio-economic effects of geothermal exploitation on the local community: a case of Suswa Geothermal Plant. Available at: http://erepository.uonbi.ac.ke/handle/11295/154486. Accessed 10 June 2025
80. Mekonnen TW, Teferi ST, Kebede FS, Anandarajah G (2022) Assessment of impacts of climate change on hydropower-dominated power system—the case of Ethiopia. Appl Sci 12(4):1954. Available at: https://doi.org/10.3390/APP12041954
81. Mhone AO, Jin J, Mhone AO, Jin J (2021) Deployment of lean six sigma and KAIZEN techniques: a case study of the concrete production plant for the 750 MW (5 * 150 MW) Kafue Gorge lower hydro power project. American J Indus Bus Manag 11(10):1052–1069. Available at: https://doi.org/10.4236/AJIBM.2021.1110064
82. Miescher SF (2021) Ghana's Akosombo Dam, Volta Lake fisheries & climate change. Daedalus 150(4):124–142. Available at: https://doi.org/10.1162/DAED_A_01876
83. Ministry of Energy (2021) National energy policy. Available at: https://www.energymin.gov.gh/sites/default/files/2023-09/2021%20ENERGY%20POLICY.pdf. Accessed 11 June 2025
84. Miskat M, Ahmed A, Rahman MS, Chowdhury H, Chowdhury T, Chowdhury P, Sait SM, Park Y-K (2021) An overview of the hydropower production potential in Bangladesh to meet the energy requirements. Environ Eng Res 26(6)
85. Mjami NB, Kimambo C, Ayeng'o S, Orozova-Bekkevold I (2025) Prospects and challenges for utilisation of low- to medium-temperature geothermal resources in Tanzania. Geothermics 129:103288. Available at: https://doi.org/10.1016/J.GEOTHERMICS.2025.103288
86. Mkabane C, Rop KV (2023) Incineration plant integration for waste management; a case of Nairobi City, Kenya, E3S Web of Conferences, 469, p 00033. Available at: https://doi.org/10.1051/E3SCONF/202346900033
87. Mlambo VH, Masuku MM (2025) Trouble in the Nile Basin: Ethiopia, Egypt, and Sudan and the Nile stalemate. Cogent Soc Sci 11(1):2491851. Available at: https://doi.org/10.1080/23311886.2025.2491851
88. Molla S, Farrok O, Rahman A, Samiul Bashir MD, Rabiul Islam MD, Kouzani AZ, Parvez Mahmud MA (2020) Increase in volumetric electrical power density of a linear generator by winding optimization for wave energy extraction. IEEE Access 8:181605–181618. Available at: https://doi.org/10.1109/ACCESS.2020.3028283

89. Mperejekumana P, Shen L, Zhong S, Gaballah MS, Muhirwa F (2024) Exploring the potential of decentralized renewable energy conversion systems on water, energy, and food security in Africa. Energy Conv Manag 315:118757. Available at: https://doi.org/10.1016/J.ENCONMAN.2024.118757
90. Mukumba P, Chivanga SY (2023) An overview of renewable energy technologies in the eastern cape Province in South Africa and the rural households' energy poverty coping strategies. Challenges 14(1):19. Available at: https://doi.org/10.3390/CHALLE14010019
91. Mwasilu F, Jung JW (2019) Potential for power generation from ocean wave renewable energy source: a comprehensive review on state-of-the-art technology and future prospects. IET Renew Power Gen 13(3):363–375. Available at: https://doi.org/10.1049/IET-RPG.2018.5456
92. Mwelwa EM (2021) Establishing the environmental flow regime for the middle Zambezi River. Establishing the environmental flow regime for the middle Zambezi River [Preprint]. Available at: https://doi.org/10.1201/9780138756543
93. Nana S, Kantanka S, Addaney M, Cobbinah PB, Akudugu JA (2023) Conflicting Rationalities: Collision of hydro-power dam construction and water availability. Available at: https://doi.org/10.21203/RS.3.RS-3173884/V1
94. NBI NBI (2022) Construction of 80MW regional Rusumo hydropower project heading to completion, was at 93% by May 2022. Available at: https://nelsap.nilebasin.org/index.php/en/newsevents/399-construction-of-80mw-regional-rusumo-hydropower-project-headingto-completion-was-at-93-by-may-2022. Accessed 10 June 2025
95. Ndiritu SW, Engola MK (2020) The effectiveness of feed-in-tariff policy in promoting power generation from renewable energy in Kenya. Renew Energy 161:593–605. Available at: https://doi.org/10.1016/J.RENENE.2020.07.082
96. Ndyamuhaki K, Ssembataya O, Balinga J (2021) Exploiting the geothermal potential of east africa for energy diversification and sustainability. 5th EAGE Eastern Africa Petrol Geosci Forum (1):1–5. Available at: https://doi.org/10.3997/2214-4609.2021605025/CITE/REFWORKS
97. Neill SP, Haas KA, Thiébot J, Yang Z (2021) A review of tidal energy—Resource, feedbacks, and environmental interactions. J Renew Sustain Energy 13(6):62702. Available at: https://doi.org/10.1063/5.0069452/15719354/062702_1_ACCEPTED_MANUSCRIPT.PDF
98. Nfaoui H, Abbou A, Sayigh A, Elmalki A (2023) Prospects for wind energy in southern morocco and Northern Mauritania, pp 403–419. Available at: https://doi.org/10.1007/978-3-031-33148-0_32
99. Noman FM, Alkawsi GA, Abbas D, Alkahtani AA, Tiong SK, Ekanayake J (2020) Comprehensive review of wind energy in Malaysia: past, present, and future research trends. IEEE Access 8:124526–124543. Available at: https://doi.org/10.1109/ACCESS.2020.3006134
100. Nøst OA, Børve E (2021) Flow separation, dipole formation, and water exchange through tidal straits. Ocean Sci 17(5):1403–1420. Available at: https://doi.org/10.5194/OS-17-1403-2021
101. Olujobi OJ (2020) The legal sustainability of energy substitution in Nigeria's electric power sector: renewable energy as alternative. Protection Control Modern Power Syst 5(1). Available at: https://doi.org/10.1186/S41601-020-00179-3
102. Omenda P, Ofwona C, Mangi P (2025) Kenya: the most successful geothermal development in Africa. Geothermal Power Gen, 863–891. Available at: https://doi.org/10.1016/B978-0-443-24750-7.00022-1
103. Osman AI, Lai ZY, Farghali M, Yiin CL, Elgarahy AM, Hammad A, Ihara I, Al-Fatesh AS, Rooney DW, Yap PS (2023) Optimizing biomass pathways to bioenergy and biochar application in electricity generation, biodiesel production, and biohydrogen production. Environ Chem Lett 21(5):2639–2705. Available at: https://doi.org/10.1007/S10311-023-01613-2
104. Parhamfar M, Sadeghkhani I, Adeli AM (2023) Towards the application of renewable energy technologies in green ports: technical and economic perspectives. IET Renew Power Gen 17(12):3120–3132. Available at: https://doi.org/10.1049/RPG2.12811;PAGE:STRING:ARTICLE/CHAPTER

105. Patel B, Patel A, Gami B, Patel P (2020) Energy balance, GHG emission and economy for cultivation of high biomass verities of bamboo, sorghum and pearl millet as energy crops at marginal ecologies of Gujarat state in India. Renew Energy 148:816–823. Available at: https://doi.org/10.1016/J.RENENE.2019.10.167
106. Perea-Moreno A-J, Kalak T (2023) Potential use of industrial biomass waste as a sustainable energy source in the future. Energies 16(4):1783. Available at: https://doi.org/10.3390/EN16041783
107. Prajapati M, Shah M, Soni B (2022) A review on geothermal energy resources in India: past and the present. Environ Sci Pollut Res 29(45):67675–67684. Available at: https://doi.org/10.1007/S11356-022-22419-9/METRICS
108. Rahman A, Farrok O, Haque MM (2022) Environmental impact of renewable energy source based electrical power plants: Solar, wind, hydroelectric, biomass, geothermal, tidal, ocean, and osmotic. Renew Sustain Energy Rev 161:112279. Available at: https://doi.org/10.1016/J.RSER.2022.112279
109. Ritcher A (2019) Ormat geothermal plant in Kenya, country's largest private power producer | ThinkGeoEnergy—Geothermal News & Insights. Available at: https://www.thinkgeoenergy.com/ormat-geothermal-plant-in-kenya-countrys-largest-private-power-producer/. Accessed 10 June 2025
110. Roga S, Bardhan S, Kumar Y, Dubey SK (2022) Recent technology and challenges of wind energy generation: a review. Sustain Energy Technol Assess 52:102239. Available at: https://doi.org/10.1016/J.SETA.2022.102239
111. Rumaherang WM, Elim HI, Latununuwe A, Huliselan EK, Lekalette JD (2025) Performance energy production with advanced design and prototype floating ducted tidal turbine for small islands. Renew Sustain Energy Rev 216:115664. Available at: https://doi.org/10.1016/J.RSER.2025.115664
112. Salah SE (2019) a study of hydroelectric power supply in the west of Algeria. Available at: http://repository.pauwes-cop.net/handle/1/351. Accessed 12 June 2025
113. Sayed ET, Wilberforce T, Elsaid K, Rabaia MKH, Abdelkareem MA, Chae KJ, Olabi AG (2021) A critical review on environmental impacts of renewable energy systems and mitigation strategies: Wind, hydro, biomass and geothermal. Sci Total Environ 766:144505. Available at: https://doi.org/10.1016/J.SCITOTENV.2020.144505
114. Seyedrahimi-Niaraq M, Nouri T (2022) Investigating the economic effects and the roadmap of developing geothermal systems to generate electricity. J Renew Energy Environ 9(3):52–64
115. Simberg-Koulumies N (2024) Just sustainabilities: lessons from the Lake Turkana Wind power project in Kenya. Local Environ 29(1):40–56. Available at: https://doi.org/10.1080/13549839.2023.2249497
116. Song F, Yu X, Li J, Ni Y, Bian C, Lv X (2020) Research and comprehensive evaluation on delivery schemes of the Grand Inga hydropower station. Global Energy Interconnection 3(6):521–531. Available at: https://doi.org/10.1016/J.GLOEI.2021.01.009
117. Srivastava S (2020) Diversification of sugar and sugarcane industry: agro-industrial alternatives, sugar and sugar derivatives: changing consumer preferences, pp 151–169. Available at: https://doi.org/10.1007/978-981-15-6663-9_10
118. Stocks C (2017) Ethiopians contribute $460 million to Renaissance Dam—international water power. International water power and dam construction. Available at: https://www.waterpowermagazine.com/news/ethiopians-contribute-460-million-to-renaissance-dam-5939460/. Accessed 10 June 2025
119. Szwaba R, Smiałek MA, Telega J, Shambira N, Mukumba P, Makaka G (2025) Assessing the wind energy potential: a case study in fort hare, South Africa, using six statistical distribution models. Appl Sci 15(5):2778. Available at: https://doi.org/10.3390/APP15052778
120. Tang Y, Dong J, Li G, Zheng Y, Chi Y, Nzihou A, Weiss-Hortala E, Ye C (2020) Environmental and exergetic life cycle assessment of incineration- and gasification-based waste to energy systems in China. Energy 205:118002. Available at: https://doi.org/10.1016/J.ENERGY.2020.118002

121. Tettey, G., Ansah, E.A. and Asante, W. (2025) The politics of renewable energy transition in Ghana: Issues, obstacles and prospects, Energy Research & Social Science, 120, p. 103939. Available at: https://doi.org/10.1016/J.ERSS.2025.103939.
122. Touimi K, Benbouzid M, Chen Z (2020) optimal design of a multibrid permanent magnet generator for a tidal stream turbine. Energies 13(2):487. Available at: https://doi.org/10.3390/EN13020487
123. Tshimbiluni HC, Tabakov PY (2019) Wind energy potential for small-scale WEC systems in port Elizabeth. IOP conference series: earth and environmental science, 342(1), p. 012014. Available at: https://doi.org/10.1088/1755-1315/342/1/012014
124. Tsui TH, Wong JWC (2019) A critical review: emerging bioeconomy and waste-to-energy technologies for sustainable municipal solid waste management. Waste Disposal Sustain Energy 1(3):151–167. Available at: https://doi.org/10.1007/S42768-019-00013-Z
125. Ugwu CO, Ozor PA, Mbohwa C (2022) Small hydropower as a source of clean and local energy in Nigeria: prospects and challenges. Fuel Commun 10:100046. Available at: https://doi.org/10.1016/J.JFUECO.2021.100046
126. Ukoba K, Jen TC, Yusuf AA (2025) Transformation of South Africa's energy landscape: Policy implications, opportunities, and technological innovations in the fourth industrial revolution. Energy Strategy Rev 59:101752. Available at: https://doi.org/10.1016/J.ESR.2025.101752
127. Verma J, Goel S (2023) State-of-the-art in bioresources for sustainable transportation. Int J Hydrogen Energy 48(10):3768–3790. Available at: https://doi.org/10.1016/J.IJHYDENE.2022.10.206
128. Wang W, Tian S, Long J, Liu J, Ma Q, Xu K, Zhang Z (2022) Investigation and evaluation of flue gas pollutants emission in waste-to-energy plant with flue gas recirculation. Atmosphere 13(7):1016. Available at: https://doi.org/10.3390/ATMOS13071016
129. Weaver T (2025) Ocean currents hold vast energy potential, particularly off Africa, study finds | Engineering and technology magazine. Available at: https://eandt.theiet.org/2025/06/06/ocean-currents-hold-vast-energy-potential-particularly-africa-study-finds. Accessed 11 June 2025
130. Wheeler KG, Jeuland M, Hall JW, Zagona E, Whittington D (2020) Understanding and managing new risks on the Nile with the Grand Ethiopian Renaissance Dam. Nature Commun 11(1): 1–9. Available at: https://doi.org/10.1038/S41467-020-19089-X; TECHMETA=119,141; SUBJMETA=134,242,2805,704,706; KWRD=DEVELOPING+WORLD,HYDROLOGY,WATER+RESOURCES
131. WHO (2023) Health data overview for the Federal Democratic Republic of Ethiopia. Available at: https://data.who.int/countries/231. Accessed 10 June 2025
132. Wilberforce T, Sayed ET, Abdelkareem MA, Mahmoud M, Olabi AG (2023) Introduction and definition of hydro energy, Renewable Energy—Volume 1: Solar, Wind, and hydropower definitions, developments, applications, case studies, and modelling and simulation, 1, pp 407–425. Available at: https://doi.org/10.1016/B978-0-323-99568-9.00008-X
133. Woldegiyorgis TA, Benti NE, Habtemicheal BA, Jembrie AA (2024) Unleashing the power of artificial neural networks: accurate estimation of monthly averaged daily wind power at Adama wind farm I, Ethiopia. Environ Res Commun 6(7):075010. Available at: https://doi.org/10.1088/2515-7620/AD592F
134. World Bank (2018) Kenya: world bank approves $180 million to support energy sector. Available at: https://www.worldbank.org/en/news/press-release/2018/04/26/kenya-world-bank-approves-180-million-to-support-energy-sector. Accessed 10 June 2025
135. Xu Y, Yang K, Zhou J, Zhao G (2020) Coal-biomass co-firing power generation technology: current status, challenges and policy implications. Sustainability 12(9):3692. Available at: https://doi.org/10.3390/SU12093692
136. Yang H, Wu Q, Li G (2024) A multi-stage forecasting system for daily ocean tidal energy based on secondary decomposition, optimized gate recurrent unit and error correction. J Cleaner Prod 449:141303. Available at: https://doi.org/10.1016/J.JCLEPRO.2024.141303
137. Yimam YM, Tesfaye N (2020) Environmental and health impacts of and challenges on Rappiekoshe waste-to-energy facility, Addis Ababa. Int. Res J Multidiscip Scope 1:9–17

138. Zhang J (2024) Energy access challenge and the role of fossil fuels in meeting electricity demand: promoting renewable energy capacity for sustainable development. Geoscience Frontiers 15(5):101873. Available at: https://doi.org/10.1016/J.GSF.2024.101873
139. Zhang X, Brown RC (2019) Introduction to thermochemical processing of biomass into fuels, chemicals, and power, thermochemical processing of biomass: conversion into fuels. Chem Power, 1–16. Available at: https://doi.org/10.1002/9781119417637.CH1
140. Zhou Y, Remón J, Pang X, Jiang Z, Liu H, Ding W (2023) Hydrothermal conversion of biomass to fuels, chemicals and materials: a review holistically connecting product properties and marketable applications. Sci Total Environ 886:163920. Available at: https://doi.org/10.1016/J.SCITOTENV.2023.163920
141. Zhu H, Sueyoshi M, Hu C, Yoshida S (2019) A study on a floating type shrouded wind turbine: Design, modeling and analysis. Renew Energy 134:1099–1113. Available at: https://doi.org/10.1016/J.RENENE.2018.09.028
142. Zulu E, Mutwale J, Zulu SL, Musonda I, Kavishe N, Moobela C (2023) Challenges, drivers and incentives to private sector participation in public-private partnership projects in developing countries: evidence from Zambia. J Eng Design Technol 23(2):386–405. Available at: https://doi.org/10.1108/JEDT-03-2023-0092/FULL/PDF
143. Falchetta G, Gernaat DE, Hunt J, Sterl S, (2019) Hydropower dependency and climate change in sub-Saharan Africa: a nexus framework and evidence-based review. J Cleaner Prod 231:1399-1417. Available at: https://doi.org/10.1016/j.jclepro.2019.05.263
144. Conway D, Dalin C, Landman, WA, Osborn TJ (2017) Hydropower plans in eastern and southern Africa increase risk of concurrent climate-related electricity supply disruption. Nat Energy 2(12):946–953. Available at: https://doi.org/10.1038/s41560-017-0037-4

Chapter 12
Building Energy Systems and Building Decarbonization

12.1 Introduction

The backbone of modern infrastructure is formed by Energy Systems providing, electricity for heating, cooling, lighting, mechanical transportation, and other critical sectors including residential, commercial, and industrial buildings [63]. Building energy systems includes equipment and infrastructure to supply and manage energy in a building to ensure the efficient use of the building. These include Heating Ventilation Air-conditioning (HVAC) systems, lighting, water heating, and other miscellaneous equipment which consumes energy [40]. HVAC systems is a key element of building energy systems which accounts for about 40–60% of the total energy consumed by the building. Additionally, lighting systems significantly increases the total cost of energy consumption by about 30% of total energy consumed in buildings. Given that these systems make up significant portions of operational energy use in buildings, optimizing their efficiency is essential reduce energy consumption and improving sustainability in buildings [64].

As urbanizations grow steadily in Africa, the need for energy efficiency building systems cannot be understated. According to the United Nations, more than half of Africa's population is expected to live in urban areas by 2025. This will lead to a substantial increase in energy consumption for buildings. Currently, energy systems in Africa mainly depend on non-renewable sources of energy which poses threat to the environment and challenging the global sustainability efforts [62]. Transitioning to more energy-efficient and renewable sources of energy for building systems is essential to reduce the carbon emissions and promote a sustainable environment. Building decarbonization is a fundamental strategy to achieve global sustainability goals. As global demand for energy increases with buildings consuming about 33% of total energy consumption and accounting for nearly 40% of greenhouse gas emission, it is imperative for buildings to be the focal point of decarbonization [66, 67]. Building decarbonization focuses on reducing the carbon emissions of building while improving energy efficiency. It is done by incorporating renewable solutions

M. N. Addy et al., *Decarbonization of Building Energy in Developing Countries*, Green Energy and Technology, https://doi.org/10.1007/978-3-032-09113-0_12

and implementing policy-driven measures aimed at reducing energy consumption of buildings [65].

Several strategies can facilitate building decarbonization. The incorporation of advanced energy-efficient technologies such as LED lighting, smart energy technologies, and the application of Internet of Things (IoT) can significantly reduce the total energy demand of buildings. Furthermore, the integration of renewable sources of energy including solar and wind power can reduce the overdependence of fossils which has adverse impact on the environment. Moreover, sustainable construction practices such as the use of eco-friendly construction materials, and adoption of circular economy principles can further reduce the embodied carbon in buildings.

12.2 Building Energy Systems

12.2.1 Ventilation and Cooling

Ventilation and Cooling is crucial for any building to improve the health and productivity of occupants. Modern infrastructure utilizes HVAC system to provide thermal comfort and high indoor air quality. In Africa, air-conditioner ownership remains low, even as high temperatures increase the need for cooling and demand for HVAC systems would increase. The high energy consumption and environmental impact of these systems is alarming. González-Torres et al. [26] stipulates that, HVAC systems alone consume about 40% of the total energy consumption of the building. Interestingly, this energy demand of these systems is high in regions with extreme climates such as Africa where constant cooling is required due to the high temperatures. However, the source of energy for this system in most African countries is generated mainly from fossil fuels which further aggravates the carbon emissions. The high demand of cooling results into substantial emission of carbon and other greenhouse gases [31]. In 2018, 8% of emission of greenhouse gases was attributed to rapid growth in the usage of air-conditioning especially in the developing countries [30].

HVAC systems further contribute to direct emission of greenhouse gases through refrigerant leakages. Air conditioners often contain Hydrofluorocarbons (HFCs) which has a very high Global Warming Potential (GWP). According to Abas et al. [1], HFCs have very high impact on the environment, and it is comparable to the embodied emission of residential buildings. Due to it high GWP, the Kigali Amendment to the Montreal protocol has established the phase out of HFCs in the coming decades [18]. Several African countries such as Cameroon have initiated a comprehensive strategy to reduce the dependence of HFCs by 30% by 2030. Andersen et al. [9] highlights the challenges of HFC's to the environment hence the need for regulatory bodies to ensure the gradual curbing of the use of HFCs. Furthermore, improper disposal of obsolete HVAC systems poses environmental threats through toxic metal contaminations [45].

12.2.2 Lighting System

Lighting systems are fundamental to every modern infrastructure. Artificial lighting plays an essential role in the daily lives of humans by enhancing productivity, safety, and quality support of life. Adequate lighting is essential in residential, commercial, and industrial settings to support daily activities. Adequate lighting further reduces accidents and improves the mental well-being and general economic growth by enhancing visibilities in various settings [13]. However, the significant amount of energy consumed by artificial lighting systems leads to carbon emissions. Traditional lighting technologies such as incandescent and fluorescent bulbs are highly energy-intensive hence consumes a lot of energy. According to the US Department of Energy, only about 10% of the energy of incandescent bulbs is used as visible light. The remainder is lost as heat. Thus a significant portion of energy is not utilized and hence, further increases the energy consumption and carbon emissions. Also, the excessive heat produced increases cooling demands hence further increasing the energy consumption for cooling [25]. Although fluorescent lamps are more efficient than incandescent bulbs, they also require considerable energy for it to operate. Fluorescent bulbs contain mercury, a hazardous substance which when disposed improperly poses a threat to the environment [60]. When these systems are powered by fossil fuel-based grid, they generate considerable carbon emissions and further impacts climate change. Other energy-efficient lighting technologies, such as light-emitting diodes (LEDs), offer substantially lower energy consumption compared to conventional lighting systems. In Africa, LED adoption has been steadily increasing, driven by falling technology costs, energy efficiency policies, and off-grid solar lighting initiatives.

12.2.3 Domestic Hot Water Systems

Domestic Hot Water (DHW) systems are essentials for residential and commercial buildings. They provide hot water for sanitation and hygiene, enhancing household and commercial usage and ensuring comfort in buildings. They include storage tanks and tankless systems which mainly rely on electricity to operate. The energy consumption of this system is largely dependent on the climatic conditions, the occupancy type and the type of building facility and user behaviour [11]. The energy consumption of DHW production for residential buildings is about 18% in the United States and 14% in the European Union [23]. However, the source of electricity is generated from fossil fuels which leads to the emission of greenhouse gases. Most especially, the conventional DHW heavily relies on non-renewable energy sources which increases the environmental impact.

12.2.4 Mechanical Movement Systems

Mechanical movement systems play important role in modern buildings typically in high rise buildings and large buildings with a lot of occupants. They form integral part of buildings providing efficiency and accessibility in vertical and horizontal transportation. They usually include elevators, escalators, moving pavements, etc. Beyond convenience, they increase mobility for people with disabilities and improve the general productivity and functionality of buildings [32]. In high rise buildings, elevators, and escalators are essential to transport people and goods from one level to the other while reducing time and effort [6]. The energy consumption of mechanical movement systems is dependent on the height of the building, the frequency of usage and the type of equipment [10]. Elevator systems consume between 5 to 15% of buildings total energy demand [57]. Conventional traction and hydraulic elevators consume significant electricity which poses significant challenges in energy efficiency [29]. Furthermore, the motors, control systems, and standby operations also contribute to high energy demands.

Since various types of mechanical movement operate using electricity produced from fossil fuels and other carbon related sources, the energy consumption indirectly results in carbon emissions. Moreover, the manufacturing and disposal of components for transportation including motors and other electronic systems contributes negatively to the environment especially when disposed inappropriately.

12.2.5 Miscellaneous Equipment

The miscellaneous equipment used in buildings include appliances and devices that contribute to the energy consumption and carbon emissions. These include pump systems, refrigerators and freezers, computers and IT equipment, electric cookers, and other appliances used in either residential, commercial, or industrial buildings. The equipment is used for diverse reasons to facilitate the use of the building. For instance, refrigerators and freezers are used for preservation of perishable substances such as food and drugs [59]. Moreover, computers, servers, data centres are crucial for modern communication. It ensures a reliable, easy, and faster transfer of information between two or more entities. Furthermore, printers, photocopier machines, security cameras, signage lights, kitchen appliances, and others all contribute to the energy consumption and carbon emission of buildings. It enhances productivity, security, and conveniency in buildings.

The energy consumption of these appliances depends largely on the efficiency of the equipment and the rate of usage. A review presented by Bansal et al. [12] showed that, refrigerators and freezer operate continuously hence resulting into high energy consumption. It accounts for about 10 to 15% of total energy use in a building. Also, adequate power is required for the operation of computers, servers, and other IT equipment. According to Pothithou et al. [50], IT equipment account for 15% of global household electricity usage. Interestingly, the energy consumption is expected to increase significantly by 250% by the year 2030. This potential increase in energy is as a result of consumer behaviour and social practices such as using multiple devices simultaneously and leaving devices on standby mode. This accounts for excessive waste of energy hence increasing the total energy consumption of buildings [50]. Due to the considerable amount of heat produced by the usage of these equipment, substantial amount of energy is used as cooling for the equipment to prevent overheating [40].

Also, the improper disposal of IT devices has led to an increase in electronic waste which contains materials such as lead and mercury [53]. Equipment such as refrigerators and freezers contain HFCs which have higher GWP than CO2 and increases about 10–15% annually [52].

12.3 Challenges

12.3.1 Lack of Climate-Responsive Design

Climate-responsive design integrates local climatic conditions into the design and construction of buildings to enhance energy efficiency and comfort of occupants [14, 35]. According to Del Pero and Martire [17], they provide sustainable approach to reduce the total energy consumption of buildings by considering climate specific conditions in the design of buildings. Without integrating climate-responsive design, buildings will depend heavily on artificial ventilation and hence increase energy consumption and carbon emissions [4]. Although climate-responsive design helps to reduce energy consumption, many buildings in Africa are designed without considering the climatic conditions. This was emphasized by Dodoo and Ayarkwa [19] in the study to examine the effect of climate change. They indicated that, current building designs neglect the climate conditions hence leads to poor indoor environmental performance and high energy consumption. Alegbe and Mtaver [5] further confirmed in his study to assess the climate resilience of building which concluded that, most existing buildings are not designed with adequate consideration of local climatic conditions hence reducing thermal comfort and increasing energy consumption. The gross neglect of climate-responsive designs can be attributed to economic and financial constraints. The higher cost of finance to implement the strategies such as climate resilient designs creates a climate investment trap in the developing countries [8]. Additionally, the lifestyle and aesthetics of occupants hinders affects the adoption of Climate-responsive designs. A study conducted by Abusafieh [2] showed that, occupants will prefer aesthetically pleasing building designs than energy saving and efficient buildings.

12.3.2 Building Energy Expertise

Building energy expertise include various aspects of sustainable designs and energy efficiency which are crucial tools to assessing the performance of buildings and optimizing energy systems, further constructing energy-efficient buildings that complies with local and global sustainability standards [28]. Unfortunately, the deficit in this specialized knowledge contributes to the low adoption of decarbonization strategies of energy systems. In Africa, the lack of building energy expertise and integrated frameworks in designs remains a significant barrier in the decarbonization of energy systems. These results in inefficient building designs that increases energy consumption and carbon emissions [15]. The lack of building energy expertise is as a result of inadequate education and training programs focused on building energy efficiency and decarbonization of building systems. Furthermore, it has increased the underutilization of energy-efficient materials and technology hence many buildings heavily rely on the conventional high energy demanding systems [38].

12.3.3 Weak Enforcement of Regulations and Unclear/Lack of Policies

Even though building decarbonization has become a major agenda globally, countries in Africa still lacks clear and comprehensive policies to regulate building decarbonization. A few African countries have developed frameworks to regulate building energy such as South African National Standard which ensures the buildings conforms to the energy specifications of South Africa [56], the Building Energy code of Morocco which is adopted to optimize energy demands of buildings [21]. In 2024, Ghana established a climate actions roadmap for buildings and construction. It is aimed at promoting low carbon, resilient and energy-efficient construction in alignment to the Paris Agreement [44]. Most countries have partial or no policies that regulate the energy use of buildings. These policies are usually outdated and do not address in-depth strategies to reduce the energy consumption and carbon emission [24].

Where policies do exist, they are poorly implemented as a result of institutional and governance challenges. These institutions lack clear blueprints for energy efficiency and decarbonization targets hence, the poor implementation. The regulatory bodies usually lack the funding, the technical capacity and skilled personnel to enforce the compliance of the building energy regulations and decarbonization standards. Hence, developers and construction firms overlook these regulations contributing to high carbon emission.

12.3.4 Technological Barriers

The adoption of advanced technologies as a decarbonization strategy has advanced in developed countries. Although energy-efficient technologies are the most pragmatic solutions to building decarbonization, the adoption of these technologies such as smart grids, renewable energy systems, and Internet of Things (IoT) have faced significant challenges in Africa. Somepalli [54] highlights that, high initial costs and lack of unskilled personnel poses significant challenges to the adoption of energy-efficient technologies. Clients, contractors and institutions usually are resistant to change and unwilling to adopt expensive technological tools without innovative financial models to support the rate of adoption. Addy et al. [3] highlights that, the absence of tax incentive initiatives, subsidies, and the perceived high initial cost of advanced technologies deter developers from adopting technological tools to enhance energy efficiency and promote decarbonization.

Hence, the effect of the slow rate of adoption of energy-efficient technologies can be classified into two namely; the high initial cost and lack of market demand. Innovative technologies often have higher installation cost about 4% of the total cost of construction. Although the total cost of usage of these technologies is reduced especially in the African region, the extra cost of installation of the technologies may increase cost overruns and increase the life cycle cost. Hence, with high initial cost perceived as a major barrier to the adoption of technologies, developers prefer to adopt the conventional technologies to reduce the total cost of construction [34].

The lack of market demand contributes to the slow adoption of technological innovations [3]. As diverse regions have diverse demands for energy-efficient technologies, it provides opportunities and threats especially for developing countries. These technological innovations are not easily accessible on a large enough scale on the market with reasonable prices. Hence, most developers, contractors, and clients are reluctant to adopt these technologies to promote decarbonization in Africa due to limited digital infrastructure in the African region and inadequate expertise on the effective operations of technological tools [34].

12.3.5 Informal Construction Practices

The informal construction practices are usually the practices considered outside the building regulations of that region. These practices do not adhere to the approved regulations for construction and energy-efficient standards. Unfortunately, these practices are rapidly growing in Africa where there is a major demand for affordable housing solutions. Individuals especially prefer to utilize the informal construction sector typically made of small firms and self-employed workers. The usage of the informal construction sector results in the heavily reliance of carbon intensive materials and lack of energy-efficient principles. It further leads to structural inefficiencies due to the absence of compliance with the building codes. The informal practices rely on conventional building materials and lack the use innovative strategies such as passive design strategies, smart technologies, and energy-efficient systems. The use of the conventional building materials and strategies increases the carbon footprint and the overall operational energy consumption. Furthermore, due to the lack of adoption of npassive design and energy-efficient strategies, the designs of the building are usually not optimal. These buildings are usually not thermally comfortable, and often require more energy for lighting, cooling and powering of other energy systems. Hence, these building do not provide adequate indoor environmental performance thereby further increasing the operational energy consumption of the building.

12.4 Pathways for Building Energy Systems Decarbonization in Africa

12.4.1 Innovative Energy Systems

Adoption of innovative energy systems is essential and an effective pathway to decarbonization and efficient energy consumption. As such, it is important to design innovative energy systems that consume less energy. For example, in lighting systems, the energy efficiency can be improved by switching to LEDs. LED lightings are more energy-efficient and consume up to 90% less energy than the traditional lighting systems, environmentally friendly and longer lifespans [41, 46]. HVAC systems can adopt the use of smart thermostats which automates the control of HVAC systems. It integrates occupancy sensing and algorithms to optimize the performance of HVAC systems [49]. This has the potential of reducing energy consumption for cooling by 20% to 25% in residential buildings [16]. Furthermore, old HVAC systems can be replaced with environmentally friendly systems that uses natural refrigerants or Hydrofluoroolefins (HFOs) to reduce the effect on the environment and energy consumption. HFOs have low impact on the environment due to its low GWP [58].

Moreover, elevators and mechanical transportation should integrate regenerative drives that recycles the energy other than losing it as heat especially during braking. The regenerative drives convert gravitational potential energy into electrical energy during descent and carrying lighter load. This electrical energy can be fed back into the electrical grids to serve as a source of power for other equipment [33]. This has the potential to significantly increase energy saving as reported in the study of [7] that, about 29 kW of regenerative power derived from a 38.6 kW motor. Due to the significant impact of energy-efficient appliances such as high energy-efficient refrigerators, kitchen equipment, and computers, it is encouraged to be adopted to reduce the operational energy and it associated impact on the environment.

12.4.2 Smart Technologies

With the world shifting into the technology age, it is essential for industries to adopt these technologies to enhance the overall performance of works. The construction industry has reacted positively towards this change. There has been development on innovative and smart technologies in the construction industry such as smart grids, Internet of Things (IoT), Artificial Intelligence (AI), Building Information Modelling (BIM), and others. These smart and emerging technologies aids in automation, real-time monitoring and have successfully reduced the carbon emission, energy consumptions and enhances the overall performance of buildings in most developed countries [37]. For instance, smart grids save energy by supplying enough energy required by the building, by utilizing real-time information hence improving energy efficiency. IoT enables automations in buildings and allows real-time tracking to reduce energy consumption of lighting and cooling of a building [51].

Africa should adopt these technologies to enhance the energy performance and consumption of buildings. Incorporating smart grids into the heavy reliance of non-renewable energy sources will balance the energy demand with the supply. Hence, excess energy can be saved for other uses. The smart technologies can be incorporated into buildings in order for them to be automated. That is, installation of sensors to control artificial lighting, infrared thermography to adjust cooling systems based on the thermal comfort of occupants to always maintain optimality. However, for this to be achieved, there should be capacity building to train individuals on the operation and installation of these technologies [27]. There should be more collaborations between the industry experts and the developers of these technologies. Furthermore, Public Private Partnership (PPP) should be encouraged to promote the adoption of these technologies.

12.4.3 Building Certifications and Standards

Building standards and certifications is a widely acceptable and effective tool for reducing carbon emissions and energy consumptions in the construction industry. There are structured frameworks which provide benchmarks and guidelines to assess and ensure that new construction or retrofit buildings are sustainable, energy-efficient, and reduce carbon emissions [55]. Several countries have adopted building certifications and standard to enhance the energy performance and reduce the carbon footprint of the construction industry. Notable among the certifications are the Leadership in Energy and environmental Design (LEED) in US which enhance sustainability and energy efficiency, Building Research Established Environmental Assessment Method (BREEAM) in Europe which assess the energy performance and carbon emissions, and Green Star in Australia that focuses on the occupancy comfort and energy performance. The adoption of these certifications has reduced the energy consumption especially in US and Europe where there has been a reduction of about

25% of energy and 34% of carbon emission and about 40% reduction of carbon emission in US and Europe, respectively [20]. Africa thereby should adopt building standards and certifications to help reduce the carbon footprint this region and meet global sustainability standards.

Several certifications have been developed in Africa such as the Excellence in Design for Greater Efficiencies (EDGE) which has helped reduce energy use by at least 20% in most African countries [39]. The Green Star in South Africa has reduced energy consumption and provided several passive design strategies to achieve energy efficiency [48], the Green Building Council of Nigeria and the Climate Roadmaps action for building and construction in Ghana. The certifications will contribute to the use of sustainable and low carbon materials, proper disposal and recycling of construction waste, promote energy-efficient systems and provides real-time information on energy usage. Hence, existing certifications should be revamped to be comprehensive and enforceable. Furthermore, the certifications should be developed to suit the climatic conditions of the region to enhance the effective performance of the certification system [14].

12.4.4 Government Policies and Incentives

Government polices play a central role in setting regulatory guidelines, enforcing standard and providing long term efficient roadmap for the reduction of carbon emission and energy usage [36]. The developed regions have capitalized on government policies to drive decarbonization and reduced energy consumption. These policies and regulations can enhance the adoption of low-energy building energy systems and provide incentives to encourage the adoption and compliance of clean energy systems. For instance, in 2023, the Renewable Energy Directive (RED) was revised to bind the European Union (EU) renewable energy target to at least 42.5% by 2030. Since the introduction of this directive, 24.5% of EU's energy consumption were attained from renewable sources [22]. Furthermore, building energy regulations will enforce the compliance to new buildings and retrofits to energy codes to improve the energy efficiency and reduce carbon emissions.

The government should promote decarbonization in Africa through financial incentives such as tax credits, low interest loans, and grants for research [47]. Providing tax credit and rebates on smart and energy-efficient technologies has the potential to attract firms and individuals to embrace low carbon solutions to enhance decarbonization [61]. Africa can invest into these strategies by enhancing its green finances through its development banks and Public private partnerships to enhance decarbonization in this region. Even though government can provide incentives to encourage decarbonization, it can utilize market-oriented techniques to control carbon emissions and energy efficiency. For instance, the Emission Trading System (ETS) in Europe aims to reduce greenhouse gas emissions in EU using the 'cap and trade' approach [43]. Furthermore, the Carbon Tax of Canada prices the carbon intensive materials to deter its usage and emission of carbon [42]. In Africa, notable among is the proposed Emission Levy of Ghana. However, this levy only applies to vehicles hence, the scope of the levy could be broadened to include the buildings to help the reduction of carbon emissions from buildings.

12.5 Summary

Building energy systems are essential in modern infrastructure. However, these systems are sometimes energy intensive with the energy source mostly relying on fossil fuels. These sources of energy increase the carbon footprint which results in global warming. As climate change grows steadily, the need to transition to low-energy building systems and sustainable building practices have become urgent. The carbon footprint of building energy systems is crucial in Africa due to the rapid urbanization growth and its associated demand for energy. Building energy systems are the largest consumers of energy in a building and largest contributor to the building operational greenhouse gas emissions. Consequently, enhancing energy efficiency through technological innovations and sustainable building design is essential to reduce the rate of energy consumption and mitigating its impact on the environment. The chapter presented a number of significant challenges especially in Africa. Pertinent among the challenges is the lack of climate-responsive building designs where building designs are not designed with the climatic conditions of the region taken into consideration. This usually results in higher energy demand for lighting and cooling. To overcome these barriers, Africa must adopt a multi-faceted approach. Policy makers should establish energy certifications and regulations such as LEED, BREEAM, EDGE to ensure sustainable construction with energy-efficient systems. Furthermore, the government should incentivize the adoption of energy-efficient systems and renewable sources of power through tax credits and rebate and subsidies of energy-efficient power systems to encourage the adoption of these towards promoting decarbonization. By adopting these strategies, Africa has the potential to effectively reduce carbon emission of building energy systems and improve energy efficiency. This will aid in the advancement of Sustainable Development Goals (SDGs) 7; Affordable and Clean Energy and goal 13; Climate Action.

References

1. Abas N, Kalair AR, Khan N, Haider A, Saleem Z, Saleem MS (2018) Natural and synthetic refrigerants, global warming: a review. Renew Sustain Energy Rev 90:557–569. https://doi.org/10.1016/j.rser.2018.03.099
2. Abusafieh S (2019) The conflict between aesthetics and sustainability: empowering sustainable architecture with aesthetics to enhance people's lifestyle and sustainable behaviour. Innov Renew Energy, 653–664. https://doi.org/10.1007/978-3-030-18488-9_52

3. Addy M, Adinyira E, Danku JC, Dadzoe F (2020) Impediments to the development of the green building market in sub-Saharan Africa: the case of Ghana. Smart Sustain Built Environ 10(2):193–207. https://doi.org/10.1108/sasbe-12-2019-0170
4. Addy MN, Ateko B, Roberts BM, Opoku D, Aigbavboa C, Kwofie TE (2024) Design optimisation for embodied carbon and thermal performance of a courtyard house in the warm–humid climate of Ghana. Arch Eng Design Manag, 1–22. https://doi.org/10.1080/17452007.2024.2442979
5. Alegbe M, Mtaver G (2023) Climate resilience and energy performance of future buildings in Nigeria based on RCP 4.5 and 8.5 scenarios. J Design Resilience Arch Planning 4(3):354–371. https://doi.org/10.47818/drarch.2023.v4i3102
6. Al-Kodmany K (2023) Smart elevator systems. J Mech Mater Mech Res 6(1):41–53. https://doi.org/10.30564/jmmmr.v6i1.5503
7. Almanda D, Ramadhan AI (2020) Analysis of the potential for savings in electrical energy consumption in lifts: case study in Indonesia. J Appl Sci Adv Technol 3(1):29–34. https://doi.org/10.24853/jasat.3.1.29-34
8. Ameli N, Dessens O, Winning M, Cronin J, Chenet H, Drummond P, Calzadilla A, Anandarajah G, Grubb M (2021) Higher cost of finance exacerbates a climate investment trap in developing economies. Nat Commun 12(1). https://doi.org/10.1038/s41467-021-24305-3
9. Andersen SO, Halberstadt ML, Borgford-Parnell N (2013) Stratospheric ozone, global warming, and the principle of unintended consequences—an ongoing science and policy success story. J Air Waste Manag Assoc 63(6):607–647. https://doi.org/10.1080/10962247.2013.791349
10. Ang JH, Yusup Y, Zaki SA, Salehabadi A, Ahmad MI (2022) Comprehensive energy consumption of elevator systems based on hybrid approach of measurement and calculation in low- and high-rise buildings of tropical climate towards energy efficiency. Sustainability 14(8):4779. https://doi.org/10.3390/su14084779
11. Arpagaus C, Bless F, Uhlmann M, Schiffmann J, Bertsch SS (2018) High temperature heat pumps: market overview, state of the art, research status, refrigerants, and application potentials. Energy 152:985–1010. https://doi.org/10.1016/j.energy.2018.03.166
12. Bansal P, Vineyard E, Abdelaziz O (2011) Advances in household appliances—a review. Appl Therm Eng 31(17–18):3748–3760. https://doi.org/10.1016/j.applthermaleng.2011.07.023
13. Boyce PR (2014) Human factors in lighting. CRC Press eBooks. https://doi.org/10.1201/b16707
14. Chang X (2025) Adapting architecture to climate: exploring climate-responsive strategies across U.S. Cities. Int J Eng Adv 2(1):1–16. https://doi.org/10.71222/qca5ng14
15. Clarke L, Gleeson C, Winch C (2016) What kind of expertise is needed for low energy construction? Constr Manag Econ 35(3):78–89. https://doi.org/10.1080/01446193.2016.1248988
16. Correia A, Ferreira LM, Coimbra P, Moura P, De Almeida AT (2022) Smart thermostats for a campus microgrid: demand control and improving air quality. Energies 15(4):1359. https://doi.org/10.3390/en15041359
17. Del Pero C, Martire M (2022) Climate-Responsive design and energy performance goals. Res Dev, 79–95, https://doi.org/10.1007/978-3-031-00284-7_4
18. Di Filippo R, Bursi OS, Di Maggio R (2022) Global warming and ozone depletion potentials caused by emissions from HFC and CFC banks due to structural damage. Energy Build 273:112385. https://doi.org/10.1016/j.enbuild.2022.112385
19. Dodoo A, Ayarkwa J (2019) Effects of climate change for thermal comfort and energy performance of residential buildings in a Sub-Saharan African climate. Buildings 9(10):215. https://doi.org/10.3390/buildings9100215

20. Eichholtz P, Holtermans R, Kok N (2019) Environmental performance of commercial real estate: new insights into energy efficiency improvements. J Portfolio Manag 45(7):113–129. https://doi.org/10.3905/jpm.2019.1.099
21. Energy Efficiency Magazine (2021) Energy efficiency law in Morocco. Available at: https://energyefficiencymagazine.com/index.php/2021/12/28/energy-efficiency-law-in-morocco/#:~:text=The%20BEC%20is%20the%20first%20regulation%20adopted%20to,services%29%20according%20to%206%20climatic%20zones%20in%20Morocco
22. European Commission (2023) Renewable Energy Directive (RED) Revision 2023. Available at: https://climate.ec.europa.eu/eu-action/renewable-energy-directive-revision-2023_en
23. Fuentes E, Arce L, Salom J (2017) A review of domestic hot water consumption profiles for application in systems and buildings energy performance analysis. Renew Sustain Energy Rev 81:1530–1547. https://doi.org/10.1016/j.rser.2017.05.229
24. Gaum T, Laubscher J (2021) Building regulatory practices in 10 african countries in the face of climate change. International conference on climate change.https://doi.org/10.17501/2513258x.2021.5104
25. Gielen D, Boshell F, Saygin D, Bazilian MD, Wagner N, Gorini R (2019) The role of renewable energy in the global energy transformation. Energ Strat Rev 24:38–50. https://doi.org/10.1016/j.esr.2019.01.006
26. González-Torres M, Pérez-Lombard L, Coronel JF, Maestre IR, Yan D (2021) A review on buildings energy information: trends, end-uses, fuels and drivers. Energy Rep 8:626–637. https://doi.org/10.1016/j.egyr.2021.11.280
27. Haliru U, Bawa N (2022) Building teachers' capacity for effective technology integration into teaching and learning in tertiary institutions in Nigeria. Int J Inf Technol Comput Eng 25:25–34, https://doi.org/10.55529/ijitc.25.25.34
28. Heidarinejad M, Srebric J (2019) Building energy modelling. Annual Rev Heat Transfer 21:1–23. https://doi.org/10.1615/annualrevheattransfer.2019030570
29. Held P, Weißhaar D, Abdeslam DO, Benyoucef D (2022) Independent and Non-Invasive State Detection for elevator based on total current consumption. 2021 International conference on computational science and computational intelligence (CSCI), pp 1274–1280. https://doi.org/10.1109/csci58124.2022.00229
30. IEA (2018) The future of cooling: opportunities for energy-efficient air conditioning. IEA, Paris https://www.iea.org/reports/the-future-of-cooling
31. Kathiravel R, Zhu S, Feng H (2024) LCA of net-zero energy residential buildings with different HVAC systems across Canadian climates: a BIM-based fuzzy approach. Energy Build 306:113905. https://doi.org/10.1016/j.enbuild.2024.113905
32. Li Y, Zheng W, Zhou Q (2024) Knowledge-driven urban innovation: dynamics of elevator installation in aging residential communities. J Knowl Econ. https://doi.org/10.1007/s13132-024-01789-0
33. Lin K-Y, Lian K-Y (2017) Actual measurement on regenerative elevator drive and energy saving benefits. International automatic control conference (CACS), pp 1–5. https://doi.org/10.1109/cacs.2017.8284249
34. Liu T, Chen L, Yang M, Sandanayake M, Miao P, Shi Y, Yap P-S (2022) Sustainability considerations of green buildings: a detailed overview on current advancements and future considerations. Sustainability 14(21):14393. https://doi.org/10.3390/su142114393
35. Looman R (2017) Climate-responsive design: a framework for an energy concept design-decision support tool for architects using principles of climate-responsive design. A+BE | Architect Built Environ 7(1):1–282. https://doi.org/10.7480/abe.2017.1.1643
36. Luo J, Jin X, Hu Y (2024) Top-down or bottom-up? Government policies, public awareness and carbon reduction. Econ Anal Policy 83:118–129. https://doi.org/10.1016/j.eap.2024.06.013
37. Ma Z, Awan MB, Lu M, Li S, Aziz MS, Zhou X, Du H et al (2023) An overview of emerging and sustainable technologies for increased energy efficiency and carbon emission mitigation in buildings. Buildings 13(10):2658. https://doi.org/10.3390/buildings13102658
38. Mahar WA, Anwar NUR, Attia S (2019) Building energy efficiency policies and practices in Pakistan: a literature review. AIP Conference proceedings. https://doi.org/10.1063/1.5115364

39. Marzouk OA (2023) Zero carbon ready metrics for a Single-Family home in the Sultanate of Oman based on EDGE certification system for green buildings. Sustainability 15(18):13856. https://doi.org/10.3390/su151813856
40. Masanet E, Shehabi A, Lei N, Smith S, Koomey J (2020) Recalibrating global data center energy-use estimates. Science 367(6481):984–986. https://doi.org/10.1126/science.aba3758
41. Masara VM (2019) The replacement of tradiocional lamps by LED in Brazil: a case of environmental and economic sustainability (short review). MOJ Civil Eng 5(1):31–33. https://doi.org/10.15406/mojce.2019.05.00146
42. McKitrick R (2016) A practical guide to the economics of carbon pricing. The School of Public Policy Publications.https://doi.org/10.11575/sppp.v9i0.42600
43. Meadows D, Vis P, Zapfel P (2019) The EU emissions trading system. Routledge eBooks, pp 66–94, https://doi.org/10.4324/9789276082569-4
44. Ministry of Works and Housing (2024) Climate action roadmap for buildings and construction—Ghana. Accra: Ministry of Works and Housing. Available at: https://www.mwh.gov.gh/
45. Naik S, Eswari JS (2022) Electrical waste management: recent advances challenges and future outlook. Total Environ Res Themes 1–2:100002. https://doi.org/10.1016/j.totert.2022.100002
46. Nair GB, Dhoble SJ (2020) General lighting. Elsevier eBooks, pp 155–176, https://doi.org/10.1016/b978-0-12-819605-2.00006-9
47. Nwani C (2022) Financing low-carbon growth in Africa: Policy path for strengthening the links between financial intermediation, resource allocation and environmental sustainability. Cleaner Environ Syst 6:100082. https://doi.org/10.1016/j.cesys.2022.100082
48. Pillay TL, Saha AK (2022) Passive, low-energy design and green star strategy for green star-rated buildings in South Africa. Energies 15(23):9128. https://doi.org/10.3390/en15239128
49. Plando NIC, Jr (2023) Advancements in smart thermostat technology for enhanced HVAC energy management. Int J Adv Res Sci Commun Technol, pp 882–887. https://doi.org/10.48175/ijarsct-12388
50. Pothitou M, Hanna RF, Chalvatzis KJ (2017) ICT entertainment appliances' impact on domestic electricity consumption. Renew Sustain Energy Rev 69:843–853. https://doi.org/10.1016/j.rser.2016.11.100
51. Poyyamozhi M, Murugesan B, Rajamanickam N, Shorfuzzaman M, Aboelmagd Y (2024) IOT—A Promising solution to energy management in smart buildings: a systematic review, applications, barriers, and future scope. Buildings 14(11):3446. https://doi.org/10.3390/buildings14113446
52. Shah N, Wei M, Letschert V, Phadke A (2019) Benefits of energy efficient and low-global warming potential refrigerant cooling equipment. https://doi.org/10.2172/1559243
53. Shehabi A, Smith S, Sartor D, Brown R, Herrlin M, Koomey J, Masanet E et al (2016) United States data Center energy usage report.https://doi.org/10.2172/1372902
54. Somepalli S (2024) Implementing smart grid technology in developing countries: challenges, opportunities, and pathways to sustainable energy solutions. J Eng Appl Sci Technol, 1–4. https://doi.org/10.47363/jeast/2024(6)e109
55. Song P, Wu L, Zhao W, Ma W, Hao J (2024) Life cycle sustainability assessment: an index system for building energy retrofit projects. Buildings 14(9):2817. https://doi.org/10.3390/buildings14092817
56. The Department of Trade, Industry and Competition (2008) National building regulations and building standards act. Available at: http://www.thedtic.gov.za/wp-content/uploads/building_standards_act.pdf
57. Thebuwena ACHJ, Samarakoon SMSMK, Ratnayake RMC (2024) Optimization of energy consumption in vertical mobility systems of high-rise office buildings: a case study from a developing economy. Energy Efficiency 17(6). https://doi.org/10.1007/s12053-024-10246-5
58. Uddin K, Saha BB, Thu K, Koyama S (2018) Low GWP refrigerants for energy conservation and environmental sustainability. Energy Environ Sustain, 485–517. https://doi.org/10.1007/978-981-13-3302-6_15

59. Vaishali, Sharma HP, Dholu U, Sharma S, Patel A (2020) Effect of freezing systems and storage temperatures on overall quality of perishable food commodities. Deleted J 9(9):114–122
60. Viana LN, Soares APS, Guimarães DL, Rojano WJS, Saint'Pierre TD (2022) Fluorescent lamps: A review on environmental concerns and current recycling perspectives highlighting Hg and rare earth elements. J Environ Chem Eng 10(6):108915. https://doi.org/10.1016/j.jece.2022.108915
61. Zhang Y, Song Y (2022) Tax rebates, technological innovation and sustainable development: evidence from Chinese micro-level data. Technol Forecast Soc Chang 176:121481. https://doi.org/10.1016/j.techfore.2022.121481
62. Agoundedemba M, Kim CK, Kim HG (2023) Energy status in Africa: challenges progress and sustainable pathways. Energies 16(23):7708. https://doi.org/10.3390/en16237708
63. Dincer, I, Abu-Rayash, A (2019) Energy sustainability. Academic Press
64. Ghanim, I. (2024) A systematic literature review on energy efficiency in buildings energy management systems. Int Inno J Appl Sci 1(2). Available at: https://doi.org/10.61856/qb4f4a29
65. Ürge-Vorsatz D, Khosla R, Bernhardt R, Chan YC, Vérez D, Hu S, Cabeza, LF (2020) Advances toward a net-zero global building sector. Annu Rev Environ Resour 45(1):227-269. Available at: https://doi.org/10.1146/annurev-environ-012420045843
66. Toyinbo O, Mensah-Attipoe J, Irankunda E, Friday IG, Berekute AK, Thulu FGD, Olawale AM, Byrne M, Huttunen K, Oluwadare T, Koala L (2025) Advancing indoor environmental quality in African countries: a call to action for awareness, research, and policy. Indoor Environments 2:100123. https://doi.org/10.1016/j.indenv.2025.100123
67. Abdul Ghani, Olabi Nabila, Shehata Usama Hamed, Issa O.A., Mohamed Montaser, Mahmoud Mohammad Ali, Abdelkareem M.A., Abdelzaher (2025) The role of green buildings in achieving the sustainable development goals. Int J Thermofluids 25:101002. Available at: https://doi.org/10.1016/j.ijft.2024.101002

Chapter 13
Building Energy Codes and Certification

13.1 Introduction

Energy savings, environmental concerns, and policy initiatives are the driving forces influencing the growing trend to achieve energy efficiency in buildings. The potential of energy saving is however the most important for this change [16, 22]. The increased energy consumption of buildings is due to increased population, migration, and lifestyles. This makes buildings key to low carbon developments. Therefore, it is generally essential to reduce the energy consumption of buildings to achieve sustainability. Building energy codes has been identified as one of the essential tools to increase energy efficiency of buildings and reduce greenhouse gases emissions [59].

Building energy codes aim to reduce energy consumption of buildings and reduce greenhouse gases emissions while maintaining or improving the integrity of the building through the establishment of minimum energy requirements for buildings [31]. Building energy codes can be used to enforce a minimum set of building energy efficiency requirements, such as envelope, heating ventilation and air-conditioning (HVAC), domestic hot water (DHW) systems, and lighting. Numerous studies have shown that building energy codes can achieve on the order of 5–20% energy savings for the building stock [16]. Elnabawi [15] revealed in his study to assess to the potential of energy savings and greenhouse emission reduction when energy codes are implemented that, about 25% to 35% of energy saving is achieved when energy codes are used. Due to this energy saving potential, it necessitates the need to certify buildings that conform to the regulations set by the building energy regulation.

Certification is used to convey how dependable a product or service is [30]. It provides the assurance that the product of service has passed the minimum threshold for customer satisfaction. Some countries have developed building energy certification systems, which are tailored to their climate and construction practices. Korea and Azerbaijan used building energy certification schemes to analyse adverse climatic conditions and energy crisis [47]. In the EU, the adoption of energy performance

M. N. Addy et al., *Decarbonization of Building Energy in Developing Countries*, Green Energy and Technology, https://doi.org/10.1007/978-3-032-09113-0_13

certificates has been one of the key drivers of the energy consumption reduction of buildings and is a key driver towards building renovations [38].

In developing countries, the adoption of building energy certifications is growing slowly. One of them is the Green Star Africa developed by the Green Building Council of South Africa (GBCSA) and Excellence in Design for Greater Efficiencies (EDGE) which has been embraced by a number the African countries such as Nigeria, Kenya, and Ethiopia. Adoption of these certifications, however, has faced significant setbacks which makes large-scale enforcement and compliance of the certifications difficult [17, 39]. Accordingly, this study fills the gap by describing developing countries' difficulties in embracing building energy certification schemes and identifying suitable avenues to raise the level of adoption.

13.2 Overview of Building Energy Certification in Developing Countries

Building energy certifications has been applied extensively by most of the developed world like US, Canada, and Europe due to its energy saving potentials and ensuring sustainability. Its energy saving potentials were demonstrated in [46] research for assessing building energy codes energy saving. It was determined in the research that commercial building energy consumption has fallen by 12% for the building that has applied the building energy codes. Li et al. [38] confirms this assertion in the building energy performance certification review. The study revealed that due to the potential for energy saving by the energy performance certificates, there is a reduction in electricity tariffs and thus increased investment in property. In addition to the potential for energy saving, it increases property value and reduces operational expense. Filippini and Obrist [62] also compared how much energy is consumed by green certified buildings and traditional buildings in Switzerland. The use of green certified building was mandated to be around 25% less than convectional buildings. Ji et al. [32] affirms that, with the utilization of building energy certifications in South Korea, there was reduced energy consumption by 33.8% to 58.1% compared to traditional buildings. This has also impacted the behaviour of occupants to be energy conscious as a result of the high energy performance of the building [37].

Estimate for rating building performance systems varies due to differences in definition and scope with some estimate about close to 1000 schemes globally for measuring and certifying energy performance and building sustainability. Interestingly, developing countries have the lowest share. Following a review of international certification schemes in developed and developing countries, ALwaer and Clements-Croome [4] concluded that their application requires contextual consideration due to distinct socio-cultural activities and climatic conditions. Developing countries are framing their certifications based on their construction practice, economies, climate, and culture. However, as at 2019 Darko [63] observed that the development and implementation of the building energy certifications are advancing at a sluggish pace.

This is because of the high initial cost and lack of government incentives to increase the uptake to these certifications with Ghana being a typical example [1]. This is also due to the lack of information regarding energy efficiency and skilled experts to design, construct, and certify energy-efficient buildings [31, 39]. In Osei-Poku et al.'s [45] study to identify factors in energy performance schemes, the implication was that Ghana has implemented no scheme to certify the energy performance of buildings. The above, evidently indicate the slow uptake of Building Energy Certification in some developing countries.

13.3 Building Energy Certification Schemes

To attain sustainability, it is crucial to pay great attention to the significance of lowering the environmental impact of the building industry. Building energy certification schemes has been identified as a valuable tool to improve building energy efficiency [38]. The schemes assess and rate the energy efficiency of the buildings and provide data on the energy consumption to the building owners, occupants, and potential buyers. The most widely used among the schemes are Building Research Establishment Environmental Assessment (BREEAM) in the UK, Leadership in Energy and Environmental Design (LEED) in the US and National Green Building Standard (NGBS) which assess the diverse environmental impact, energy efficiency and indoor environmental quality of a building [21].

In the late 1980s, North America and Europe were introduced to mandatory energy-efficient design requirements for buildings due to concerns about climate change and excessive energy consumption in buildings. The earliest schemes, such as the BREEAM and LEED, concentrated on the whole range assessment of building sustainability that included energy performance as a key component. Over the past three decades, building energy certification schemes have thrived globally, with numerous countries and regions developing their own programmes tailored to local contexts and policy priorities. Thus, there are a wide variety of certification schemes targeting at different climates; others use different rating systems and these programmes have different criteria.

Building energy certification adopts various methodologies to provide ratings for the energy performance of the building. Common among these methodologies are the prescriptive-based methods, performance-based methods and hybrid methods [42]. In the prescriptive-based approach, a benchmark is set to serve as the minimum baseline for specific building components such as requirements for insulations, windows and doors, HVAC systems, lighting, and many other [34]. It is usually preferred due to its simplicity in adhering to the requirements [42]. Park et al. [47] indicated that, prescriptive approach was used to set compulsory design specifications for the main section of the Building Energy Conservation code in the study to determine the potential of building energy certifications. The performance-based approach focuses on the meeting the benchmark requirements but does not focus on how it is attained [28]. The hybrid method combines both the prescriptive approach and the performance approach. This is evidenced in the study of Bernardi et al. [5] to understand the main differences between the certifications and its design. Diverse certification schemes use various methodologies and approach to rate building energy performance tailored to their specific regions.

Due to the success stories in Europe and America, several developing countries began similar efforts in the 1990s. Countries such as South Africa, Nigeria, Rwanda, and others are adopting energy certifications to promote energy-efficient and sustainable construction. Notable among these certifications is the Green Star South Africa which was established by the South African Green Building Council to serve as the focal regulatory framework for assessing the environmental impact of building in South Africa. The certification evaluates the performance of buildings in nine (9) different areas including energy, emission, and innovation [26].

Similarly, the Kenya Green building Society endorse building energy certifications in Kenya aiming to enhance the energy efficiency and reduce carbon emissions [9]. The most prominent of these is the International Finance Corporation's (IFC) EDGE certification, which targets a minimum of 20% improvement in energy efficiency, water efficiency, and embodied energy in materials. EDGE promotes affordable green building practices and provides accessible tools that supports large-scale adoption of sustainable construction in emerging markets. Moreover, the Nigerian Building Energy Efficiency Code was developed to reduce the energy consumption and determine standards to promote sustainability; and sustainable building materials; and method in Nigeria through air-conditioning systems and lighting systems [20]. Furthermore, the Ghana Green Building Council (GHGBC) ensure the implementation of green certification systems through international agencies such as Excellence in Design for Greater Efficiency (EDGE). Ghana has six out of eight certified green building in the region bearing an EDGE certification [1].

13.4 Challenges to Building Energy Certification

The efficiency of building energy codes and certifications lies on its implementation. Unfortunately, the implementation of these codes and certifications have encountered complex web of challenges that demands careful considerations especially in Africa. There is a significant gap between the standards and real-world practices [48]. Since its emergence, building energy certifications have been plagued by low adoption rates especially amongst developing countries [38].

Although occupant behaviour has significant impact on the energy consumption of a building, the certification is not occupant oriented thus the occupant are not the subject of the certification. The certification and codes do not certify the building's use and construction but focuses on the design of the building. The energy certificates depend on simulations rather than actual energy demand from the appliances and loads of equipment as well as users' behaviour [11]. This is upheld by Herrando et al. [25] that, there is about a 30% difference between the simulated consumption of energy and the actual building's energy use. The variability is due to energy consumption during the operations of equipment and individual behaviour. This leads to uninformed choices and acts as a setback to achieve the energy efficiency targets. In a study to assess the barriers to energy certifications, non-user-friendliness, authentic data, and quality was also emphasized as a key barrier to the implementation of energy certifications [50]. Additionally, startup cost is also a significant obstacle to energy building certification.

13.5 Initial Startup Cost

The expense and intricacy of certification processes may be a barrier to adoption on a large scale. Several studies have also reported high initial cost as the main reason for slow adoption of building energy certification in Africa. Research conducted by Dwaikat and Ali [13] in Malaysia indicated that, green certified buildings have higher life cycle cost than the conventional building but uptake is low due to the high initial cost of such certifications. Simpeh et al. [53] concluded that, one of the greatest challenges that constrain the uptake of building energy certification in South Africa is high initial cost.

However, in most instances, the construction industry in Africa functions under a very limited budget thereby, low or inadequate funds always exist to establish energy certification schemes. Others in the industry contend that obtaining these energy certifications would cost an additional 20% of the total project expense [8]. This is so because installing an environmentally friendly building technology and an environmentally friendly building material may prove to be costly compared to installing conventional technologies, materials, and building practices. Besides, the effective use of energy certification schemes would be hindered by a premium cost of procuring expert experts in Africa to install and certify [64].

13.6 Experts and Skilled Technicians

Zuo and Zhao [61] outline that, due to the complexity nature of building energy certifications, experts and skilled technicians are required. Unfortunately, these skilled personal are not readily available especially in Africa hence, affects the overall quality of construction and the adoption of building energy certification schemes. When there is the absence of certified building energy professionals, it impedes the implementation of building energy codes and certifications.

In Ghana, Agyekum et al. [2] emphasized that, the successful implementation of building energy certifications is dependent of the availability of skilled and trained personnel. However, the shortage in skilled professional in Ghana hampers the adoption and implementations of building energy certifications. Similarly, Ganyiu et al. [19] found it necessary to train and develop local professionals on the concepts of building energy certifications to fill the gap on skilled expertise on building energy certifications in Nigeria to enhance the rate of adoption. In South Africa, there is evidence on the shortage of skilled labour and qualified tradesmen [58]. How will sustainable practices be adopted in such a region? The absence of certified professionals is as a result of formal education on the certification processes. Construction firms in Africa are not willing to invest in human resources. This is also as a result of lack of government incentives and support for the industries. Hence, results in the slow adoption of building energy and certification tools in Africa.

13.7 Level of Awareness and Understanding of Building Energy Certifications

The level of awareness and understanding of building energy certifications is a major challenge to building energy certifications. Amecke [65] posits that, building owners are not entirely abreast with the significance of energy certifications and how it can be attained. This results in the reluctance of adopting energy certifications. This is further confirmed by Keleş et al. [33] in the study to determine the level of awareness of green building in Turkey. Most of the participants do not have sufficient information and understanding on the energy certification schemes. Watts et al. [57] further emphasizes that, low level of awareness and lack of understanding is a critical challenge limiting the adoption of energy performance certificates.

Similarly in Africa, clients and construction professionals are not conversant with the full importance of building energy efficiency hence usually reluctant in pursing certifications. This unawareness is as a result of the absence of accessible information on existing energy-efficient buildings. Construction professionals in Ghana, Nigeria, Tanzania, and several other African countries are not well-informed on building energy efficiency. This has retarded the adoption of building energy certification in Africa [29].

13.8 Building Certification Processes and Building Permit Issues

In many African countries, it is always hectic, bureaucratic with long delays to acquire building permits for construction works to proceed. While Australia and many U.S. jurisdictions allow fast-tracking of permitting procedures for above-code buildings [66], applicants of building permits in Africa usually need to visit several offices to complete various procedures in completing building permit applications. Ghana typically has one of the slowest building permit issuances. It usually takes up to six months or more for clients and contractors to receive feedback of permit applications [23]. This is as a result of inadequate human and capital resources for the specified institution. Hence, individuals usually commence construction without the permit to avoid financial losses due to inflation [7]. Likewise, building permit acquisition has faced significant challenges such as lack of transparency in the issuance hence causing lags and delays in Nigeria [43]. Kenya also experiences long delays in building permits acquisition and can take up to several months to over a year or more for building permits to be issued. These inefficiencies in acquiring building permits deter developers from pursuing other certifications such as building energy certifications [8].

13.9 Lack of Policy Direction and Clear Regulations on Energy Certification

Unlike in developed countries where building energy certifications such as BREEAM and LEED have been integrated into building energy codes; there are no mandatory policies and clear regulations in most African countries that require buildings to be issued energy-efficient certifications. This is due to a fragmented coordination among key stakeholders with different stakeholders having diverse expectations. Developers and builders are motivated to comply when they face meaningful penalties for non-compliance. Where there are regulations to ensure energy efficiency, they are weakly enforced due to low institutional capacity for supervision and enforcement. South Africa has relatively advanced regulatory framework through the South African National Energy Development Institute (SANEDI), however, there are inconsistencies in the enforcements [58]. The lack of institutional capacity to enforce the adoption of energy certified buildings discourages several developers from investing in energy certifications. Countries such as Ghana and Nigeria have national building codes but they do not mandate the compliance of building energy certifications. They typically do not have building energy certification systems hence developers are not mandated to pursue it [2]. When there are no clear enforced rules, developers may also import materials where foreign analysis conducted on these materials may not be compatible with domestic requirements [16].

13.10 Pathways to Building Energy Certification

Several studies have explored pathways for reducing energy and carbon emissions in buildings. For instance, the study of Odubiyi et al. [52] which sought to measure the energy-saving potential of buildings as well as analyze the technical progress and utilization of renewable energy made recommendations that for energy to be conserved, there is a need for governmental officials to update their existing standards to cater for advanced methods to improve energy conservation and the utilization of renewable energy. The study also found the deployment of technology as a gateway to improving energy conservation. Similarly, Leibowicz et al. [35] concluded that having building energy efficiency standards and carbon reduction measures was a sure way to steer the built environment towards energy efficiency.

In addressing the issue of low adoption and utilization rates of energy performance certifications, Li et al. [38], identified some problems of the existing energy certification system and proposed measures to improve them. From their study, improvements need to be made first and foremost to the data acquisition process where technological devices and Building Information Models (BIM) could be employed to speed up the process and save cost. According to Hasan and Defer [67], emerging technologies enable continuous monitoring and when implemented and enhance decision making. The use of advanced technologies is recommended to improve energy efficiency, reduce carbon emissions, and improve comfort levels of occupants in the study of De Almeida et al. [12]. Shen et al. [51] further emphasizes the importance of adopting and implementing innovative and smart technological tools to improve the efficiency of buildings. With the ability of machine learning to analyse large data and AI to optimize building energy consumption when in use, Ma et al. [40] suggests that, capitalizing on these tools is essential to improve energy efficiency and building energy certifications. Furthermore, by incorporating BIM into building certifications which provides a visual representation of the building, it helps stakeholder to make informed decisions on the energy efficiency potentials of the building [40]. On the contrary, the Africa region is maximizing the full potential of emerging technologies. This is due to high initial cost, little or no incentives and unavailability on the large market scale. Ejidike and Mewomo [14] advocates for a comprehensive understanding on emerging technologies to enhance its full utilization in Africa.

Additionally, enhancing public awareness on energy efficiency, developing reliable databases for easy policymaking, and improving indicator sets were amongst some advancements that could be made to improve the implementation of energy certifications [38]. It was noted that transitioning to low energy building codes could rather be challenging especially in the case of Africa due to high initial cost, lack of skilled labour, poor regulatory bodies, etc. To alleviate these difficulties, recommendations encompassing state and local complementary policies are an indispensable pathway for implementation of building energy certifications. Policy makers and energy experts are to develop a framework recommending minimum design criteria to achieve a high energy efficiency. For instance, in the European Union, Energy Performance Certificates (EPC) provides a list of measures for retrofitting of buildings to improve the energy efficiency. Furthermore, it emphasizes on the development of several methodologies to simulate the energy consumption of the building before issuing energy performance certificates [38]. The EU further mandated the use of Building performance certificates to improve the energy efficiency [49]. The authorization of building energy certification regulates the prescriptive compliance of occupants and prioritizes energy efficiency and sustainability [27]. Some African countries have developed building energy codes but it is usually not exhaustive. Hence, these codes should be improved to cover diverse scenarios of the region.

Furthermore, economics pathways are effective approaches to foster the adoption of building energy certifications. According to Suh et al. [55], certified buildings usually have a high market value than that of the traditional buildings. There is a higher demand for buildings which are either LEED or energy certified. This is further confirmed by Odubiyi et al. [44] who argues that, adopting green concepts which includes smart technologies and increasing incentives to the building owners in the construction of residential buildings raises the property value of the building. This reduces the total lifecycle cost of the building. However, since affordability is always a concern in developing countries than the impact of the construction on the environment, the initial cost for adopting energy certifications in building construction is always high. Hence, innovative policies to incentivize the implementation of these certifications is in the right direction. Incentives, tax reductions and subsidies should be made available for building owners to serve as encouragement to invest into energy efficiency. Buildings with certification from an accredited institution usually have high energy efficiency. This reduces cost on energy consumption [10]. However, these results might vary due to difference in number of occupants and construction periods [68]. Cashflow is generally increased when green building certifications are adopted due to increased rents, reduced operational cost, and reduced risk [36].

Furthermore, technical pathways have been identified as one of the integral approaches towards reducing energy use in both new and existing buildings through building energy certifications. The Building America programme strives to ensure that by 2020, there is a reduction in energy use within the existing home stock by 20–30% [24]. This programme is made up of computer modelling, field demonstrations, and long-term monitoring working in support of integrated approaches for energy reductions [24]. Green building certification strategies can be categorized into practices that give private gains through efficiency investments and those that result in public benefits [18]. Energy efficiency can be achieved through a number of ways, including green building rating systems, policy use, technology implementation, and strategy application among others as demonstrated by Evans et al. [16]. Low-carbon transition pathways for buildings are integrated with techniques on energy saving, clean energy supply, and flexible demand response [60]. Such approaches, along with energy and carbon trading systems, provide a total impetus toward sustainable and energy-efficient buildings.

The government and policy makers play a significant role in the development and implementation of building regulations and standards. The effective government oversight for building constructions includes setting clear and goal-oriented building codes, enforcing building codes, ensuring monitoring, and compliance of the building codes. For this to be achieved, government should liaise with private organizations to provide training programmes to help build local capacity of expertise on energy certifications. This will create job opportunities while ensuring environmental protection. Also, financial incentives and grants can be awarded to encourage clients and building developers to adopt energy certifications. Moreover, the full benefit of building energy codes and certifications lies in its implementation from the design stage to the construction and usage stage [16]. To enhance its implementation, building energy certifications should be developed based on the climatic characteristics of this region. Also, imported materials should be based on the acceptable standards of the energy certification. Several studies have emphasized on the need to implement building energy codes and certifications. Hence, government should collaborate with third party support systems such as the EDGE and Energy Star to enhance and enforce energy certification of buildings.

The third-party support systems are usually made up of outside agencies, professionals, and organizations that aid in the provision of energy certifications to building owners and developers. They make sure that, building owners and developers comply with the building energy codes and standards, provide advisory and consultancy services regarding building energy efficiency achievement, and conduct energy performance audits. Typically, the building design is inspected to determine if it meets the energy codes design specification before the construction process. Inspections are conducted during construction to ensure developers adopt sustainable construction practice [16]. These services are critical to guaranteeing the effectiveness of building energy certifications. In most developing countries where government agencies lack the ability to enforce strict compliance with the energy standards, third party systems guarantee that certifications are extensively utilized. They have specialized knowledge and provide guidelines for building design and construction to make buildings environmentally friendly and operates at high performance with maximum power usage [41]. They also provide capacity building training that enhances the competency of the stakeholders involved in the construction industry on building energy certifications.

In South Africa, the Green Building Council of South Africa (GBCSA) collaborate with third party systems to certify buildings under the Green Star South African (SA) rating system. This has enhanced long-term savings on operation cost through enhancing energy efficiency and has significantly increased the number of certified green buildings in SA. Nigeria has also incorporated third party verification systems that ensure compliance with the Building Energy Efficiency (BEE) guideline. Additionally, the application of EDGE certifications has been facilitated by third party systems that ensure energy efficiency of buildings and compliance with the energy-efficient standards.

Social Pathways is a growing strategy in building energy certifications tackling both the environmental and social sustainability. Even though most certification systems do not factor occupants' behaviour, a study by Stender and Walter [54] suggested that, building assessment certifications systems should incorporate social sustainability indicators. Tinarwo et al. [56] argue that, human influence is an integral aspect of building energy performance and should not be ignored in the development of frameworks to promote energy efficiency. In order to reduce performance gap between the estimated energy use and actual energy consumption, Ebuy et al. [69] maintain that, the occupants' behaviour should not be neglected. Building energy certifications should have indicators to incorporate the active and passive behaviour of occupants. In Africa, most part of the region is faced with high solar gains hence, building energy certifications should be developed to meet the climatic conditions of this region in order to enhance implementation.

The construction industry in most developing countries is characterized with unprotected and unregulated employees. In Africa, about 80% of the construction labour force typically in countries such as Ghana, Nigeria, and South Africa are in the informal sector [6]. Hence, there is lack of adherence to building codes and standards leading to significant challenges in implementing building energy certifications. Hence, formalizing the informal sector is crucial for the widespread adoption of building energy certifications. Training and skill development programmes are essential to formalize the construction industry. Training is particularly important to implement performance-based codes. Many countries are switching to performance-based codes; such codes can be more complex, requiring more training and analytical understanding to demonstrate compliance via whole-building simulation. The need for effective training is emphasized in a study in United States which found a significant level of confusion and lack of understanding in energy codes among the code officials and field inspectors [70]. Although the use of software is correlated to higher compliance rate, training ensures the efficient usage of this software. This underscores the importance of training in the industry. Policies should be established to enhance the transition from the informal practices to the formal practices through the provision of vocational education, certification training, and internship programmes in order for construction workers to increase their understanding on the intricacies on the effect of building of the environment [3]. This will enhance the adoption and implementation of building energy certifications.

13.11 Summary

This chapter illustrates that, energy saving and environmental concerns are primary drivers for energy-efficient buildings. With the gross energy consumption of buildings and large carbon emissions, it is imperative to develop innovative techniques to reduce the energy consumption and carbon emission of building in order to promote sustainability in the construction industry. The evolution of Building Energy Certification has become an effective instrument for improving the energy efficiency and reducing emissions from buildings by putting in minimum energy requirements for particular parts of the building, like heating, ventilating, air-conditioning systems, and lighting. Diverse studies have indicated that, there is huge energy saving potentials when building energy certifications are adopted. Adopting building energy certifications have resulted in about 5% to 35% energy savings. This calls for buildings that have complied with such regulations to be certified, thus giving assurance that they have met the minimum energy standards. The building energy certifications has been developed in many developed countries tailored to the climate conditions and methods of construction. Most developing countries are still in the early stages of establishing regulatory frameworks for building energy certifications however, the rate of adoption is slow.

Although Building energy certifications have high prospects of energy saving and a roadmap to decarbonization, developing countries are faced with various barriers to the adoption of these certifications. Majority of these barriers are; high initial cost and lack of government incentives such as tax remission, subsidies and grants offered to individuals that adopt energy certifications; and unavailability of skilled professionals who have comprehensive knowledge on energy savings and certifications. Also, there is lack of skilled professionals and low awareness among stakeholders. Furthermore, due to occupants' behaviour, there is a discrepancy between actual energy consumed and estimated energy. This heavily misinforms the decision making and hinders the attainment of energy efficiency goals. These schemes vary according to the priorities of policies and local contexts. The schemes further adopt several methodologies such as prescriptive, performance-based, and hybrid method that give energy performance ratings. Since the effectiveness of these certifications lie on its implementation, the adoptions of these energy certifications have faced considerable challenges such as lack of awareness and reluctance of occupants, high initial cost and complexity with certifications. This creates a huge gap between the standards and real-world practices.

To address these concerns, this chapter proposes five key pathways to improve the implementation of building energy certifications in developing countries. These pathways include policy reforms, financial incentives, capacity building, technological innovations, and stakeholder engagement including private sector players, government, developers, and clients.

References

1. Agyekum K, Adinyira E, Ampratwum G (2020) Factors driving the adoption of green certification of buildings in Ghana. Smart Sustain Built Environ 9(4):595–613. https://doi.org/10.1108/sasbe-02-2019-0017
2. Agyekum K, Adinyira E, Baiden B, Ampratwum G, Duah D (2019) Barriers to the adoption of green certification of buildings. J Eng Design Technol 17(5):1035–1055. https://doi.org/10.1108/jedt-01-2019-0028
3. Ahadzie DK, Debrah Y, Ofori G (2021) Formality and informality in Sub-Saharan Africa and the Ghanaian construction industry. Routledge eBooks, pp 103–132. https://doi.org/10.4324/9780429431135-6-6
4. ALwaer H, Clements-Croome DJ (2010) Key performance indicators (KPIs) and priority setting in using the multi-attribute approach for assessing sustainable intelligent buildings. Build Environ 45(4):799–807. https://doi.org/10.1016/j.buildenv.2009.08.019
5. Bernardi E, Carlucci S, Cornaro C, Bohne R (2017) An analysis of the most adopted rating systems for assessing the environmental impact of buildings. Sustainability 9(7):1226. https://doi.org/10.3390/su9071226
6. Boadu EF, Wang CC, Sunindijo RY (2020) Characteristics of the construction industry in developing countries and its implications for health and safety: an exploratory study in Ghana. Int J Environ Res Public Health 17(11):4110. https://doi.org/10.3390/ijerph17114110
7. Botchway EA, Afram SO, Ankrah J (2014) Building permit acquisition in Ghana: The situation in Kumasi. Developing Country Studies 4(20):11–22
8. Chan APC, Darko A, Olanipekun AO, Ameyaw EE (2017) Critical barriers to green building technologies adoption in developing countries: the case of Ghana. J Clean Prod 172:1067–1079. https://doi.org/10.1016/j.jclepro.2017.10.235
9. Chelang'at FN, Lawrence R (2024) Always with the best intentions? Interrogating the use of sustainable building assessment systems in developing countries: Kenya. Sustainability 16(9):3868. https://doi.org/10.3390/su16093868
10. Clay K, Severnini E, Sun X (2023) Does LEED certification save energy? Evidence from retrofitted federal buildings. J Environ Econ Manag 121:102866. https://doi.org/10.1016/j.jeem.2023.102866

11. Cozza S, Chambers J, Brambilla A, Patel MK (2020) Energy Performance Certificate for buildings as a strategy for the energy transition: Stakeholder insights on shortcomings. IOP Conference series. Earth and Environmental Science, Vol 588 No 2, p 022003. https://doi.org/10.1088/1755-1315/588/2/022003
12. De Almeida MO, Vieira JC, Oliveira RD, Reis ED (2022) From efficient to smart: transitioning from certified high-performance commercial buildings. Brazilian J Dev 8(11):74673–74692. https://doi.org/10.34117/bjdv8n11-262
13. Dwaikat LN, Ali KN (2016) Green buildings cost premium: a review of empirical evidence. Energy Build 110:396–403. https://doi.org/10.1016/j.enbuild.2015.11.021
14. Ejidike CC, Mewomo MC (2023) Benefits of adopting smart building technologies in building construction of developing countries: review of literature. SN Appl Sci 5(2). https://doi.org/10.1007/s42452-022-05262-y
15. Elnabawi MH (2021) Evaluating the impact of energy efficiency building codes for residential buildings in the GCC. Energies 14(23):8088. https://doi.org/10.3390/en14238088
16. Evans M, Roshchanka V, Graham P (2017) An international survey of building energy codes and their implementation. J Clean Prod 158:382–389. https://doi.org/10.1016/j.jclepro.2017.01.007
17. Falcone PM (2023) Sustainable energy policies in developing countries: a review of challenges and opportunities. Energies 16(18):6682. https://doi.org/10.3390/en16186682
18. Flowers ME, Matisoff DC, Noonan DS (2018) For what it's worth: evaluating revealed preferences for green certification. J Environ Planning Manage 62(5):843–861. https://doi.org/10.1080/09640568.2018.1447444
19. Ganiyu A-Y, Mayowa RW, Taibat AR, Orire AM, Soliu I, Eluwa SE (2020) Barrier factors affecting adoption of green building technologies in Nigeria. Built Environ J 17(2):37. https://doi.org/10.24191/bej.v17i2.8860
20. Geissler S, Österreicher D, Macharm E (2018) Transition towards energy efficiency: developing the Nigerian building energy efficiency code. Sustainability 10(8):2620. https://doi.org/10.3390/su10082620
21. Giama E, Papadopoulos AM (2012) Sustainable building management: overview of certification schemes and standards. Adv Build Energy Res 6(2):242–258. https://doi.org/10.1080/17512549.2012.740905
22. Hafez FS, Sa'di B, Safa-Gamal M, Taufiq-Yap YH, Alrifaey M, Seyedmahmoudian M, Stojcevski A et al (2023) Energy efficiency in sustainable buildings: a systematic review with taxonomy, challenges, motivations, methodological aspects, recommendations, and pathways for future research. Energy Strategy Rev 45:101013. https://doi.org/10.1016/j.esr.2022.101013
23. Hammah NK (2015) Streamlining of building permit approval processing of town and country planning department in Ghana. Cogent Soc Sci 1(1). https://doi.org/10.1080/23311886.2015.1060730
24. Hendron R (2006) Building America performance analysis procedures for existing homes. https://doi.org/10.2172/883018
25. Herrando M, Cambra D, Navarro M, De La Cruz L, Millán G, Zabalza I (2016) Energy performance certification of faculty buildings in Spain: the gap between estimated and real energy consumption. Energy Convers Manage 125:141–153. https://doi.org/10.1016/j.enconman.2016.04.037
26. Hoffman D, Huang L-Y, Van Rensburg J, Yorke-Hart A (2020) Trends in application of Green Star SA credits in South African green building. Acta Structilia 27(2). https://doi.org/10.18820/24150487/as27i2.1
27. Hoxha V, Lecaj V (2022) Regulatory barriers to implementing sustainable buildings in Kosovo. Prop Manag 42(2):293–309. https://doi.org/10.1108/pm-01-2022-0008
28. Hui S (2002) Using performance-based approach in building energy standards and codes
29. Ikingura A, Grabiec AM, Radomski B (2025) Examining key barriers and relevant promotion strategies of green buildings adoption in Tanzania. Energies 18(5):1081. https://doi.org/10.3390/en18051081

30. ISO (2022) ISO—Certification, available at: https://www.iso.org/certification.html. Accessed 22 July 2024
31. Iwaro J, Mwasha A (2010) A review of building energy regulation and policy for energy conservation in developing countries. Energy Policy 38(12):7744–7755. https://doi.org/10.1016/j.enpol.2010.08.027
32. Ji C, Hong T, Kim H, Yeom S (2022) Effect of building energy efficiency certificate on reducing energy consumption of non-residential buildings in South Korea. Energy Build 255:111701. https://doi.org/10.1016/j.enbuild.2021.111701
33. Keleş AE, Önen E, Górecki J (2022) Determination of green building awareness: a study in Turkey. Sustainability 14(19):11943. https://doi.org/10.3390/su141911943
34. Lee WL (2013) A comprehensive review of metrics of building environmental assessment schemes. Energy Build 62:403–413. https://doi.org/10.1016/j.enbuild.2013.03.014
35. Leibowicz BD, Lanham CM, Brozynski MT, Vázquez-Canteli JR, Castejón NC, Nagy Z (2018) Optimal decarbonization pathways for urban residential building energy services. Appl Energy 230:1311–1325. https://doi.org/10.1016/j.apenergy.2018.09.046
36. Leskinen N, Vimpari J, Junnila S (2020) A review of the impact of green building certification on the cash flows and values of commercial properties. Sustainability 12(7):2729. https://doi.org/10.3390/su12072729
37. Li P, Froese TM, Brager G (2018) Post-occupancy evaluation: state-of-the-art analysis and state-of-the-practice review. Build Environ 133:187–202. https://doi.org/10.1016/j.buildenv.2018.02.024
38. Li Y, Kubicki S, Guerriero A, Rezgui Y (2019) Review of building energy performance certification schemes towards future improvement. Renew Sustain Energy Rev 113:109244. https://doi.org/10.1016/j.rser.2019.109244
39. Liu F, Meyer AS, Hogan JF (2010b) Mainstreaming building energy efficiency codes in developing countries. World Bank Working Paper, https://doi.org/10.1596/978-0-8213-8534-0
40. Ma Z, Awan MB, Lu M, Li S, Aziz MS, Zhou X, Du H et al (2023) An Overview of emerging and sustainable technologies for increased energy efficiency and carbon emission mitigation in buildings. Buildings 13(10):2658. https://doi.org/10.3390/buildings13102658
41. Mazur Ł, Resler M, Koda E, Walasek D, Vaverková MD (2023) Energy saving and green building certification: case study of commercial buildings in Warsaw, Poland. Sustain Energy Technol Assess 60:103520. https://doi.org/10.1016/j.seta.2023.103520
42. Nicholson C (2023) Performance versus Prescriptive Path", Energy Diagnostics, 1 May, available at: https://energydiagnosticsinc.com/performance-vs-prescriptive-path/#:~:text=The%20prescriptive%20energy%20code%20path%20is%20a%20prescriptive%2Dbased%20approach,lighting%2C%20and%20other%20building%20components
43. Nwanekezie OF, Nwanguma RW (2020) Examination of the challenges and impact of building plan approval procedure on real estate development in Uyo, Nigeria. Archit Res 10(1):15–20
44. Odubiyi TB, Abidoye RB, Aigbavboa CO, Thwala WD, Ademiloye AS, Oshodi OS (2024) Impact of green features on rental value of residential properties: evidence from South Africa. Real Estate 1(1):65–79. https://doi.org/10.3390/realestate1010005
45. Osei-Poku G, Koranteng C, Amos-Abanyie S, Botchway EA, Gyimah KA (2023) A review of frameworks for the energy performance certification of buildings and lessons for Ghana. Springer eBooks, pp 63–80, https://doi.org/10.1007/978-3-031-25998-2_6
46. Papineau M (2017) Setting the standard? A framework for evaluating the cost-effectiveness of building energy standards. Energy Econ 64:63–76. https://doi.org/10.1016/j.eneco.2017.02.011
47. Park D, Yu K, Yoon Y, Kim K, Kim S (2015) Analysis of a building energy efficiency certification system in Korea. Sustainability 7(12):16086–16107. https://doi.org/10.3390/su71215804
48. Romano S, Riediger N (2019) BIM as a tool for Green building certifications: an evaluation of the energy category of LEED, BREEAM and DGNB. J Phys: Conf Ser 1425(1):012162. https://doi.org/10.1088/1742-6596/1425/1/012162

49. Rotimi A, Bahadori-Jahromi A, Mylona A, Godfrey P, Cook D (2017) Estimation and validation of energy consumption in UK existing hotel building using dynamic simulation software. Sustainability 9(8):1391. https://doi.org/10.3390/su9081391
50. Sesana MM, Salvalai G (2018) A review on building renovation passport: potentialities and barriers on current initiatives. Energy Build 173:195–205. https://doi.org/10.1016/j.enbuild.2018.05.027
51. Shen W, Tang W, Siripanan A, Lei Z, Duffield CF, Hui FKP (2019) Understanding the green building industry in Thailand. Green energy and technology, pp 161–180. https://doi.org/10.1007/978-3-030-24650-1_9
52. Shi J, Chen W, Yin X (2016) Modelling building's decarbonization with application of China TIMES model. Appl Energy 162:1303–1312. https://doi.org/10.1016/j.apenergy.2015.06.056
53. Simpeh EK, Smallwood JJ, Ahadzie DK, Mensah H (2021) Analytical taxonomy of challenges to the implementation of green building projects in South Africa. Int J Constr Manag 23(2):286–296. https://doi.org/10.1080/15623599.2020.1863172
54. Stender M, Walter A (2018) The role of social sustainability in building assessment. Build Res Inf/Build Res Inf 47(5):598–610. https://doi.org/10.1080/09613218.2018.1468057
55. Suh MJ, Pearce AR, Song Y, Kwak YH, Kim JI, Zhang Y (2019) The impact of Leed-energy star certified office buildings on the market value of adjoining buildings in new York City. J Green Build 14(1):31–52. https://doi.org/10.3992/1943-4618.14.1.31
56. Tinarwo B, Rahimian F, Ghanem DA (2023) Towards a trajectory for sustainable policies and market strategies governing building lifecycle energy performance. Smart and sustainable built environment.https://doi.org/10.1108/sasbe-01-2023-0024
57. Watts C, Jentsch MF, James PA (2011) Evaluation of domestic energy performance certificates in use. Build Serv Eng Res Technol 32(4):361–376. https://doi.org/10.1177/0143624411404486
58. Windapo AO (2016) Skilled labour supply in the South African construction industry: the nexus between certification, quality of work output and shortages. SA J Hum Resour Manag 15. https://doi.org/10.4102/sajhrm.v14i1.750
59. Xie Y, Tyler M, Huckett J, Bartlett R, Chen Y, Salcido V, Mendon V et al (2023) A Review of the evaluation of building energy code compliance in the united states. Energies 16(16):5874. https://doi.org/10.3390/en16165874
60. Zhou Y (2023) Worldwide carbon neutrality transition? Energy efficiency, renewable, carbon trading and advanced energy policies. Energy Rev 2(2):100026. https://doi.org/10.1016/j.enrev.2023.100026
61. Zuo J, Zhao Z-Y (2014) Green building research–current status and future agenda: a review. Renew Sustain Energy Rev 30:271–281. https://doi.org/10.1016/j.rser.2013.10.021
62. Filippini M, Obrist A (2022) Are households living in green certified buildings consuming less energy? Evidence from Switzerland. Energy Policy 161:112724
63. Darko A (2019) Adoption of green building technologies in Ghana: Development of a model of green building technologies and issues influencing their adoption. PhD thesis, The Hong Kong Polytechnic University
64. Fernandes BA (2023) Analysis of financing schemes targeting energy efficiency and energy poverty mitigation case studies: European Union, UK, Australia, and New Zealand. Master's thesis, Universidade NOVA de Lisboa (Portugal)
65. Amecke H (2012) The impact of energy performance certificates: a survey of German home owners. Energy policy 46:4-14
66. Pitt S (2014) National energy efficient building project. Available at: https://www.pittsh.com.au/assets/files/Projects/NEEBP-final-report-November-2014. pdf
67. Hasan OA, Defer D (2019) The role of new technologies in understanding the building energy performance: a comparative study. Int J Smart Grid Clean Energy 8(4):397-401. Available at: https://doi.org/10.12720/sgce.8.4.397-401
68. Al-Zubaidy MSK (2015) A literature evaluation of the energy efficiency of leadership in energy and environmental design (LEED)—certified buildings. Am J Civ Eng Archit 3(1):1-7. Available at: https://doi.org/10.12691/ajcea-3-1-1

69. Ebuy HT, El Haouzi HB, Benelmir R, Pannequin R (2023) Occupant behavior impact on building sustainability performance: a literature review sustainability 15(3):2440. Available at: https://doi.org/10.3390/su15032440
70. Gowri K, Williams J (2014) Streamlining the energy code compliance process to reduce enforcement challenges and harvest real energy savings. 2014 ACEEE summer study on energy efficiency in buildings. Available at: http://aceee.org/files/proceedings/2014/data/papers/4-393.pdf

Chapter 14
Refurbishment and Retrofit Guidelines and Appendages

14.1 Introduction

Africa stands at a critical crossroad of unprecedented urban expansion, rapid population growth, and mounting climate change pressures. Characterized with about 77% of its population below 35 years, the continent is experiencing one of the highest urbanization rates globally, as urban population has been projected to triple by 2050, reaching about 1.34 billion people [1]. According to the 2020 World Cities Report, urban sprawl in African regions have not only been difficult to manage, but have also been predicted to geometrically increase by a factor of 3.7 by 2050 [2]. This surge in population, coupled with accelerating industrialization, has triggered an insatiable demand for building—be it residential, commercial, or industrial. As these new construction projects proliferate across the continent, the already existing conventional building stock, which is characterized by high CO_2 emissions and poor energy performance, requires swift attention in view of the global decarbonization agenda.

The African built environment accounts for about 56% of the continent's total final energy consumed and contributes significantly to greenhouse gas emissions [3, 79]. The utilization of energy during the construction, demolition, disposal, and for upstream activities like the production of building materials like steel and cement have been well documented as significant consumers and emitter of energy and CO_2, respectively [4]. The current state of energy intensity continues to worsen as a result of inadequate building codes, inadequate regulatory frameworks, financial constraints, and a strong emphasis for upfront construction costs in lieu of lifecycle energy performance among many others [5–7]. In addition to these, the widespread adoption of modern construction techniques (e.g. glazed buildings) and over reliance on active systems like heating ventilation air-conditioning (HVAC) further implicate the degrading situation [8]. Consequently, a number of studies have concluded that developing countries, particularly those across the African continent, will disproportionally experience the adverse impacts of climate change. This assertion is based

M. N. Addy et al., *Decarbonization of Building Energy in Developing Countries*, Green Energy and Technology, https://doi.org/10.1007/978-3-032-09113-0_14

on rising temperature and heat wave, increasing water insecurity, drought, famine, rising sea levels, and others prevailing in the continent [7, 9, 10].

It is worth noting that these highlighted issues continues to persist to date within the continent even though a significant part of the region continue to battle with access to electricity [6, 11]. According to the Mo Ibrahim Foundation report on Africa's energy deficit for instance, [12] iterate that over 600 million of people in the African continent have not access to electricity. This further begs the question: what will happen, should the entire continent have access to electricity given its inefficient managerial approach to energy and response to building demand? Considering all of these, it is imperative to commence tackling issues from the sustainability stance. And the onus lies with the architectural engineering and construction (AEC) industry, as a major consumer of resources and contributor to environmental issues, to inculcate certain sustainable protocols in its activities. Among which the issue of energy use is highlighted as most expedient [13]. Basically, there are two ways to attend to this: (1) addressing energy efficiency of new building and, (2) retrofitting and refurbishing already existing building structures.

In light of the former approach, the industry in collaboration with academic and research institutions have begun to explore several options including Circular Economy, Green buildings using geogenic and biogenic materials, vernacular construction concepts, and strategies, among many others in the attempt to engender energy-efficient buildings in the region [14–18]. While these options are more focused on new construction projects, both [19, 20] contend that existing buildings could benefit from sustainable retrofitting and refurbishment practices. This is mainly because between 2030 and 2050, they will form a substantial part of the built environment [21], even though most of them have been built without energy efficiency standards and considerations [22]. Building retrofitting and refurbishment strategies present the architectural engineering and construction industry with compelling opportunities for energy decarbonization. Both concepts are characterized by distinct procedures.

Completely demolishing a building that has failed to meet expectation or has come to the end of its service life [23], and redeveloping a new one either in its place or in a different locality, comes at a considerable high cost, significant environmental degradation, and energy consumption [24]. The construction procedure needed to revive or cure the building from its state of dilapidation and obsolescence is considered as refurbishment [23]. To refurbish a building means to renovate, restore, or improve a building to return it to a good condition or make it appear fresh and modern [19]. Fahlstedt et al. [25] view refurbishment as a way of ameliorating the features and functionality of a building, to the end that it aligns with prevailing standards. Typically, such corrective works encompass, but not limited to, improving heating, ventilation, and air-conditioning (HVAC) systems, façades, structural components, and building materials [25].

Retrofitting, on the other hand, means refers to the application of conversion techniques or modification activities, which encompasses the additions, deletions, and replacement of a part or multiple parts of a building [26, 27]. It can also be described as the "process of replacing and upgrading systems and technology in existing building to address its technological or environmental obsolescence" [28], p. 1458]. Although building retrofitting aims to reduce water use and improve indoor environmental quality Davoudi et al. [24, 29] opine that the need for energy efficiency is most paramount. In view of this, retrofitting can be delineated as the "installation of individual or multiple energy efficiency measures to an existing building" [24], p. 19]. Retrofitting interventions meticulously consider passive design improvements (enhanced insulation, natural ventilation, and sun shading), active system upgrades (efficient HVAC systems, renewable energy integration), and behavioural modifications [30, 31]. Briefly put, [29] categorized retrofitting technologies into two types: "fit and forget" interventions and those that require monitoring, control, and maintenance.

Overall, engaging in refurbishment and retrofitting activities in the built environment transcends their immediate practical necessity. Rather, these activities have become essential means to engender sustainability—addressing economic, social, and environment issues—particularly in developing regions like Africa, where many infrastructure projects were constructed without energy efficiency standards and techniques. According to United Nations Environment Programme (UNEP) [32], refurbishment and retrofitting programmes have the potential to reduce carbon emissions by 40%. Ohene et al. [31] also demonstrate a potential 47–54% reduction in energy consumption. In addition, Harvey [33] highlights that optimizing HVAC systems and lighting in commercial spaces can reduce energy consumptions by 25% to 50% and 30% to 60%, respectively. According to Camarasa et al. [9], retrofitting and refurbishing buildings in some parts of Europe resulted in an 85% and 89% cut in final energy consumption and carbon emissions respectively.

For the African continent, this serves as a clarion call to rise above the status quo. As the continent predicted to drive global demand for energy in the coming years, these findings present significant opportunities to remain resolute in enhancing climate resilience activities, create employment while developing more sustainable infrastructure without following the high-carbon development pathways of developed countries. The practice of retrofitting and refurbishing, like many sustainable construction practices, are at their nascent stages in the region. Nevertheless, advancing on this development trajectory is the viable option the built environment must embrace in order to engender a future of sustainability within the continent.

Having established the fundamental concepts of retrofitting and refurbishment, it is crucial to examine their specific role in decarbonization efforts in the built environment, particularly in the African continent. Construction and building operations demand for nearly one-third of the world energy, subsequently resulting to very significant carbon emission into the environment [34, 35]. The quest to provide high-performance building reveals building retrofitting and refurbishment as viable pathways to transform the existing building stocks into less carbon-intense and more energy-efficient structures. This section explores the decarbonization potentials of building retrofitting and refurbishment activities, examining both the quantifiable benefits supported by recent research and the implementation challenges that must be addressed to scale these practices effectively, particularly in Africa where the need and opportunity are enormous.

14.2 The Role of Building Retrofitting and Refurbishment for Decarbonization

14.2.1 Environmental Sustainability

A primary motivation for building retrofitting and refurbishment is to improve the structural performance by optimizing energy use and implementing construction strategies that reduce carbon emissions. Building retrofitting and refurbishment presents a significant opportunity to address environmental challenges and enhance climate actions by lowering carbon footprint through the use of energy-efficient technologies [36]. It also reduces demolition waste and landfill pressure. In their environmental impact analysis between building demolition and retrofitting simulations, Dragonetti et al. [37] demonstrated that carbon gases are significantly reduced in renovated buildings as compared to demolished ones since energy consumption is also lowered. Circular economy concepts like design for disassembly techniques encourages the reuse or recycle of building materials [37, 38]. In developing regions, where many buildings have been constructed without energy efficiency considerations, retrofitting offers a particularly valuable opportunity as it aims to enhance energy efficiency, while reducing reliance on non-renewable energy sources. Ohene et al. [31] and Organ [39] both agree that the use of Photovoltaic technologies as renewable energy sources to offset buildings' operational energy. These interventions directly translate to substantial carbon emission reductions for Africa since it has a huge opportunity to harness solar and other renewable energy technologies.

14.2.2 Preservation of Cultural and Historical Heritage

Building retrofitting and refurbishment enable the preservation of cultural and historical heritage while advancing decarbonization objectives. Historic buildings represent irreplaceable cultural assets that embody architectural styles and historical narratives central to community identity [40]. The demand for heritage tourism is estimated to increase 15% annually [41]. Engaging in building retrofits and refurbishment keeps the building in good shape, which further influences the living conditions of the communities [42]. Another significance of cultural and historic structures is in its contribution to sustainable urban development by enhancing vibrancy while maintaining low-carbon neighbourhood. In their review, Ameen et al. [43] demonstrated that retrofitting of historic buildings can achieve energy savings of 15–69% while preserving their architectural significance. In the same, Loli and Bertolin [42] found that integrating modern energy-efficient technologies with traditional building systems can substantially reduce operational carbon emissions without compromising historical integrity. Apart from making heritage preservation climate-friendly,

Ameen et al. [43] and Organ [41] noted that the indoor conditions of historic structures are enhanced by incorporating climate-responsive features that corroborates with modern sustainability goals.

14.2.3 Economic Benefits

Building retrofitting and refurbishment offer substantial economic advantages while making significant contribution to decarbonization efforts. Energy-efficient retrofits reduce operational cost through the installation of modernized and efficient building subsystems such as air-conditioning and lighting. According to their longitudinal study in Hong Kong, Zheng and Lai [44] found that building retrofits that focus on air-conditions can reduce energy consumption and carbon emissions by 32.3% and 42.9%, respectively, creating significant long-term savings for investors, building owners and users. These savings also offer stakeholders attractive return on investments (ROI). Beyond this point, retrofitted buildings are usually relatively cheaper and less environmental degrading than new construction. Organ [41] mentioned that avoiding more invasive and costly repair works on buildings, the cost of renovation can be moderated. Retrofitting and refurbishment works requires skills labour. This enhances job security and fosters employment opportunities, which are imperatives for determining life satisfaction (i.e. social sustainability) [41, 45].

14.2.4 Job Creation and Skills Development

Building retrofitting and refurbishment creates job opportunities and a niche of work within the supply chain of the architectural engineering and construction industry. This is because renovating building demands a set of skills and knowledge [41], and the inadequacy thereof could further aggravate building conditions. This job subsequently requires academics, engineers, technicians, and local artisans, particularly in Africa, to arise to the occasion. To this end, construction stakeholders must willingly undergo training programmes to enhance their knowledge and skills in conducting sustainable building refurbishment and retrofitting. According to Loli and Bertolin [42], stakeholders in Scandinavian regions are undertaking studies on traditional buildings in order to establish guidelines for their refurbishment and maintenance. For the Africa continent, this offers an opportunity to collaborate and share knowledge. Though architectural concepts may vary, researchers could still learn from various settings within African region or beyond and implement novel materials and techniques.

14.2.5 Improved Building Performance and Indoor Environmental Quality

Through the use of technological advancements in simulations, several retrofits tools have been designed to explore how buildings operations can be optimized [46]. Building retrofit and refurbishment offers the chance to improve resilience against climate-induced phenomena such as extreme heat and flooding. Extant studies indicate that building users spend a very significant amount of their time within indoor environments [47, 48]. This makes the concept of building performance a dire one to the built environment. The indoor environment created impacts the health, well-being, and productivity of occupants. Accordingly, building retrofitting and refurbishment offers the opportunity to enhance indoor environmental parameters like thermal comfort and indoor air quality with minimal negative implication on the environment [44]. Additionally, the lifespan of existing structure is further elongated as old building systems are replaced with modernized and more efficient one, enhancing sustainability within the community.

14.3 Challenges of Building Retrofitting and Refurbishment for Decarbonization

Building retrofit and refurbishment in the African context presents unique challenges due to the acceleration of climate change and its related implications and urban population growth. The potential benefits of transforming existing building stock through retrofitting and refurbishment could influence the attainment of Africa's sustainable development agenda. This section examines the multifaceted barriers militating against practitioners' undertaking of retrofit projects, illuminating both visible obstacles and subtler impediments that often remain unacknowledged in policy discourse. Literature reveals these challenges encompass technical, market and supply chain, financial and economic, organizational, social and behavioural, regulatory and policy issues.

14.3.1 High Initial Investment Cost

Building retrofitting and refurbishment projects for decarbonization usually require significant upfront monetary commitment, creating a substantial barrier to their implementation in many parts of the world. The cost of financing building retrofits would typically include capital, operational energy, replacement, and maintenance cost [40]. This usually leads to an incremental implementation of building retrofitting, such that renovation works are carried out over a number of years [36]. Several factors significantly influence the total cost of renovating buildings, including the scope of retrofitting, location, material selection, installation costs, building type, and its prevailing condition. Initial financing is a particularly crucial barrier to the integration of advanced energy-efficient technologies in developing countries [31]. Furthermore, the limited presence of funding agencies within these region poses a significant challenge for building owners seeking to engage in retrofit practices. Securing loans from financial institutions represents a dire hurdle for many building owners due to uncertainties regarding return on investment and the lack of standardized evaluation metrics [49]. Ardron and Allwood [50] further emphasize that the decarbonization agenda for domestic buildings primarily lies on landlords, making it expensive for many property owners. The affordability challenge is further exacerbated by "energy efficiency gap"—that is the difference between actual and cost-effective energy use. This gap emerges because building owners, particularly in developing countries, tend to undervalue the prospect of future energy savings relative to upfront costs, leading to systematic underinvestment in energy efficiency measures [51].

14.3.2 Technical and Structural Challenges

Retrofitting existing buildings presents numerous technical and structural challenges that can complicate or even prevent its successful implementation. In a region where many existing buildings are likely to remain in use for several decades, but lack energy efficiency considerations from their design stage, retrofitting becomes particularly challenging. Many existing structures were not designed to support modern retrofitting methods, complicating renovation activities since any work needed now would have to be deep and technically demanding, and sometime structurally unfeasible [45]. Traditional buildings present unique challenges as retrofitting activities are often faced with methodological limitations. For instance, Loli and Bertolin [42] mentioned that wooden historic buildings risk being attacked by insects and other biological phenomena, which can cause decay among other things. To improve such structures, there is a need for specialized skills and customized solutions, which further aggravate retrofitting cost [26, 36]. In the same vein, building-specific properties like wall thickness, roof strength, and HVAC systems create challenging situations. The resultant effect of such a situation is a limitation on the scalability and standardization of retrofit methodologies. In addition, the scope of retrofitting work may require more than one specialist, often leading to communication and coordination challenges, increased cost and suboptimal performance [36]. The unavailability of experts, lack of accurate building documentation for renovation works, and creation of inconveniences—since occupants may have to consider temporal lodging arrangements—can further impact willingness to under building retrofitting and refurbishment.

14.3.3 Limited Awareness and Expertise

A significant knowledge gap among stakeholders is another concern evident in the implementation process of building retrofitting and refurbishment. Many building owners, property managers, and even construction professionals lack comprehensive understanding of available technologies, potential benefits, and implementation processes for deep energy retrofits. According to Jagarajan [26], unawareness of the benefits of this sustainable practice stems from a lack of information. This knowledge deficit in the industry directly influences decision-making during a building's lifecycle. The unavailability of knowledge in retrofitting is further revealed in stakeholders' inability to accurately assess the energy efficiency gap of a building. They often undervalue return of investment and other non-energy benefits like thermal comfort while overestimating payback periods [46, 52]. Inaccurate initial financial assessments often result in retrofit projects being deemed financially unfeasible. In furtherance, the inadequate integration of retrofitting and refurbishment concepts in academic curricula presents another challenge. The built environment would benefit significantly if its players were adequately expose to such practices. The lack of education and exposure worsens the issue of skills shortage in the market and increases energy consumption in the long run. Updating academic curricula could act as a pipeline through which circularity concepts like design for disassembly, reusing, and recycling would diffuse among parties in the sector [38]. In a field experiment, Stojanovski et al. [53] highlighted two important things: (1) energy consumers in developing countries are eager to receive actionable information; (2) accessibility to information is critical in fostering energy-efficiency awareness in developing countries. However, little is done to raise public awareness and education on energy-efficient building practices. The fragmentation of construction works often lead contractors to specialize in a specific areas of building work, meaning that multiple professionals are required for comprehensive retrofitting [45]. Though such diversity can be beneficial in some ways, it is likely to cause delays in project implementation, increase costs, and create communication and coordination problems. Furthermore, since retrofitting activities can be very complex, the scarcity of professional expertise in integrated energy renovation approaches would obviously discourage clients' interest.

14.3.4 Displacement and Socioeconomic Issues

Building retrofits can trigger significant socioeconomic issues, especially within the rental housing subsector of the built environment. The "split incentive" issue—a situation where building owners bear retrofitting cost while building users enjoy related benefits like thermal comfort and reduced energy utility bill—would typically deter many property owners from engaging in renovating procedures [50]. However, in situations where owners increase rental costs following retrofits, tenants could be displaced from properties if rent surpasses their affordability threshold. In other words, low-income residents are often displaced from renovated buildings, creating a form of "green gentrification" [54]. In addition, retrofit programs are usually very disruptive to occupants, especially when the work under consideration is extensive. To the aged, disabled individuals, and families with very young children, these disturbances can be particularly significant if there are no alternative lodging arrangements available [55]. Furthermore, due to lack of communication among stakeholders, including the community, retrofits solutions are sometimes considered irrelevant because they do not conform to certain cultural practices or lifestyle requirements [26]. This misalignment between technical solutions and social context further reduces the effectiveness and acceptance of retrofitting programs in various communities.

14.3.5 Challenges in Achieving Energy Savings

After the installation of, say, a new HVAC system, building users tend to lodge complaints against the renovation works that have been carried out. When the projected benefits of engaging in retrofits do not align with the actual energy performance of the buildings, investors, building owners, and users begin to lose confidence in the program. According to their study, Ardron and Allwood [50] found that building users or owners are not always content with renovation works. This phenomenon can be referred to as the "performance gap". Lee et al. [46] pointed out that building users who do not conduct comprehensive assessments before retrofitting buildings, but choose to rely on rule-of-thumb procedures and inexperienced professionals, often encounter such problems. While engaging in energy retrofitting should ideally make buildings use energy more efficiently and incorporate renewable energy technologies, this is not always achieved in practice. Building owners who may want to cut down cost, may resort to focusing on parts of the building with little to no implication for energy efficiency, such as building aesthetics, or employ local artisans using conventional construction methods that do not optimize energy performance. Additionally, retrofitted buildings that are still depend solely on national power grids, which are usually unstable and inefficient in many regions, may experience little change in energy utility expenses during operational stages. This discourages stakeholders from committing to such programs. Likewise, the prevailing inconsistent energy pricing policies in some parts of the world also discourage building users from adopting efficiency improvements techniques like building retrofitting and refurbishing [53]. The absence of pricing structures that properly value energy savings significantly diminishes financial incentives for undertaking retrofitting.

14.3.6 Climate and Weather-Related Constraints

Building retrofits face increasingly complex challenges related to climate change conditions like heat wave, flooding, drought, and other extreme weather events. With the rise in global temperature and change in weather patterns, current retrofit techniques may become ineffective over time. Webb [40] projects that buildings, particularly traditional buildings, can be susceptible to higher risks when future climate changes make implemented retrofits potentially inappropriate. Seasonal variations limit the implementation window for certain retrofit measures, particularly in regions with harsh weather conditions. These seasonal constraints create bottlenecks in the construction sector during viable working periods and extends overall project timelines. Additionally, variations in climate complicate testing and commissioning of building systems, potentially hiding performance issues until extreme conditions reveal them. The wide range of climate variations in the African continent exacerbates the challenge of developing standardized documentation for building retrofits. Having generalized standardized approaches can impede retrofit adoptions since some buildings may require additional resilience measures specific to their local climatic conditions.

14.3.7 Supply Chain and Material Availability

The building renovation sector often faces supply chain constraints that hinder widespread implementation of decarbonization retrofits. High-performance building materials and advanced energy systems frequently experience limited availability, extended lead times, and premium pricing due to insufficient production capacity and distribution networks [56]. Manufacturing capacity for specialized retrofit components such as heat pumps, high-performance windows, advanced insulation materials remain insufficient to meet the demand required for rapid decarbonization of the building stock. Manufacturing companies would need to upscale their rate of production of building components to meet growing demand. The supply chain is also highly fragmented, with poor integration between material suppliers, equipment manufacturers, design professionals, and installation contractors [56]. This fragmentation leads to coordination problems and increased cost of transaction cost, among other issues. Geographical disparities and demand for locally available specialized materials and equipment further create variations in retrofit feasibility and cost across different regions.

It is clear that while building retrofitting and refurbishment present clear both energy and non-energy related benefits—including sustainability, cost efficiency, job creation, and higher building performance—significant barriers such as high costs, lack of technical expertise, fragmentation in supply chain, and policy gaps must be addressed for widespread adoption in Africa. A strategic approach that includes financial incentives, workforce training, enhanced collaboration, and regulatory reforms could maximize the impact of retrofitting efforts across the continent.

14.3.8 Energy Efficiency

The issue of energy consumption is crucial in decarbonizing the built environment. Energy efficiency looks at sustainable ways to manage and restrain the consumption of energy [57]. According to Alnuaimi and Natarajan [47], unnecessary energy consumption not only increases carbon emissions but further implicate electrical systems. In their opinion, reducing high energy peaks in developing countries is a step in the right direction, and it will further be encouraged by adopting climate-friendly practices. Retrofitting and refurbishment of building allows the replacement of old building systems, since they age and become inefficient over time [44]. Harvey [33] demonstrates that installing new HVAC and light systems in commercial spaces can reduce energy consumptions by 25% to 50% and 30% to 60%, respectively. Apart from new installations, retrofitted buildings benefit from the installations of renewable energy technologies like the photovoltaic solutions, biomass systems, and green roof [31, 58]. According to Abdalazeem et al. [59], integrating these into buildings can achieve a double reduction in energy demand.

14.4 Pathways to Building Retrofit and Refurbishment for Decarbonization

This section highlights three critical pathways through which building retrofit and refurbishment can contribute to the decarbonization agenda: technical pathways, regulatory and policy frameworks, and economic instruments and financial mechanisms.

14.4.1 Technical Pathway

This pathway highlights the technical strategies and approaches aimed at promoting the implementation of retrofitting and refurbishment in existing buildings to reduce their carbon footprint, improve their energy efficiency and enhance sustainability. According to the Climateworks [60], the renovation subsector requires comprehensive data on retrofitting techniques for different scenarios across varied climatic regions. Developing a robust information pool also necessitates transparency and accessibility. Building retrofitting is a complex engagement which encompasses physical, technological, and social components. Without transparency and access to databases, stakeholders will remain ignorant about the advantages of sustainable renovation practices [61]. In Sweden, for instance, von Platten et al. [62] demonstrated how Google Street View integrated with machine learning can be used in creating and improving the accuracy of building databases for energy retrofitting purposes. Sharing critical information on retrofitted building performance and its impacts subsequently shapes policy-makers' decisions as well as industrial practices.

Another factor to enhance retrofitting is having a collaborative industry [63, 64]. After creating databases and making them accessible, Buchner [65] opine that these databases can be improved by collecting and harmonizing information from various works and enhancing interoperability of software and simulation tools. Going further, the absence of an collaboration between the many sectors in the building industry worsens the case for building retrofitting [66]. Renovating existing structures, similar to constructing new ones, involve different activities which require diverse skillsets and technical personnel. However, Bell et al. [63] mention the presence of sectoral silos in the built environment supresses the advancement of practices like building retrofitting. So, in order to encourage innovation adoption, while achieving an energy-efficient industry in decades to come, it will be highly important to ensure collaboration and co-operation among stakeholders [61].

Prioritizing education and organizing workshops to improve awareness of renovation practices is also key in advancing retrofitting in sub-Saharan Africa. Education and training do not only increase awareness among professionals, but normalizes the implementation of modern construction techniques and sustainable practices. Professional bodies should encourage continuous development programs for their members, keeping them updated about technologies, new breakthroughs, and methodologies such as industrialized approaches like prefabrication and serial renovation. Funding for educational institutions and research programs is imperative, particularly for exploring diverse vernacular retrofitting strategies across different climatic regions [42, 67]. These investigations have the potential of discovering continent-specific building materials that could minimizing waste through reuse, enhance building envelopes, lighting, and HVAC systems among many others.

An emerging strategy involves the integration of Building Renovation Passports (BRPs) and the adoption of data-driven decision-making technologies, such as artificial intelligence (AI), to optimize retrofit strategies. BRPs provide property owners with personalized, long-term renovation roadmaps, offering insight into a building's lifecycle and enabling informed, phased upgrades [68]. Several countries, particularly those in the EU have already implemented BRPs. For example, France's *Passeport Efficacité Énergétique* (Energy Efficiency Passport), Germany's *Individueller Sanierungsfahrplan* (Individual Renovation Roadmap or iSFP), and Belgium's *Woningpas* (Dwelling ID). These tools assist homeowners in planning sequential retrofits while maintaining consistency with overarching energy performance goals [69].

14.4.2 Regulatory and Policy Frameworks

Mandatory building performance standards serve as potent regulatory instrument for accelerating retrofit actions. The prominent example is the EU's Energy Performance of Building Directive (EPBD), which established minimum energy performance standards (MEPS) that building are required to meet within defined deadlines [70]. The goal of designing such codes is meant to systematically raise the energy efficiency of the building stock and support broader decarbonization targets. Africa's capacity to mobilize experts and establish a centralized regulatory body dedicated to overseeing energy efficiency and building renovation efforts would represent a major step towards effective energy performance management in the built environment. Statutory bodies must review construction activities with standardized benchmarks. These codes must be performance-oriented instead of being simply prescriptive, allowing for innovation and cost-effectiveness while ensuring buildings meet emission targets. When retrofitting standards are designed to be more than prescriptive, it nullifies the notion of the "all-size-fits-all" approach, making allowance for professionals to apply distinct renovation practices to different properties as the need arises. Developing standards must not only take into account contextual factors of the entire region, but must harmonize them [71].

In addition, building owners and tenants can be mandated to submit building energy performance certifications. Recent policy developments across some African nations signal growing recognition of building decarbonization. Countries like South Africa, Morocco, Ghana, and Kenya have begun incorporating building energy efficiency into their climate action plans, though implementation remains challenging [32, 72]. To maximize the potential of building retrofitting, disclosure policies must be accompanied by robust compliance mechanisms, standardized assessments strategies, and clear communication of benefits to property owners and building occupants. This disclosure further enhances information asymmetry within the property markets by highlighting energy efficiency-related benefits to potential building buyers and tenants. The Australia's Commercial Building Disclosure program and the EU's Energy Performance Certificates (EPCs) for instance, demonstrate how transparency can influence property values and rental premiums, creating market incentives for retrofits [71, 73, 74].

14.4.3 Economic Instruments and Financial Mechanism

This pathway encompasses the strategic use of financial tools, incentives, and funding mechanisms to support and accelerate decarbonization of the built environment through retrofitting and refurbishment. In the context of the African continent, several key components of this pathway warrant critical attention, including green financing and low-interest credit schemes. Establishing green banks and dedicated financing channels is essential for driving the transition from inefficient, high-emission buildings to energy-efficient alternatives [73]. Given that many property owners in Africa struggle accessing loans or lack adequate financial support, there is an urgent need to develop dedicated financial vehicles that facilitate affordable credit acquisition for retrofitting. The high initial cost of renovation continues to deter property owners and potential investors. To address this, financial institutions should offer options such as green bonds and climate financing [75], accompanied by sufficient public education to enhance awareness of the long-term profitability and environmental benefits of retrofitting. This, in turn, implies that banking agents upgrade their knowledge and skills, particularly in areas such as green accounting and sustainable investment advisory.

Deploying economic incentives offer another critical pathway to fostering decarbonization in Africa. Economic incentives represent those instruments designed to motivate building owners and occupants to undertake retrofitting for energy efficiency. These instrument include but are not limited to energy savings performance contracting, tax exemptions, capital subsidies, grants, and subsidized loans [76]. Carbon pricing is another promising economic tool to consider in the African built environment. It creates economic signals that favour low-carbon retrofit solutions. This can be either in the form of carbon taxes or emission trading system. Pricing carbon emission can help skew property owners and building occupants to embrace sustainable practices, while enabling cost–benefit analyses in favour of energy efficiency investments [77, 78]. To close substantial funding gaps that persists in many part of the continent, it is crucial for international financial institutions, such as the Green Climate Fund and the African Development Bank's Sustainable Energy Fund for Africa, to intensify investments targeted at transforming the built environment (Barbara [65]). These institutions, along with dedicated non-governmental bodies and academic institutions must work collaboratively with governments to align funding with national energy efficiency goals.

14.5 Conclusion

Retrofitting and refurbishing Africa's existing building stock represent a critical strategy for decarbonization, energy efficiency enhancement, and climate resilience. This paper identifies three interconnected pathways that must be implemented cohesively to transform the continent's-built environment. The technical dimension emphasizes Building Renovation Passports and data-driven technologies to optimize retrofit planning and execution. Regulatory and policy framework incorporating performance-based standards and codes, and centralized oversight provide essential governance structures. Economic and financial mechanisms, including innovative funding instruments and international climate financing, address resource constraints that currently inhibit implementation. Research findings indicate that successful retrofitting initiatives in the African context require a holistic approach that integrates these dimensions while remaining sensitive to local socioeconomic conditions. Overall, this study contributes to sustainable development scholarship by demonstrating how cross-sectoral collaboration can unlock the inherent potential of retrofitting towards the reduction of emissions while delivering socioeconomic co-benefits in rapidly urbanizing African communities.

References

1. World Economic Forum (2020) The children's continent: keeping up with Africa's growth. World Economic Forum. Available at: https://www.weforum.org/stories/2020/01/the-children-s-continent/. Accessed 27 Feb 2025
2. UN Habitat (2020) World Cities Report 2020. No HS/045/20E, UN Habitat, pp 1–418
3. Miller E, Buys L (2008) Retrofitting commercial office buildings for sustainability: tenants' perspectives. J Prop Invest Financ 26(6):552–561. https://doi.org/10.1108/14635780810908398.
4. Anzagira LF, Duah D, Badu E, Simpeh EK, Amos-Abanyie S, Marful A (2022) Application of green building concepts and technologies for sustainable building development in Sub-Saharan Africa: the case of Ghana. Open House Int 47(3):408–427. https://doi.org/10.1108/OHI-02-2022-0054
5. Addy MN, Adinyira E, Dadzoe F, Opoku D (2022) The market for green buildings in Sub-Saharan Africa: experts perspective on the economic benefits in Ghana. J Constr Dev Ctries 27(1):173–188
6. Debnath R, Bardhan R, Sunikka-Blank M (2019) How does slum rehabilitation influence appliance ownership? A structural model of non-income drivers. Energy Policy 132:418–428. https://doi.org/10.1016/j.enpol.2019.06.005
7. IPCC (2022) Climate change 2022: impacts, adaptation and vulnerability. contribution of working Group II to the sixth assessment report of the intergovernmental panel on climate change. Intergovernmental Panel on Climate Change, Cambridge University Press, Cambridge, UK and New York, USA
8. Koranteng C, Simons B, Gyimah KA (2021) Potential measures towards the reduction of cooling loads of office buildings in Ghana. Front Eng Built Environ 1(2):161–172. https://doi.org/10.1108/FEBE-03-2021-0016
9. Camarasa C, Mata É, Navarro JPJ, Reyna J, Bezerra P, Angelkorte GB, Feng W et al (2022) A global comparison of building decarbonization scenarios by 2050 towards 1.5–2 °C targets. Nat Commun 13(1):3077. https://doi.org/10.1038/s41467-022-29890-5
10. Dodoo and Ayarkwa (2019) Effects of climate change for thermal comfort and energy performance of residential buildings in a sub-saharan African climate. Buildings 9:215. https://doi.org/10.3390/buildings9100215
11. Dioha MO, Abraham-Dukuma MC, Dato P (2025) Unearthing the reality of 'Zombie energy systems' in Africa's energy transition. Environ Res: Energy 2(1):013001
12. Chandler B (2022) Addressing Africa's energy defict: climate change, renewables, and gas. Mo Ibrahim Foundation
13. Santamouris M, Vasilakopoulou K (2021) Present and future energy consumption of buildings: challenges and opportunities towards decarbonisation. E-Prime 100002. https://doi.org/10.1016/j.prime.2021.100002
14. Adetooto J, Windapo A, Pomponi F, Companie F, Alade K, Mtya A (2024) Strategies to promote the acceptance of sandbag building technology for sustainable and affordable housing delivery: the South African case. J Eng, Des Technol 22(5):1505–1522. https://doi.org/10.1108/JEDT-06-2022-0290
15. Agyekum K, Amudjie J (2024) Challenges to implementing circular economy: empirical evidence among built environment firms in Ghana. Int J Constr Manag 24(3):281–297. https://doi.org/10.1080/15623599.2023.2222993

16. Ebekozien A, Aigbavboa C, Samsurijan MS (2023) Social sustainability under threat: a case of two collapsed buildings in Lagos, Nigeria. Prop Manag 41(3):431–453. https://doi.org/10.1108/PM-10-2022-0072
17. Hafez H, El-Mahdy D, Marsh ATM (2023) Barriers and enablers for scaled-up adoption of compressed earth blocks in Egypt. Build Res Inf 51(7):783–797. https://doi.org/10.1080/09613218.2023.2237133
18. Rademaekers K, Smit TAB, Artola J, Koehler J, Hemkhaus M, Ahlers J, Van Hummelen S et al (2021) Circular Economy in the Africa-EU Cooperation: Continental Report. European Commission, Luxembourg, pp 1–147
19. Le ATH, Park KS, Domingo N, Rasheed E, Mithraratne N (2021) Sustainable refurbishment for school buildings: a literature review. Int J Build Pathol Adapt 39(1):5–19
20. Lin Y, Cui C, Liu X, Mao G, Xiong J, Zhang Y (2023) Green renovation and retrofitting of old buildings: a case study of a concrete brick apartment in Chengdu. Sustainability 15(16):12409. https://doi.org/10.3390/su151612409
21. Kristl Ž, Temeljotov Salaj A, Roumboutsos A (2020) Sustainability and universal design aspects in heritage building refurbishment. Facilities 38(9/10):599–623. https://doi.org/10.1108/F-07-2018-0081
22. Mapfumo ET, Emuze F, Smallwood J, Ebekozien A (2024) Appraising challenges facing Zimbabwe's building retrofitting for energy efficiency using structural equation model approach. Int J Build Pathol Adapt 42(7):76–92. https://doi.org/10.1108/IJBPA-05-2024-0105
23. Shah Ali A, Nizam Kamaruzzaman S, Salleh H (2009) The characteristics of refurbishment projects in Malaysia. Facilities 27(1/2):56–65. https://doi.org/10.1108/02632770910923090
24. Okorafor C (2019) Retrofitting to Reduce Carbon Emissions from Existing Buildings in Bloemfontein, South Africa. PhD Thesis, Central University of Technology, Bloemfontein
25. Fahlstedt O, Temeljotov-Salaj A, Lohne J, Bohne RA (2022) Holistic assessment of carbon abatement strategies in building refurbishment literature—a scoping review. Renew Sustain Energy Rev 167:112636
26. Jagarajan, R., Abdullah Mohd Asmoni, M.N., Mohammed, A.H., Jaafar, M.N., Lee Yim Mei, J. and Baba, M. (2017) Green retrofitting—a review of current status, implementations and challenges. Renew Sustain Energy Rev 67:1360–1368. https://doi.org/10.1016/j.rser.2016.09.091
27. Sanvido VE, Riggs LS (1991) Managing retrofit projects. Technical Report No. 25. Department of Civil Engineering, University of Texas at Austin, pp 1–84
28. Che Husin S, Mohd Zaki NI, Abu Husain MK (2019) Implementing sustainability in existing building through retrofitting measures. Int J Civ Eng Technol 10:1450–1471
29. Davoudi S, Zhao P, Brooks E (2014) Retrofitting cities for low-carbon urban futures in Europe and China: Introduction to special issue. disP—Plan Rev 50(3):6–10. https://doi.org/10.1080/02513625.2014.979036
30. Juliardi RD, Misnan MS, Khalid AG, Haron L (2019) Recognizing of building components to achieve green performance for renovation and retrofitting works. IOP Conf Ser: Earth Environ Sci 353(1):012017. https://doi.org/10.1088/1755-1315/353/1/012017
31. Ohene E, Hsu S-C, Chan APC (2022) Feasibility and retrofit guidelines towards net-zero energy buildings in tropical climates: a case of Ghana. Energy Build 269:112252. https://doi.org/10.1016/j.enbuild.2022.112252
32. United Nations Environment Programme (2021) 2021 Global status report for buildings and construction: towards a zero-emission, efficient and resilient buildings and construction sector. United Nations, Nairobi, pp 1–105
33. Harvey LDD (2013) Recent advances in sustainable buildings: review of the energy and cost performance of the state-of-the-art best practices from around the world. Annu Rev Environ Resour 38(38):281–309. https://doi.org/10.1146/annurev-environ-070312-101940
34. Gavu EK, Peiser RB (2024) Embodied carbon and the nuances in office-to-residential conversions. Sustainability 16(7):2711. https://doi.org/10.3390/su16072711
35. Plörer D, Hammes S, Hauer M, van Karsbergen V, Pfluger R (2021) Control strategies for daylight and artificial lighting in office buildings—a bibliometrically assisted review. Energies 14(13):3852. https://doi.org/10.3390/en14133852

36. Regnier C, Sun K, Tianzhen H, Piette MA (2017) Quantifying the benefits of a building retrofit using an integrated system approach: a case study. Energy Build 159:332–345. https://doi.org/10.1016/j.enbuild.2017.10.090
37. Dragonetti L, Papadaki D, Mazzoli C, Monacelli A, Assimakopoulos M-N, Ferrante A (2025) Circular deep renovation versus demolition with reconstruction: Environmental and financial evaluation to support decision making in the construction sector. Energy Build 115610. https://doi.org/10.1016/j.enbuild.2025.115610
38. Al-Obaidy M, Carens H, Campain C, Giannasi E, Mori M, van Vliet M, Attia S (2025) Expert system for building disassembly potential evaluation and inspection. J Build Eng 103:112148. https://doi.org/10.1016/j.jobe.2025.112148
39. Parracho DF, Nour El-Din M, Esmaeili I, Freitas SS, Rodrigues L, Poças Martins J, Corvacho H et al (2025) Modular construction in the digital age: a systematic review on smart and sustainable innovations. Buildings 15(5):765
40. Webb AL (2017) Energy retrofits in historic and traditional buildings: a review of problems and methods. Renew Sustain Energy Rev 77:748–759. https://doi.org/10.1016/j.rser.2017.01.145
41. Organ S (2019) The opportunities and challenges of improving the condition and sustainability of a historic building at an international tourist attraction in the UK. Int J Build Pathol Adapt 38(2):329–355. https://doi.org/10.1108/IJBPA-09-2018-0076
42. Loli A, Bertolin C (2018) Towards zero-emission refurbishment of historic buildings: a literature review. Buildings 8(2):22. https://doi.org/10.3390/buildings8020022
43. Ameen A, Bahrami A, El Tayara K (2022) Energy performance evaluation of historical building. Buildings 12(10):1667. https://doi.org/10.3390/buildings12101667
44. Zheng L, Lai J (2018) Environmental and economic evaluations of building energy retrofits: case study of a commercial building. Build Environ 145:14–23. https://doi.org/10.1016/j.buildenv.2018.09.007
45. Ma Z, Cooper P, Daly D, Ledo L (2012) Existing building retrofits: Methodology and state-of-the-art. Energy Build 55:889–902. https://doi.org/10.1016/j.enbuild.2012.08.018
46. Lee SH, Hong T, Piette MA, Sawaya G, Chen Y, Taylor-Lange SC (2015) Accelerating the energy retrofit of commercial buildings using a database of energy efficiency performance. Energy 90:738–747. https://doi.org/10.1016/j.energy.2015.07.107
47. Alnuaimi AN, Natarajan S (2020) The energy cost of cold thermal discomfort in the global south. Buildings 10(5):93. https://doi.org/10.3390/buildings10050093
48. Elnaklah R, Alnuaimi A, Alotaibi BS, Topriska E, Walker I, Natarajan S (2021) Thermal comfort standards in the Middle East: current and future challenges. Build Environ 200:107899. https://doi.org/10.1016/j.buildenv.2021.107899
49. Bertoldi P, Economidou M, Palermo V, Boza-Kiss B, Todeschi V (2021) How to finance energy renovation of residential buildings: review of current and emerging financing instruments in the EU. Wiley Interdiscip Rev: Energy Environ 10(1):e384. https://doi.org/10.1002/wene.384
50. Ardron TG, Allwood JM (2025) Exploring time and cost optimization in energy retrofit programmes of traditionally constructed, owner-occupied UK dwellings. Build Res Info, pp 1–19. https://doi.org/10.1080/09613218.2024.2446949
51. Fowlie M, Greenstone M, Wolfram C (2018) Do energy efficiency investments deliver? Evidence from the weatherization assistance program. Quart J Econ 133(3):1597–1644
52. Ürge-Vorsatz D, Khosla R, Bernhardt R, Chan YC, Vérez D, Hu S, Cabeza LF (2020) Advances toward a net-zero global building sector. Annu Rev Environ Resour 45(1):227–269. https://doi.org/10.1146/annurev-environ-012420-045843
53. Stojanovski O, Leslie G, Wolak F, Wong JEH, Thurber M (2018) Promoting energy efficiency in emerging economies through consumer education: results from a field experiment in Mexico. SSRN
54. Quinton J, Nesbitt L, Sax D (2022) How well do we know green gentrification? A systematic review of the methods. Prog Hum Geogr 46(4):960–987. https://doi.org/10.1177/03091325221104478
55. Khairi M, Jaapar A, Yahya Z (2017) The application, benefits and challenges of retrofitting the existing buildings. In: IOP conference series: materials science and engineering, vol 271. IOP Publishing, p 012030

56. Brocklehurst F, Morgan E, Greer K, Wade J, Killip G (2021) Domestic retrofit supply chain initiatives and business innovations: an international review. Build Cities 2(1). https://doi.org/10.5334/bc.95
57. Martínez-Molina A, Tort-Ausina I, Cho S, Vivancos J-L (2016) Energy efficiency and thermal comfort in historic buildings: a review. Renew Sustain Energy Rev 61:70–85. https://doi.org/10.1016/j.rser.2016.03.018
58. Alsaied HMH, El Madawy AET, El Sayyad NAEH (2024) Life cycle impact assessment methodology for building envelope retrofits using photovoltaic systems in Egypt. Civ Eng Arch 12(2):850–865. https://doi.org/10.13189/cea.2024.120214
59. Abdalazeem ME, Hassan H, Asawa T, Mahmoud H (2024) Enhancing energy efficiency in hot climate buildings through integrated photovoltaic panels and green roofs: an experimental study. Sol Energy 270:112419. https://doi.org/10.1016/j.solener.2024.112419
60. Climateworks Centre (2022) Renovation Pathways. Climateworks Centre. Available at: https://www.climateworkscentre.org/project/renovation-pathways/. Accessed 2 Apr 2025
61. UKGBC (2024) Building the case for net zero: Retrofitting office building. UK Green Building Council, pp 1–56
62. von Platten J, Sandels C, Jörgensson K, Karlsson V, Mangold M, Mjörnell K (2020) Using machine learning to enrich building databases—methods for tailored energy retrofits. Energies 13(10):2574. https://doi.org/10.3390/en13102574
63. Bell S, Bush J, Lara-Hamilton E (2023) Retrofitting cities: challenges and opportunities in Australia. Melbourne Centre for Cities, pp 1–19
64. Zhai J, LeClaire N, Bendewald M (2011) Deep energy retrofit of commercial buildings: a key pathway toward low-carbon cities. Carbon Manage 2(4):425–430. https://doi.org/10.4155/cmt.11.35
65. Buchner B, Naran B, Padmanabhi R, Stout S, Strinati C, Wignarajah D, Miao G et al (2023) Global landscape of climate finance 2023. Climate Policy Initiative (CPI), pp 1–56
66. Rauland V, Newman P (2015) Making it work. In: Decarbonising Cities. Springer International Publishing, Cham, pp 205–233. https://doi.org/10.1007/978-3-319-15506-7_13
67. Chung-Camargo K, González J, Chen Austin M, Carpino C, Mora D, Arcuri N (2024) Advances in retrofitting strategies for energy efficiency in tropical climates: a systematic review and analysis. Buildings 14(6):1633. https://doi.org/10.3390/buildings14061633
68. Global Alliance for Buildings and Construction and UNEP (2021) The building passport: a tool for capturing whole life data in construction and real estate—practical guidelines. United Nations Environment Programme
69. Buchholz, M. and Lützkendorf, T. (2023), "European building passports: developments, challenges and future roles", *Buildings & Cities*, Vol. 4 No. 1, https://doi.org/10.5334/bc.355.
70. Joint Research Centre (European Commission), Maduta C, Tsemekidi-Tzeinaraki S, Castellazzi L, Dagostino D, Melica G, Paci D et al (2025) Updates on the energy performance of buildings directive implementation in the EU member states. Publications Office of the European Union
71. World Green Building Council (2023) Energy Performance Certificates (EPC): Implementation of the EU Taxonomy in the Built Environment. World Green Building Council and European Public Real Estate Association, pp 1–9
72. Ministry of Works and Housing (2024) Climate Action Roadmaps for Buildings and Construction, Ghana. Ministry of Works and Housing, UN Environment Programme, Global Alliance for Building and Construction, UN-Habitat, Ghana, pp 1–84
73. Green Building Council of Australia (2025) Commercial building disclosure scheme. Available at: https://new.gbca.org.au/archive/commercial-building-disclosure-scheme/. Accessed 4 Apr 2025
74. Sesana MM, Salvalai G, Della Valle N, Melica G, Bertoldi P (2024) Towards harmonising energy performance certificate indicators in Europe. J Build Eng 95:110323. https://doi.org/10.1016/j.jobe.2024.110323
75. MacRae H, Tozer L (2024) The use of green bonds in financing energy retrofits in buildings. Energy Res Soc Sci 112:103500. https://doi.org/10.1016/j.erss.2024.103500

76. Allouhi A, El Fouih Y, Kousksou T, Jamil A, Zeraouli Y, Mourad Y (2015) Energy consumption and efficiency in buildings: current status and future trends. J Clean Prod 109:118–130. https://doi.org/10.1016/j.jclepro.2015.05.139
77. Boyce JK (2018) Carbon pricing: effectiveness and equity. Ecol Econ 150:52–61. https://doi.org/10.1016/j.ecolecon.2018.03.030
78. Klenert D, Mattauch L, Combet E, Edenhofer O, Hepburn C, Rafaty R, Stern N (2018) Making carbon pricing work for citizens. Nat Clim Chang 8(8):669–677. https://doi.org/10.1038/s41558-018-0201-2
79. IEA (2023) Africa energy outlook 2022. Int Energy Agency 1–250

Chapter 15
Urbanization and Decarbonization

15.1 Introduction

Cities have become the heart of human civilization, with more than half of the world population now living in urban area—a figure projected to rise to nearly 70% by 2050 [1]. The significance of cities extends beyond their sheer population size (whose current estimation is about 50% of the world population); they are hubs for opportunities, offering greater access to jobs, education, healthcare, and connectivity than rural areas [2]. According to the International Energy Agency, urban centres serve as engines or systems that drive global economies, contributing to about 80% of global Gross Domestic Product (GDP) [1]. This notwithstanding, challenges encompassing inequality, environmental strains, and the need for sustainable development among others present a critical concern in shaping a sustainable future for all. Notably, the periods between 2015 and 2020 witnessed a geometric surge in urban populations, with a significant majority—over 90%—of this growth concentrated in Asia and Africa [2, 3] (Fig. 15.1).

Today, Asia stands at the forefront of this urbanization, boasting megacities like Tokyo, Delhi, Mumbai, and Shanghai, each accommodating tens of millions and still growing at an unprecedented rate. While many of these regions are actively implementing measures to manage population growth and its associated challenges, Africa presents a distinct scenario. As the second fastest-urbanizing region [4, 5], it grapples with substantial infrastructure gaps, particularly in critical areas like energy, housing, and other socio-technical domains. As Wang et al. [6] assert, urbanization is "as an inevitable process of human progress", and as such, the African continent's pursuit of its economic potential is a natural consequence. Nevertheless, it is imperative to recognize that such advancement is not dissociated from its potential impact on sustainable development. This fundamental truth positions the continent as a critical global actor and a focus point for global attention in the foreseeable future. As the choices and activities hold the power to significantly shape the world's environmental and socio-economic future, for better or worse.

M. N. Addy et al., *Decarbonization of Building Energy in Developing Countries*, Green Energy and Technology, https://doi.org/10.1007/978-3-032-09113-0_15

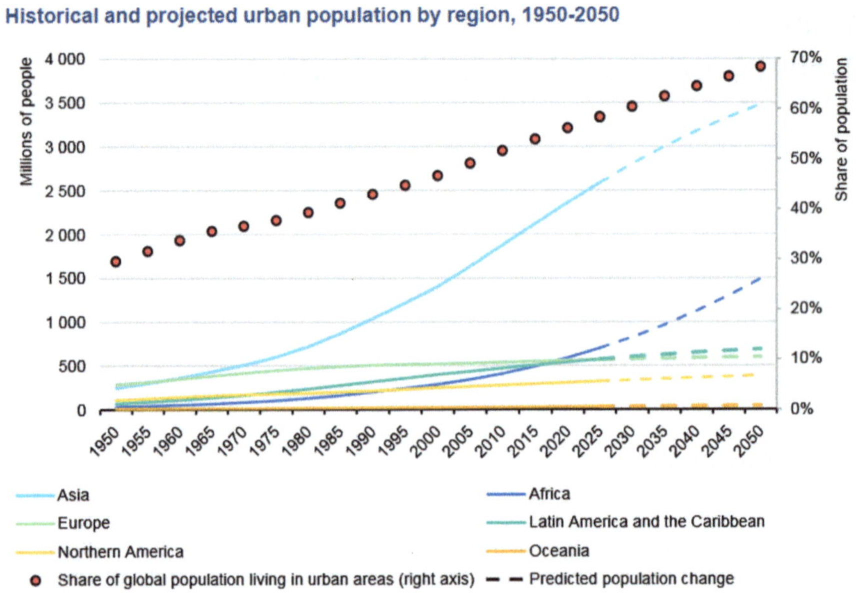

Fig. 15.1 Continental urban population projectio. *Source* Adopted from the International Energy Agency [1]

Projections indicate that Africa's population is set to double by 2050 and triple by 2100 [7]. Compounding this demographic expansion, a recent report from the United Nations Economic Commission for Africa [8] anticipates that about 50% of Africa's population will reside in urban centres by 2050. This presents a significant implication, considering that cities consume over 60% of final energy and contribute more than 75% of global carbon emissions [9], rendering them as critical venues in the global fight against climate change.

Standing at a pivotal moment in its development trajectory, Africa faces an urgent imperative for building decarbonization and climate action. Concerns with respect to climate change has become a predominant theme in almost every discourse, dictating how the future trajectory of human civilization and social production must proceed [10]. This unprecedented rate of urbanization calls for the urgent need for building decarbonization and climate actions, necessitating the necessary transformative measures across all sectors [11, 12].

The sheer magnitude of Africa's impending expansion cannot be overstated. Over the next three decades, the population growth will be accompanied by a surge in demand for infrastructure amenities, including housing units, communication networks, and industrial facilities [10]. The residential shift—from rural to urban areas already underway and set to continue—will significantly reshape the physical, social, and economic landscapes of the region [4]. This presents both unexampled opportunities and formidable challenges for sustainable development, with particular attention on the built environment. As a crucial player in satisfying these burgeoning needs while simultaneously striving to attain global climate objectives, the built environment's role is indispensable.

Currently, the built environment is under intense scrutiny due to its conventional practices, which are often considered as exploitative and environmentally destructive [13, 14]. The global construction sector is a most resource-intensive sector, consuming over 50% of global resources like aggregates, trees, and water among others in large quantities [15],Circle [16]. Despite this, its efficiency in resource utilization and management remains a significant concern, as highlighted by [17]. Whether through the construction of new buildings or the retrofitting of existing ones, this sector is responsible for about 45% of final energy consumption [13] and contributes about 34% of global carbon emission [8]. Furthermore, the ineffective handling of construction and demolition (C&D) waste contributes to environmental degradation, leading to excessive landfilling and the leaching of harming substances into soil and water systems [18, 19]. Figure 15.2 illustrates the built environment's dependence on raw, processed, or semi-processed materials, and how factors throughout a project's lifecycle can compromise or advance the principles of sustainability.

Although the concepts of urbanization and decarbonization are seemingly distinct, an integrated framework is essential, since ignoring them would further exacerbate climatic conditions. The benefits of addressing urbanization and decarbonization concurrently extend beyond emission reduction to encompass broader socio-economic and environmental gains. Consequently, it lies in the interest of the built environment to effectively address this situation. Beyond carbon emissions reduction, the construction of energy-efficient buildings enhances indoor environmental

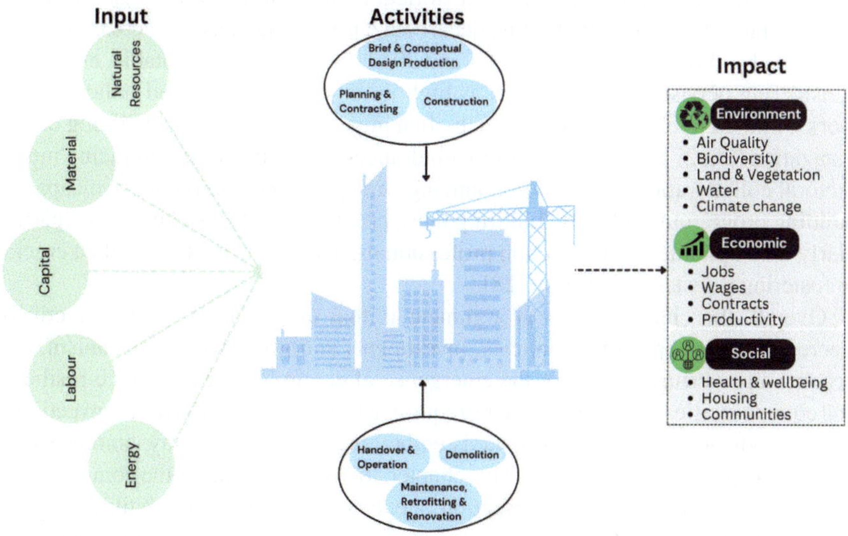

Fig. 15.2 Demands, activities, and impacts of the built environment. *Source* Adapted from [15]

quality, strengthens building resilience, and reduces energy poverty [11, 20]. Moreover, integrated urban planning, which promotes compact, walkable neighbourhood, and robust public transportation, directly contributes to improved urban liveability and emissions [9, 21].

This notwithstanding, the intricate nexus of urbanization and decarbonization in Africa presents unique complexities, stemming from the continent's distinctive urban growth patterns. A juxtaposition of Africa's urbanization against that seen in developed economies historically reveals a crucial distinction: African urbanization predominantly emerges from informal and peri-urban areas, wherein conventional planning and regulatory frameworks often possess limited capacity to effectively operate [5, 9, 22]. This situation necessitates innovative approaches to building decarbonization that can function efficiently within informal settings while simultaneously addressing critical issues of affordability and social equity [7, 50]. The decisions made today regarding urban planning, new construction projects, retrofitting and renovating existing structures, building standards, infrastructure systems, and policies frameworks will determine whether Africa's urbanization becomes a catalyst either for sustainable development or worsening prevailing environmental and social challenges [2].

This represents a significant socio-technical challenge that demands the creation of inclusive pathways to low-carbon development, ensuring that no segment of the population is left behind. In view of this, recent development witnessed across the region are indicative of a gradual embrace of innovative solutions in the pursuit of decarbonization. These initiatives include the African Carbon Market Initiative [23], the Africa Green Hydrogen Alliance [24], and the Africa Climate Risk Insurance Facility for Adaptation [25]. Furthermore, a report by the Climate Policy Initiative [4] revealed that despite the region's historical inability to meet the scale of investment required for thorough building decarbonization, it managed to increase climate financing by about 48% between 2021 and 2022. This progress is notable, especially considering persistent barriers such as high initial costs, fragmented policy frameworks, limited public awareness, which often impede the progress of building decarbonization [26, 27]. Overcoming these hindrances will necessitate embracing multisectoral collaboration, innovation financing strategies, and comprehensive capacity building programmes [1, 22]. Tailoring these programs for key stakeholders, particularly urban planners, construction professionals, and policy makers, will be crucial in fostering a sustainable future [28].

Overall, the chapter's central argument is that urbanization, when effectively steered through appropriate policies, investments and governance mechanism, can become a powerful catalyst for the emergence of energy-efficient, climate-resilient, and equitable cities across Africa. This requires a fundamental shift from convectional "take-use-dispose" approaches towards embracing circular economy, among others sustainable practices [17, 27, 29]. This chapter, therefore, examines the complex relationship emerging from the intersection between urbanization and building decarbonization in Africa. Drawing from recent case studies, policy developments, and emerging best practices across the continent, it explores the pathways, challenges, and opportunities for achieving scalable, inclusive, and affordable low-carbon building development.

15.2 Urbanization Through the Lens of Time

Urbanization, at its core, describes the transformative demographic and spatial shift from rural to urban areas. This phenomenon depicts both an increase in the proportion of a population residing in urban settlements and the consequential physical expansion and socio-economic metamorphosis of cities and towns [30]. While urbanization inherently reflects population redistribution and often growth, the concept itself is not exclusively defined by population growth. Instead, it can be explained as transitions in industrialization and economic development. [31], for instance, posits urbanization as a natural consequence of economic progress, particularly when societies transition beyond agricultural dependency of economic venture towards modern industry. Furthermore, the term "urbanization" can delineate the outward expansion of urban centres and/or the formation of novel settlements. As [32] articulate, urbanization is a phenomenon resulting from the outward growth of existing cities and the creation of new urban areas, especially when former villages expand or evolve to satisfy national requirements for urban status. Therefore, the definition of urbanization can be summarized in terms of population size, settlement density or spatial expansion, and industrialization [33].

Historically, the earliest forms of urbanization emerged with the Neolithic Revolution, around 10,000 BCE. These nascent urban areas, characterized by monumental architecture, centralized governance, and complex economic systems, flourished notably in Mesopotamia and the Egyptian regions around the Nile [34, 35]. These civilizations flourished due to their dependency and advancements in horticultural and pastoral activities [36, 37], thereby establishing the city as the centre of power, trade, and cultural exchange.

The period between the fifth and fifteenth centuries constitutes the Classical and Middle Age eras. The classical era witnessed the rise of massive urban networks in the European continent, particularly during the Roman and Greek empires—in cities such as Rome and Alexandria serving as the administrative capitals and unprecedented infrastructure development [38, 39]. The Middle Age on the other hand, saw prominent urban growth among Islamic regions and in certain parts of Asia, including cities like Baghdad and Cordoba. [40] assert that Islamization fundamentally sponsored the expansion of cities, fostering agricultural innovation and trade, and opening land and soil routes to numerous regions.

The dawn of the eighteenth century marked the onset of the Industrialized Revolution, characterized by the invention of complex machinery such as steam engines, lightbulbs, and telephones [41, 42]. This era instigated a shift in the socio-economic structure of societies. The widespread demand for labour to operate and maintain machinery in the emerging factories prompted a significant migration of people away from the agrarian lifestyle towards urban centres. Subsequently, this industrial transformation has evolved into a global phenomenon, basically reconfiguring many economies from agriculture-based systems to those dominated by manufacturing and services provision [42].

The pursuit of better employment opportunities and improved infrastructure constitutes some of the factors that influence individual's decisions to migrate from their "undeveloped" villages to cities [43]. Subsequently, the population growth resulting from this migration is further heighten by high birth rates and low death rate within urban settings [44, 45]. The spatial-cycle model, developed by Van den [46], illustrates demographic growth in urban populations in four distinct stages:

- **Urbanization:** This initial phase depicts a continued rural–urban population rift, currently more pronounced in developing nations, driving growth in cities. During this stage, the unit of population change typically increases, consistent with Van den [46] model.
- **Suburbanization:** This stage illustrates the spilling effects of expansion experienced in urban regions into surrounding areas, leading to the formation of suburban areas.
- **Disurbanization/ Counter-Urbanization:** This stage is characterized by a decline in population within central cities, as residents migrate towards rural areas.
- **Re-urbanization:** This final stage depicts the regeneration and revitalization or cities, attracting new residents and businesses back into urban regions.

Contemporary urbanization is a global phenomenon, with characteristics significantly varying across different regions in the world. While countries in the Global North may be experiencing re-urbanization, economies in the Global South, particularly in Asia and Africa, are undergoing unprecedented rates of urban expansion [33, 42]. Viewing urbanization through the lens of time reveals a complex, multi-stage progression, far from a singular, monolithic event. Instead, it connotes a dynamic situation that is characterized by different historical phases, with each driven by distinct socio-economic, technological, and cultural forces. This quintessential human phenomenon is not merely a modern occurrence but a profound historical transformation of human societies on a global scale [47].

15.3 Urbanization in Africa

Africa is currently undergoing an unprecedented urban expansion as a result of rapid growth in population. Projections show the African continent will host a population of about 1.4 billion by 2050 (see Figs. 15.3 and 15.4) [48]. While historically the most rural continent [49], Africa has become the fastest-urbanizing region in the world, with cities growing at an average rate of 3.5% per year [50]. Presently, Egypt's Cairo, the Democratic Republic of the Congo's Kinshasa, Nigeria's Lagos, Ivory Coast's Abidjan, and Tanzania's Dar es Salaam are among the five most populous cities in Africa. [49] mention that out of a sample of 100 urban dwellers in present times, are

13 people who are probably Africans. He added that, within a period of three decades, the figure will change to about 22. With over 60 cities characterized by more than 1 million population, cities in Africa has been predicted to also increase to about 159 by 2050 [50]. Individually, urbanization in each of the 54 states vary considerably. For instance, Gabon, Libya and Djibouti in 2023 recorded over 90%, 82%, and 79% of their population dwelling in urban centres respectively [51]. Meanwhile, Malawi, Rwanda, Niger, and Burundi record about 20 percent of their population dwelling in urban areas [51]. Despite this development, Bujumbura in Burundi for instance, is set to witness a geometric increase—of about 123%—in urban–rural drift by 2035.

[52] highlight the uneven nature of urbanization in individual African countries and sub-regions. This is as a consequence of the distinct prevailing situations, challenges and opportunities, and different intervention mechanisms as touching issues of urbanization. The Northern sub-region of Africa, made of Algeria, Egypt, Libya, Morocco, Tunisia among others, generally exhibit higher levels of urbanization relative to those in the eastern, central, and western parts of Africa [49]. The level of urbanization in this region of Africa is about 51% [52, 53], with projections indicating that by 2043, this percentage would have increased to 61%, with about 270.5 million people [54].

Ghana, Cote d'Ivoire, the Gambia, Togo, Burkina Faso and Nigeria, are some countries in Western Africa. Presently, it is considered as the second fastest urbanizing region characterized by an annual growth rate of 3.7% after Central Africa [48]. According to [49], urbanization in the West is about 48%, and is projected to increase drastically from this percentage in the next decade. Baeumler et al. [30] named Lagos as one of the four mega-cities that will be in Africa by 2030 to host

Fig. 15.3 Urban growth in African sub regions. *Source* [49]

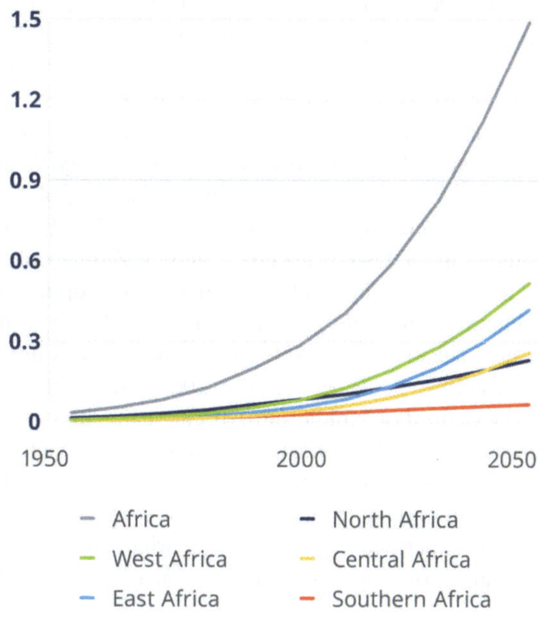

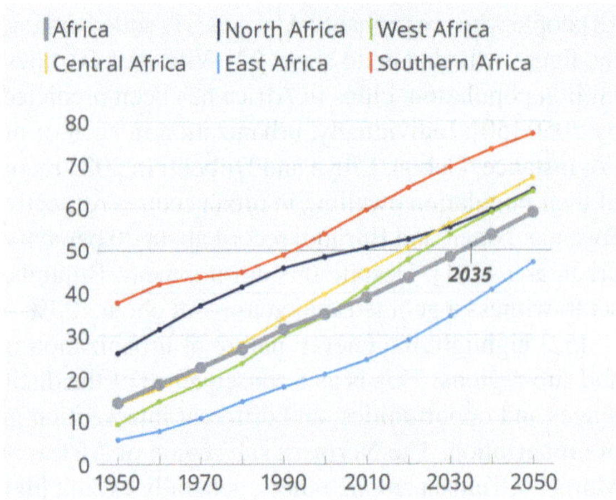

Fig. 15.4 Percentage of African urban population level. *Source* [49]

more than 10 million people; taking Cairo's place in the process [55]. This prospect makes the region a hub for innovation and growth if managed well. [54] asserts that the region's inevitable population explosion will enlarge its economy, overtaking Northern Africa's by 2029.

The Eastern sub-region of Africa, made of Uganda, Kenya, Tanzania, Rwanda, and Burundi, has witnessed the least rate of urbanization among the other sub-regions. Statistics from [30, 59, 113] suggest that urbanization level in this sub-region ranges from 23 to 29%. In spite of this, the urban population within this region is projected to be about five times more than that of its population of 2010 in the next decade [56]. Between 2020 and 2025, reports indicate the Eastern sub-region of Africa will record an average urban population growth rate of 4.32% per year, vis a vis the 1.84% from the Southern sub-region within the same year span [57, 58].

Southern Africa, commonly referred to as the Southern African Development Community (SADC), is similar to the Northern sub-region, represents one of the most urbanized regions on the continent. Notably, the urban development level in the sub-region, led mainly by South Africa and Angola in recent years, ranges between 60 and 65 percent of the population [53, 58]. Despite increasing records of city residents in Malawi, Tanzania, Zambia, Zimbabwe, among others—in the region due to economic shift and infrastructure development—projections reveal a gradual deceleration in South Africa's urban population in the coming years [48, 53]. In the next decade (i.e. 2020 to 2030), the SADC's urban population growth is set to decrease from 3.8% per year (i.e. 2000 and 2020) to 3.7% per year (UN [59]).

The Central or Middle sub-African region houses countries including Cameroon, Central African Republic, Gabon, and others. Like Eastern sub-region of Africa, this region represents one of the least urbanized zones on the continent, with urban population level ranging from about 40 to 48% [52, 53, 60]. Yet, projections show that the region is experiencing an increasing urban growth rate. As a consequence, the level of urbanization is set to increase to about 66% by 2050 [47, 48]. Democratic Republic of Congo's Kinshasa is projected to be Africa's second-largest city after Nigeria's Lagos by 2035. Furthermore, other cities like Douala and Yaoundé in Cameroon will also be transformed into major cities, characterized by huge population sizes.

15.4 Characteristics of Urbanization in Africa

Africa's rapid urbanization is occurring at a historically unprecedented. [61] notes the widespread growth in urban population as "unpredictable and uncontrollable". Projections reveal that within the next two decades, an additional billion people will be added to the continent's population [50]. Studies indicate that a significant part of this population expansion will be in urban centres, placing immense pressure on governance and amenities [11, 87].

Unlike the urbanization seen in the Global North, where urban centres' expansion was in tune with industrial development [61, 62], Africa's urbanization is treading a different trajectory. This is based on several key distinctions: (1) a significant portion of the labour force remaining in the agricultural sector; (2) the absence of major manufacturing industries to drive economic activities; and (3) the predominant informal nature of the urban economy [114]. As [63] explained, meeting food requirements is an imperative before a workforce can shift towards industrial works. The Africa context, therefore, challenges the conventional notion that urbanization is primarily influenced by agricultural push and industrial pull factors [33, 63]. What truly differentiates Africa's urbanization journey from that of the Global North is the intricate interplay of forces including natural increase, evolving economic opportunities, infrastructure and services availability, and climate change. This complex dynamics, driving urbanization within African cities, gives rise to a set of distinct characteristics that delineate the influx of people into urban centres.

15.4.1 Rapid Growth and Uneven Urban Expansion

Africa's urbanization is characterized by unprecedented speed and scale. The continent's urban population is set to grow from 376 million people in 2015 to about 1.4 billion people by 2050 [30, 34]. According to the [57], more than 60% of Africa's population will be urban centres dwellers by 2050. Similarly, another report stated that the projected population growth will be massive to the extent that two-thirds of Africans will be living in cities [48]. This growth is as a result of both natural population increase and rural–urban migration. [64] consider the inevitable urban population surge on the continent as a symbol of modernity and advancement of the society and the economy. This implies that the ongoing spatial expansion is considered as a fundamental driver and an indispensable functionary in Africa's pursuit of socio-economic transformation and modernity. However, there is a widespread of regional variations in this development. While urban population continues to expand, this growth often results in a disproportionate concentration of people and activities within urban areas, leading to footprints and population densities within cities [53]. This situation of unequal concentration of population and economic activities within only a few primate African sub-regions and cities can be referred to as spatial inequalities or uneven urban expansion. A glance at what is happening in the various sub-regions, several factors account for the continent's unequal development and urban footprint. For instance, North Africa's industrialization-led economy, South Africa's post-apartheid spatial policies, and Central Africa's weak state capacity as a result of war and conflicts either deepen or ameliorate the spatial inequalities in the continent [53, 57, 65].

15.4.2 Dominance of Primate Cities

Primate cities' dominance represents one of the most pronounced characteristics of urban centres in the Global South, particularly in Africa [66]. It reflects a spatial concentration pattern where a city is described as being at least twice as large as the next largest city and more than twice as important, pertaining to economic activities, population, and administrative stronghold [67], effectively meaning it outranks the next city [66]. Capital cities in African countries like Egypt's Cairo, Nigeria's Lagos, Ethiopia's Addis Ababa exemplify this primacy due to their extreme concentration of people, resources, and power. This pattern is often traced to the colonial and post-colonial development trajectory, which concentrated infrastructure and administrative functions in single city [55]. Nigeria's Lagos city, for instance, projected to host about 18 million residents by 2025, dwarfs other cities like Kano and stands as the largest urban agglomeration on the African continent. Similarly, Cairo dominates settlement in Egypt, just as Kinshasa and Dakar dominate other cities in Congo and Senegal respectively. The prevalence of urban primacy across the continent of Africa is particularly striking, as the region hosts about 11 out of the 31 identified primate cities globally. Over the years, the concept of primacy has evolved, since some dominant cities have grown to unprecedented scale, giving way to terminology like "megacity".

15.4.3 Proliferation of Informal Settlement

As African cities experience rapid growth and urbanization, and in the absence of structure policies and infrastructure, the most visible manifestation of this expansion in population is seen in the proliferation of informal settlements. Statistics show that more than 50% of Africa's urban population live in informal settlement, characterized by insecure tenure, poor housing quality, lack of basic services like water, sanitation, and electricity [50]. In these structures, residents tend to rely on expensive, inefficient and polluting fuels like charcoal and kerosene. The unplanned nature of such structures leads to irregular communication networks, single-layered buildings with limited functions, and neglect of ecological considerations, which impede sustainable urban development [68]. [69] observed four variants of informal settlements across the continent namely:

1. Former peri-urban or traditionally held lands that have transitioned into urban neighbourhoods as cities have expanded;
2. Residential developments initially intended for regulated occupancy that evolved into overcrowded zones through incremental informal modifications and unmonitored tenancy;
3. Residential enclaves formed on state or privately held land acquired through irregular means, frequently outside the bounds of statutory planning frameworks;
4. Temporarily authorized settlements by government bodies that have evolved into long-standing communities that have evolved into long-standing communities in the absence of formal recognition, while informal market actors continue to fill critical service and land provision gaps left by inadequate state intervention.

15.4.4 Youthful Urban Population

Africa has the world's youngest population, and this youth bulge is increasingly urban [70, 71]. A significant portion of Africa's urban development, mainly in countries like Rwanda, Ethiopia, among others are due to internal dynamics within the cities, where natural increase defines population growth. In tandem, some developed cities tend to attract young people seeking better education, jobs, including other essential services creating human mobility between rural and urban centres [72]. According to [73], the youthful demography is mainly driven by the former in lieu of the later. This demographic surge requires to be proactively managed and equipped with knowledge and skills in order to adequately harness its potential in the fight against unsustainable developmental practices [74]. Meanwhile, the failure to attend to the needs of such a formidable force has and will continue to impose immense pressure on housing, services, and employment within the continent.

15.4.5 Climate Vulnerability

Although the African continent contributes minimally to global emissions (i.e. less than 4%), the continent remains the most vulnerable to its implications, particularly climate change [75]. The region has a significant part of economies benefitting from the agrarian work structure, which heavily relies on climate-sensitive sectors like rain-fed farming. Therefore, exposure to vulnerabilities manifests in different but significant ways, for instance, sea-level rise threatens in coastal cities, increased flooding affects informal settlements, droughts strain water supplies for hydropower and cooling, and rising temperatures increase the need for cooling systems [55].

This inherent situation has made development in Africa a major concern to all stakeholders. The existing infrastructure, which lack energy efficiency principles, adopting foreign building concepts, reliance on mechanical systems, together with political, financial, market, and regulatory shortfalls in the continent worsen expectations, especially in the advent of massive population increase.

15.4.6 Weak Urban Planning and Governance

Weak urban planning and governance characterize African urbanization, leading to challenges such as poor infrastructure, housing deficits, and environmental degradation [76]. [72] maintain urban planning is an integral concern to adopting sustainable practices, especially in such times where urbanization is projected to significantly increase in the region. Many African cities struggle with fragmented governance structures, limited financial resources, lack of technical capacity, corruption, and outdated planning frameworks (Burdett and [77, 78]). This makes coordinated, long-term planning for infrastructure, land use, along with the provision of crucial services like water, sanitation, and energy systems, extremely difficult.

15.4.7 Infrastructure Deficits and Service Gaps

The African continent exhibits a substantial infrastructure gap in fundamental urban infrastructure, encompassing transportation, water, sanitation, solid waste management, and energy systems. This is exemplified by approximately 600 million people lacking access to reliable energy [79]. As [80] emphasize, existing infrastructure is already inadequate for current demands, much less satisfy future needs. Rapid urban growth in cities such as Freetown and Kampala consistently outpaces formal urban planning and infrastructure development, overwhelming existing structures and hindering the development of new ones. Additionally, deficient maintenance practice significantly exacerbate infrastructure declining services and quality [81, 82]. These persistent gap in infrastructure and services has not only hindered economic progress, but has also implicated socio-economic inequalities, particularly among vulnerable demographic groups.

15.4.8 Limited Industrialization

While urbanization often precedes industrialization historically, Africa's path has been considered different. African's urbanization is characterized by limited industrialization. Despite high urban population growth, many cities maintain high agricultural workforce, without a parallel rise in formal industrial manufacturing jobs.

This disconnect manifests critically in urban labour markets: the informal workforce constitutes the largest percentage of labour across much of the continent [114], reflecting the underdeveloped state of many African cities [55]. This leads economies to rely heavily on informal sectors and non-farming jobs [72]. Consequently, the limited presence of industries has made many countries import dependent, as domestic industries fail to meet local demand for goods.

15.5 Implications for Building Energy

Building energy sustainability is inextricably linked to three interconnected pillars: economic, environmental, and social structures. African urban centres play a pivotal role in shaping economic development, consequently, influencing the resilience of these pillars [74]. While African cities generate approximately 50% of the continent's gross domestic product (GDP) [83], they are also responsible for about 75% of energy-related emission [22]. Therefore, in pursuit of the UN Sustainable Development Goals (SDGs), particularly SDG 11 (Sustainable Cities and Communities), it is imperative to inculcate energy-efficient strategies into building construction. In addition, Africa's rapid and uneven urbanization trend profoundly impacts the built environment's energy landscape, necessitating the presence of strategic energy-efficient policies for the built environment.

Concurrent surges in demand for residential, commercial, and industrial building stock is likely to amplify building-related energy consumption. A crucial challenge emerges: the current energy systems which is lacking in capacity to both meet the growing demand for new infrastructure and drive substantial transformation needed for decarbonization [62, 73]. Consequently, the embodied energy of new construction and the operational energy required for appliances, lighting, and cooling systems particularly in residential and industrial buildings mandate a dramatic expansion of energy supply, which will significantly escalate the carbon footprint of buildings in the region.

In furtherance, in the bid to modernize the outlook of existing structures while constructing new ones in cities, many building owners have skewed towards adopting western building designs, which are usually energy-intensive, demanding high energy since energy-efficiency considerations are less prioritized [84]. A lack of a strategic intervention in building designs, materials, and energy systems places Africa's growing urban areas at risk of being carbon-intensive, undermining efforts towards climate change mitigation and SDGs [62].

The heightened climatic vulnerability of African cities mandates building designs that are both energy-efficient and climate-resilient [83]. A report from the African Development Bank identifies all the 100 fastest growing African cities as extremely vulnerable to climate change [83]. As a tropical region, the impact of climate change will significantly impact anthropogenic activities, as both present and future weather conditions will worsen [85]. This vulnerability poses a direct threat to building energy security and sustainability. Elevated temperature, intensified heatwaves, and urban heat island effects significantly increase energy demands for cooling rooms [86, 87]. Beyond the fundamental challenges with energy accessibility and reliable system coverage, extreme hydro meteorological conditions including flooding and severe storms frequently disrupt existing power supply infrastructure, compelling energy users to depend on inefficient backup generation systems [79].

Additionally, the proliferation of informal settlements, a situation occurring due to the absence of structured infrastructure systems, oftentimes lead to access restriction to roads networks, portable water and other essential services including energy (or grid connection). There are about 600 million people who lack access to reliable electricity in the region [88], who tend to rely on alternative fuel sources like firewood and biomass for their anthropogenic activities. The alternatives people rely on are mostly inefficient, costly and environmentally degrading sources of energy [69, 89]. The consequences of this trend leads to deforestation, creates low indoor environmental quality—which impacts productivity and comfort, and endangers the lives of building users—and so forth [69].

Africa's burgeoning youthful population represents an enormous potential workforce, yet they often lack access to comprehensive training and quality education. The region's import-driven economies have resulted in fewer manufacturing industries, limiting job opportunities for this demography [9]. As industrial expansion inevitably occurs, it will lead to a concurrent surge in demand for residential and commercial structures—a trend already contributing to informal settlements. This situation creates an opportunity for stakeholders including governments to investments massively into infrastructure and buildings, which will subsequently increase both embodied and operational energies. Ultimately, this creates a vicious cycle of population growth, rural–urban migration, escalating infrastructure demands, and strained energy systems, all of which are likely to exacerbate climatic conditions through increased greenhouse gas emissions.

15.6 Slum Developments

The need for infrastructure to accommodate the growing population in Africa's urban regions continues to remain high, with affordable housing and communication networks being the priority [90]. Houses are human-made structures put up to provide comfort, shade building users from the elements of the weather while keeping them healthy [91]. Apart from being considered as man-made environment, buildings play an enormous role in the lives of people, impacting their health and well-being, the surrounding environment and even resources. Building, as an investment, represents a link that influences the economic development, as well as contribute to the formation of fixed capital and employment [92]. To this end, it encapsulates a formidable instrument against poverty, particularly in developing countries [51]. Due to this, the expectation is to have housing units at low cost, creating an equitable environment for everyone to acquire one. However, the inability to address the demand–supply gap—as a result of high demand, rising pricing, and lack of interest in high-rise buildings, limited technologies, and the lack of data—for affordable housing units tend to influence homelessness and urban slum developments [96, 128, 129], be it in a developed or developing country. Currently, approximately 1.6 billion people are compelled to live in such conditions [57].

[92] mention that housing deficit is able to promote settlements that are characterized by an environment where crime and diseases are rampant, escalating prices of accommodation, and poor living standard. Such settlements can be considered as slums or informal settlements. These informal settlements represent affordable neighbourhood where rural–urban migrants, low-income earners, and spill-over from urban areas get to settle in. Developing countries like Nigeria, Mozambique and DRC are characterized by high urban densities, which take the form of overcrowded informal settlements instead of planned dwelling centres [53]. Most informal settlements across the region have been as a result of spontaneous self-organized housing production, without formal planning and legal consent [93]. By implication, these settlements are not considered during the sharing of developmental quotas and social amenities.

Given time and in the absence of proper urban planning in the face of urbanization, under-resourced sub-region administrative offices, and the absence of housing policies, people are likely to experience high poverty level as the rate of unemployment also rises [57]. These conditions are able to create a favourable environment for the proliferation of slums. A slum encompass a household or a group of households that lack one or more of the following conditions: access to portable water, access to enhanced sanitation systems, adequate living space, and durable housing structure [94]. [95] described a slum as an area marked by issues including defective infrastructure, high population density, limited space, and lack of access to essential services like portable water, sanitation systems, and reliable energy. Such environments are usually located in hazardous areas and are referred to as "ghettos"—a place where criminals activities flourish—characterized by the depleting state of many of the structures [96].

As discourse on sustainable urban development in Africa gains prominence, persistent challenges in urban planning and resource allocation remain particularly acute amid rapid urbanization. With an estimated 51% of urban populations dwelling in slum in 2020 [58], Africa maintains the highest concentration of slum settlements globally [57]. Given that Africa's urbanization is marked by inadequate infrastructure, widespread urban poverty, and high unemployment, critical questions emerge about equitable allocation of resources to informal settlements. As a consequence, slum residents are prone to plethora of challenges including limited access to quality healthcare and education [95]. As the hazardous and unkept environment of slums negatively impact public health, women tend to be disproportionately affected. Their quality of life encompassing their health, economic stability, and social welfare are compromised by inadequate sanitation [97, 98]. In addition to these issues are pervasive social problems including substance abuse, teenage pregnancies, and gender-based violence—all consistent with Africa's slum [99]. Some prominent cities in Africa that host significant slum settlements include Lagos in Nigeria, Nairobi in Kenya, Accra in Ghana, Cape Town in South Africa, Cairo in Egypt, and so on.

Regarding continental building decarbonization efforts, the proliferation of slums and informal settlements across Africa presents both a formidable challenge and a substantial opportunity requiring innovative solutions. These settlements present a significant challenge as they house approximately half of Africa's urban population—a demography that is too significant to overlook in the pursuit of building decarbonization [58]. Marked by poverty, slums complicate decarbonization efforts through their dependence on high-carbon energy sources and inefficient construction methods [57]. Primarily, residents rely on polluting fuels like charcoal, kerosene, firewood, along with unauthorized electricity connections, creating significant carbon footprints despite low individual consumption levels. The informal nature of these settlements often restricts access to cleaner energy alternatives including renewable sources like the Solar Photovoltaic and Biomass, worsening the problem [100]. Alarmingly, a significant funding is required to address decarbonization issues targeting conventionally constructed buildings, hence the urgent need to address practices in these rapidly expanding informal communities [28, 101].

Despite these challenges, African slums also present opportunities for sustainable decarbonization approaches that could potentially outdo conventional development models. Transitioning from makeshift structures to green building materials and passive design principles could significantly influence environmental sustainability [26, 102]. Furthermore, the inherent population density in slums makes them ideal for efficient district-level energy systems and community-based energy initiatives. Without deliberate intervention, slums will continue to constrain the transformative potential of Africa's cities. [103] emphasize strategic investments in infrastructure and services could yield significant improvements if governments implement inclusive policies, foster public–private partnerships, and prioritize community empowerment through education and public awareness programmes. Sustainable urban development does not only require judicious use of land and spatial planning, but also considers how efficiently services, particularly energy, can be used in a sustainable way [56]. Policy makers need to also initiate slum upgrading programmes, in order to improve infrastructure quality, though energy-efficient techniques, access to services including clean energy.

15.7 Pathway Towards a Decarbonized Urban Future in Africa

Africa faces a pivotal moment in urban development as rapid urbanization creates both immense challenges and unique opportunities for sustainable urban development. The urgency of establishing decarbonization pathways is now more critical than ever, especially as African cities have been projected to accommodate nearly one billion more residents by 2050 [8, 48]. Simultaneously, a staggering 600 million people currently lack access to reliable electricity, underscoring the continent's profound energy deficit [57]. As home to the world's youngest and fastest-growing population, there is a necessity to meticulously balance urban development with environmental sustainability. Recognized as the world's most underdeveloped continent, Africa is characterized by widespread informal settlements and slums. These communities often situated in precarious environments, highly prone to recurring floods, poor sanitation, disproportionate exposure to violence and abuse, while simultaneously faced with low quality medical care and education. It is therefore expedient to consider innovative approaches to maximize the potential of Africa's burgeoning youth demographic, while concurrently establishing measures to enhance both global climate stability and local prosperity.

Geographically situated within the topics and marked by hot climates, Africa's rapid population growth will inevitably require significant infrastructural expansion for residential, commercial, and industrial purposes. This, in turn, will heighten the demand for robust electrification frameworks and cooling solutions, imposing massive pressure on the already strained energy generation systems. In spite of these, the continent regrettably remains largely locked into carbon-intensive development patterns, rendering immediate interventions less effective. While Africa contributes the least (i.e. less than 4%) to global greenhouse gas emissions [104], it suffers disproportionate impacts including extensive heat waves, flooding, droughts, and irregular rainfall patterns due to climate change [62].

Given the complex challenges of rapid urbanization, energy deficits, and climate vulnerability in Africa, it becomes unequivocally clear that the sustainable development of the continent's urban centres demand strategic yet imperative undertakings. The pathways towards decarbonizing both existing metropolises and emerging urban settlements necessitate comprehensively innovative approaches. These approaches must meticulously integrate Africa's distinctive demographic trend and development context to foster pioneering sustainable low-carbon models. A growing number of African cities are progressively leading this transformative shift, proving that tailored approaches can deliver both significant climate benefits and an improved quality of life, thereby ensuring inclusive progress towards a sustainable urban future. To this end, the following actionable pathways can catalyse this transformation while addressing Africa's urban realities:

15.7.1 Promoting Energy-Efficient Building Design and Practices

Integrating energy-efficient building designs and principles with construction practices is an imperative to managing the burgeoning energy demand. Passive design strategies are structures that leverage natural elements to fulfil a desired goal without depending on electricity [105, 106]. By integrating passive design considerations into buildings, occupants may want to shield parts of the buildings, reduce thermal transmittance, improve ventilation and daylight among others [105]. Subsequently, without the use of electricity, a building's operational energy consumption drastically reduces. Key strategies include building orientation and position, strategically positioning rooms to maximize natural light and minimize heat gain, and incorporating shading devices for large windows [106]. Besides this, the built environment in Africa needs to adopt bioclimatic building techniques, where construction practices are based on the local context [84]. African sub-regions, due to their different climatic and geographic characteristics, can initiate vernacular designs while using readily available geogenic and biogenic materials, which are not only cost effectively, but are also characterized by low embodied energy as compared to conventional materials [107].

With projections indicating a significant urban growth, developing building codes and energy performance standards specific to all the regions in Africa is a non-negotiable for a sustainable urban future. These tools serve as references, offering guidance as to how buildings should be constructed, which can help reduce inefficient operational energy. As demand for infrastructure increase, there is a need to develop and mandate the use of green building certification programs like LEED, Green Star, and EDGE [108]. Report show that about 38 African countries are yet to implement some of these strategies [62]. Even though, the individual's consumption is low, it is crucial to adopt energy-efficient appliances like air-conditioning units and lighting systems like LEDs in urban centres. There is also the need to complement this with attitudinal change. Africans would have to adopt simple yet impactful behaviours like opening/closing windows, turning off lights when not in use, and so on. This requires some form of orientation and mental shift, which can happen through the provision of education purposely for the youth and creating public awareness. As a region characterized by hot climates and tidal resources, it is necessary to integrate comprehensive renewable energy sources including solar and wind. A typical situation is seen in South Africa, where at least half of the quantity of hot water used in a new house must be solar-heated [109]. The aim of all of these is substantially reduce energy demand.

While these design principles and technologies offer clear benefits, their widespread adoption is often hampered by socio-economic and technical factors. For instance, the perceived high upfront costs for green building solutions deters developers and homeowners [108], while the use of some bioclimatic principles lack technical expertise [110]. Another is the redirection of investments in energy-measures due to low electricity prices. To overcome these barriers, policy interventions must combine regulatory frameworks with targeted financial incentives and comprehensive awareness campaigns.

15.7.2 Leveraging Decentralized Energy Solutions

Africa possesses immense renewable energy potential. It is estimated that region could lead renewable energy generation in the world by 2030 by 310 GW [10]. It also has the potential to harness 10 TW of solar energy, 350 GW of hydro energy, 110 GW of wind and 15GW of geothermal energy [10]. Despite these potentials, the continent is currently under exploiting these alternatives. For example, the region has less than 2% of solar PV installations, even though it is the world's largest solar resource [79]. The establishment of community-based programmes like decentralized energy infrastructure, including photovoltaic mini-grids networks and independent systems, presents an affordable, expandable, and dependable approach to clean electricity access [111, 112]. Implementing these initiatives eliminates reliance on expensive and often delayed grid expansion. Such approaches prove particularly viable for informal settlements where a significant portion of energy-deprived population resides [113].

The deployment of these interventions addresses both safety and environmental concerns, particularly in informal settlements, by reducing fire risks and smog pollution linked to carbon-intense fuels and illegal electricity connections, while concurrently creating employment pathways and income-generating opportunities [113, 114]. Furthermore, these solutions will enhance energy reliability, bypassing the vulnerability inherent with centralized grid systems. Some prominent case studies include the BIPVGREEN (Building Integrated Photovoltaic with Greenry) project in Kenya and the Umbane projects in Qandu-Qandu in South Africa demonstrate the transformative potential of solar initiatives in informal settlements [115, 116]. Ultimately, using sustainable energy interventions needs to be considered as catalysts for a holistic regional development, offering multi-faceted benefits beyond technical solutions.

15.7.3 Strengthening Governance and Policy Frameworks

Effective governance is an essential component to Africa's urban transformation, permitting the formulation and implementation of comprehensive, context-specific National Urban Policies (NUP) and National Development Policies (NDP) as catalysts for progressive change [117, 118]. These frameworks demand multi-tiered governance structures characterized by explicit objectives and decentralization commitments that empower local authorities to manage urban growth effectively. Strengthening the relevance, capacity, and speed of urban planning and policies in Africa has become an imperative to ensure alignment with local realities and evolving challenges. Developing and effectively implementing building codes and energy performance standards are crucial approaches to ensure new constructions are energy-efficient from the inception stage. Countries like Morocco, South Africa and Tunisia, with their mandatory building regulations, serve as examples of such a framework [62]. Concurrently, regulatory restructuring—particularly regarding cost-of-service electricity pricing mechanisms—emerges as a priority for establishing equitable and sustainable energy markets.

15.7.4 Fostering Local Sustainable Material Production

Africa possesses abundant indigenous, sustainable building materials, encompassing adobe, laterite, timber, stone, bamboo, and dry vegetation [107, 119]. The strategic utilization of these local resources with traditional construction techniques, including rammed earth, sun-dried bricks, wattle-and-daub, and woven thatch techniques, can significantly reduce the need for energy-intensive air-conditioning, minimize the long-range transport of building materials, and lower greenhouse gas emissions associated with construction [110, 120]. A deliberate transition towards green industrialization, emphasizing low-carbon manufacturing and value-added processing, has the potential to transform the continent's reliance on raw material exportation to domestic processing of critical minerals. To this end, leveraging on local construction materials and techniques is able to offer numerous co-benefits: employment generation, economic growth, reduced negative environmental impacts, cultural heritage preservation, and enhanced housing quality and climate resilience.

15.7.5 Integrating Climate Resilience

The design and construction of climate-resilient residential infrastructure represents an essential component for urban environments to endure, recover from, and withstand the adverse effects of climate change [83, 121]. Core mitigation strategies encompass identifying low-risk construction sites to minimize the impact of hazards, employing raised foundations to protect against flooding, and utilizing sustainable materials like Durabric—a stabilized compressed earth block technology that reduces embodies carbon by approximately 85% compared to fired bricks which offer an affordable and environmentally friendly alternative to traditional bricks [101, 122]. Urban planning needs to prioritize passive design principles, liked shaded walkways green spaces, to effectively address urban heat island effects. Policy initiatives, such as Nigeria's 2021 Climate Change Act, South Africa and Ghana's Green Star certification, and Kenya's Green Building code (requiring affordable housing projects to meet global green certification requirements), prove that important steps are being taken towards embedding climate considerations into urban development [108, 123, 124]. Despite the merit of these national efforts, the various sub-regions on the continent would benefit substantial advantage from developing their own green building certification frameworks tailored to their context. This would ensure a more equitable and standardized approach to sustainable and climate-resilient urban development across Africa, ensuring comprehensive inclusion of all sub-regions in the sustainability transition.

15.7.6 *Green Financing and Partnerships*

Africa is poised to become a renewable energy powerhouse, especially in solar, yet a critical barrier remains: securing sufficient affordable funding [125]. This is where international bilateral and multilateral institutions, like the International Monetary Fund (IMF), World Bank, and the African Development Bank, become crucial. They are vital sources for attracting large-scale investments in climate-resilient housing and sustainable energy solutions by offering de-risking financial products [4]. Financial de-risking essentially involves external support, either through financial institutions or national and international policy that reduces the risks involved with investments. To bridge the existing financial gap, a combination of public and philanthropic capital with private investment (also known as blended finance) is highly recommended to reduce risks [125, 126]. Unlocking access to finance and developing local technical expertise are identified as key action areas for sustainable energy development [4]. Public–private partnerships are vital for promoting the widespread adoption of renewable energy and overcoming financial barriers to green building initiatives. These collaborative models can aid to co-create and scale sustainable building and energy solutions across the continent [64, 127].

15.8 Summary

This chapter examines Africa's unprecedented urban transformation, driven by both natural population and rural–urban migration, which is set to double the continent's urban double by 2050. This makes Africa the fastest-urbanizing continent. While cities are economic engines with manifold opportunities, Africa's urbanization remains distinct: it is characterized by informality and marked by a massive infrastructure gap, particularly in energy and housing. The built environment is a critical player in meeting these burgeoning needs, yet its conventional practices are inherently resource-intensive and high in carbon emissions.

Ignoring the nexus of urbanization and decarbonization poses significant risks; therefore, an integrated approach is essential mitigate these impending challenges. This chapter highlighted how energy-efficient buildings can improve indoor quality and alleviate energy poverty, while integrated urban planning fosters more liveable, low-emission cities. Nonetheless, Africa's urbanization, predominantly occurring in informal and peri-urban areas, necessitates innovative, affordable, and socially equitable decarbonization strategies. The continent's youthful population offers immense potential, yet it also presents daunting challenges with respect to housing and employment that, if not proactively managed, could worsen existing problems.

Addressing these issues requires a multi-pronged approach. The chapter emphasizes the importance of effective governance to formulate and implement context-specific National Urban and Development policies, fostering multi-tiered governance and empowering local authorities. It further advocates for strengthening urban planning, developing comprehensive building energy codes (like those in Morocco and Tunisia), and robust green building certification frameworks (such as Green Star certification program), while concurrently reforming electricity pricing. While national green building certifications represent a significant leap in the right direction, it is acknowledged that the development of tailored sub-regional certification programs is necessary to ensure a standardized urban development across the continent's diverse contexts.

Furthermore, the chapter underscores the imperative to utilize Africa's expansive local geogenic and biogenic materials and construction techniques, including passive designs principles, to reduce energy consumption, minimize transport, and lower emissions. This trend not only present numerous environmental benefits but also engender employment opportunities, stimulate economic growth, and fosters cultural preservation. Crucially, the chapter maintain that addressing the proliferation of slums and informal settlements involve innovative, community-centric upgrading rather than forced evictions. These strategies should prioritize in-situ improvements, tenure security, and enhanced access to basic services.

Finally, the chapter addresses the critical challenge of green financing. Attracting sufficient funding remains a critical challenge, despite Africa's renewable energy potential. It underscores the integral role of international financial institutions in providing de-risking financial products such as blended finance, refinancing, portfolio approaches, and currency risk mitigation. Similarly, it advocates for strategic public–private partnerships, and quality education to develop local technical expertise. These elements are salient to accelerating the adoption of renewable energy and green building initiatives, ultimately facilitating the scaling of sustainable solutions across Africa.

In essence, the chapter posits that urbanization in Africa can be a powerful catalyst for energy-efficient, climate-resilient, and equitable cities. However, this demands a fundamental shift from conventional practices towards sustainable practices, innovative financing, robust governance, and inclusive, context-specific urban planning that prioritize the well-being of all communities.

References

1. IEA (2024) Empowering urban energy transitions: smart cities and smart grids. International Energy Agency
2. Smith T (2024) Accelerating investment into clean energy in cities across Africa. ESI-Africa.Com. Available at: https://www.esi-africa.com/industry-sectors/future-energy/accelerating-investment-into-clean-energy-in-cities-across-africa/. Accessed 5 Jun 2025
3. Adeodun A (2024) African cities lead charge in clean energy revolution. Energy News Africa Plus. Available at: https://energynews.africa/2024/05/06/african-cities-lead-charge-in-clean-energy-revolution/. Accessed 5 Jun 2025
4. CPI (2024) Landscape of climate finance in Africa. Climate Policy Initiative
5. Joshua A, Precious A, Mathew O, Usman I, Oduwole M (2023) Unleashing sustainable development potential in Africa. Int J Soc Sci Manag Res

6. Wang S, Gao S, Li S, Feng K (2020) Strategizing the relation between urbanization and air pollution: Empirical evidence from global countries. J Clean Prod 243:118615. https://doi.org/10.1016/j.jclepro.2019.118615
7. Chandler B (2022) Addressing Africa's energy deficit: climate change, renewables, and gas. Mo Ibrahim Foundation
8. UNECA (2025) State of urbanization in Africa: digital transformation for inclusive, resilient and sustainable urbanization. United Nations Economic Commission for Africa
9. Duan X, Li X, Tan W, Xiao R (2022) Decoupling relationship analysis between urbanization and carbon emissions in 33 African countries. Heliyon 8(9):e10423. https://doi.org/10.1016/j.heliyon.2022.e10423
10. Revi A, Rosenzweig C (2013) The urban opportunity: enabling transformative and sustainable development. Background paper for the high-level panel of eminent persons on the post-2015 development agenda. Prepared by the sustainable development solutions network-thematic group on sustainable cities, pp 131–138
11. Jaber JO, Marahleh G, Al-Lubani S, Khrisat M (2025) Energy efficiency retrofitting for old buildings in a low-income district of Amman, Jordan. Acad Green Energy 2(1). https://doi.org/10.20935/AcadEnergy7487
12. Lin Y, Cui C, Liu X, Mao G, Xiong J, Zhang Y (2023) Green renovation and retrofitting of old buildings: a case study of a concrete brick apartment in Chengdu. Sustainability 15(16):12409. https://doi.org/10.3390/su151612409
13. Emere CE, Aigbavboa CO, Thwala WD, Akinradewo OI (2024) A principal component analysis of sustainable building construction features for project delivery in South Africa. J Eng, Des Technol. https://doi.org/10.1108/JEDT-01-2024-0015
14. Rauf A, Attoye DE, Crawford RH (2024) Evaluating the impact of material service life on embodied energy of residential villas in the United Arab Emirates. Eng, Constr Arch Manag 31(13):244–270. https://doi.org/10.1108/ECAM-05-2023-0514
15. Linde L, Lewis AN, Sanchez F (2022) Towards a sustainable global construction and buildings value chain. Leadership Group for Industry Transition
16. Economy C (2023) The circularity gap report 2023. Circle Economy, Amsterdam, pp 1–64
17. Shooshtarian S, Hosseini MR, Kocaturk T, Arnel T, Garofano N (2023) Circular economy in the Australian AEC industry: investigation of barriers and enablers. Building Research & Information 51(1):56–68. https://doi.org/10.1080/09613218.2022.2099788
18. Frempong-Jnr EY, Ametepey SO, Cobbina JE (2022) Impact of stakeholder management on efficient construction waste management. Smart Sustain Built Environ 12(3):607–634. https://doi.org/10.1108/SASBE-08-2021-0147
19. Hao JL, Ma W (2023) Evaluating carbon emissions of construction and demolition waste in building energy retrofit projects. Energy 281. https://doi.org/10.1016/j.energy.2023.128201
20. Abdulgader M, Lashhab F (2021) Energy-efficient thermal comfort control in smart buildings. In: 2021 IEEE 11th Annual Computing and Communication Workshop and Conference (CCWC). In: IEEE 11th annual computing and communication workshop and conference (CCWC), pp 0022–0026. https://doi.org/10.1109/CCWC51732.2021.9376175
21. Cartwright A (2015) Better growth, better cities: rethinking and redirecting urbanisation in Africa. In: The new climate economy working paper. New Climate Economy Washington, DC, USA
22. Soares F, Silva MC, Azevedo I (2025) Urban decarbonization policies and strategies: a sectoral review. Renew Sustain Energy Rev 215:115617. https://doi.org/10.1016/j.rser.2025.115617
23. ACMI (2023) Africa carbon markets initiative (ACMI) Overview. Africa Carbon Markets Initiative
24. GH2 (2022) Africa's green hydrogen potential. Green Hydrogen Organization
25. ACRIFA (2023) The Africa climate risk insurance facility for adaptation—ACRIFA, African Development Bank Group.
26. Addy M, Adinyira E, Danku JC, Dadzoe F (2021) Impediments to the development of the green building market in sub-Saharan Africa: the case of Ghana. Smart Sustain Built Environ 10(2):193–207

27. Wang K, Costanza-van Den Belt M, Heath G, Walzberg J, Curtis T, Berrie J, Schröder P, et al (2022) Circular economy as a climate strategy: current knowledge and calls-to-action, No. NREL/TP-6A20-84141, 1897625, MainId:84914, p. NREL/TP-6A20-84141, 1897625, MainId:84914. https://doi.org/10.2172/1897625
28. UNEP (2025) Global status report for buildings and construction 2024/2025: not just another brick in the wall—the solutions exist. Scaling them will build on progress and cut emissions fast. United Nations Environment Programme, Paris. https://doi.org/10.59117/20.500.11822/47214
29. Kurita A, Espuny M, Campos TLR, Kazançoğlu Y, Kandsamy J, de Oliveira OJ (2023) Drivers for circular economy development: making businesses more environmentally friendly. Environ Sci Pollut Res Int 30(33):79553–79570. https://doi.org/10.1007/s11356-023-28048-0
30. Tisdale H (1942) The process of urbanization. Soc Forces 20(3):311–316. https://doi.org/10.2307/3005615
31. Henderson V (2003) The urbanization process and economic growth: the so-what question. J Econ Growth 8(1):47–71. https://doi.org/10.1023/A:1022860800744
32. McGranahan G, Satterthwaite D (2014) Urbanisation concepts and trends
33. Felix Meier zu Selhausen (2022) Growing cities: urbanization in Africa. In the history of African development: an online textbook for a new generation of African students and teachers. African Economic History Network, pp 1–18
34. Childe VG (1950) The urban revolution. Town Plan Rev 21(1):3. https://doi.org/10.3828/tpr.21.1.k853061t614q42qh
35. Rai S (2019) Concept of urbanization in ancient society: a review
36. Davis K (1955) The origin and growth of urbanization in the world. Am J Sociol 60(5):429–437
37. Kleniewski N, Thomas AR (2019) Cities, change, and conflict: a political economy of urban life, 5th ed. Routledge, New York. https://doi.org/10.4324/9780429022920
38. Neiberg MS (2015) Warfare in history. In: Wright JD (ed) International encyclopedia of the social & behavioral sciences, 2nd Ed. Elsevier, Oxford, pp 414–419. https://doi.org/10.1016/B978-0-08-097086-8.96033-1
39. Weber M (1958) Translated and edited by D. Martindale and G. Neuwirth
40. Jedwab R, Johnson ND, Koyama M (2022) Medieval cities through the lens of urban economics. Reg Sci Urban Econ 94:103598. https://doi.org/10.1016/j.regsciurbeco.2020.103598
41. Berry BJL (2008) Urbanization. In: Marzluff JM, Shulenberger E, Endlicher W, Alberti M, Bradley G, Ryan C, Simon U et al (eds) Urban ecology: an international perspective on the interaction between humans and nature. Springer US, Boston, MA, pp 25–48. https://doi.org/10.1007/978-0-387-73412-5_3
42. Khan F (2024) Impact of industrial revolution on urbanization patterns in 19th-century Europe. Eur J Hist Res 3(3):49–60. https://doi.org/10.47672/ejhr.2465
43. Ren Y, Li H, Shen L, Zhang Y, Chen Y, Wang J (2018) What is the efficiency of fast urbanization? A China study. Sustainability 10(9):3180
44. Kang H, Fu M, Kang H, Li L, Dong X, Li S (2024) The impacts of urban population growth and shrinkage on the urban land use efficiency: a case study of the northeastern region of China. Land 13(9):1532. https://doi.org/10.3390/land13091532
45. Samways D (2022) Population and sustainability: Reviewing the relationship between population growth and environmental change. J Popul Sustain 6(1):15–41
46. van den Berg L, Drewett R, Klaassen LH, Rossi A, Vijverberg CHT (eds) (1982) The costs of urban growth (CURB) Project. In: A study of growth and decline. Pergamon, p ii. https://doi.org/10.1016/B978-0-08-023156-3.50001-2
47. Das S, Choudhury MR, Chatterjee B, Das P, Bagri S, Paul D, Bera M et al (2024) Unraveling the urban climate crisis: exploring the nexus of urbanization, climate change, and their impacts on the environment and human well-being—a global perspective. AIMS Public Health 11(3):963–1001. https://doi.org/10.3934/publichealth.2024050
48. OECD, African Development Bank, Cities Alliance, and United Cities And Local Governments Of Africa (2025) Africa's urbanisation dynamics 2025: planning for urban expansion. OECD Publishing. https://doi.org/10.1787/2a47845c-en

49. Bos J-M (2023) Africa drives global urbanization—DW—05/17/2023. Dw.Com. Available at: https://www.dw.com/en/africa-drives-global-urbanization/a-65653428. Accessed 11 Jun 2025
50. Africa Center for Strategic Studies (2025) Africa's Unprecedented urbanization is shifting the security landscape. Africa Center. Available at: https://africacenter.org/spotlight/africa-urban-growth-security/. Accessed 11 Jun 2025
51. Galal S (2024) Africa: urbanization rate by country 2023. Statista. Available at: https://www.statista.com/statistics/1223543/urbanization-rate-in-africa-by-country/. Accessed 13 Jun 2025
52. Cobbinah PB, Erdiaw-Kwasie MO, Amoateng P (2015) Africa's urbanisation: Implications for sustainable development. Cities 47:62–72. https://doi.org/10.1016/j.cities.2015.03.013
53. Gambe TR, Turok I, Visagie J (2023) The trajectories of urbanisation in Southern Africa: a comparative analysis. Habitat Int 132:102747. https://doi.org/10.1016/j.habitatint.2023.102747
54. Aikins ER (2023) North Africa. ISS African Futures
55. Zoma V, Sawadogo Y (2022) Main characteristics of African cities. Quest JS J Res HumIties Soc Sci 10(4):11–17
56. Jiang S, Ren H, Zhang Z (2025) Monitoring and assessing the growth law of urban land using multi-source data: a case study of five East African countries. Land Use Policy 153:107530. https://doi.org/10.1016/j.landusepol.2025.107530
57. UN-Habitat (2022) World Cities Report 2022: Envisaging the future of cities. United Nations Human Settlements Programme
58. Weimann A, Smit W, Magidi M, Safougne B (2022) The urban governance implications of African urbanisation. AMALI. Available at: https://amalicities.africa/the-latest-trends-the-urban-governance-implications-of-african-urbanisation/. Accessed 16 Jun 2025
59. UN DESA (2019) World urbanization prospects the 2018 revision. No. ST/ESA/SER.A/420, United Nations, New York
60. World Bank Group (2025) Western & Central Africa: development news, research, data. World Bank. Available at: https://www.worldbank.org/en/region/afr/western-and-central-africa. Accessed 16 Jun 2025
61. Parves Rana M (2009) Sustainable city in the global North and South: goal or principle? Manag Environ Qual: Int J 20(5):506–521
62. Gollin D, Jedwab R, Vollrath D (2013) Urbanization with and without structural transformation. George Washington University, Mimeograph
63. Kriticos S (2019) Urbanisation and structural transformation in Africa. International Growth Centre. Available at: https://www.theigc.org/blogs/urbanisation-and-structural-transformation-africa. Accessed 18 Jun 2025
64. Kong Y, Donkor M, Jin C, Musah M, Nkyi JA (2024) Do financial development, urbanization, economic growth and renewable energy promote the emission mitigation agenda of Africa? Evidence from models that account for cross-sectional dependence and slope heterogeneity. Front Environ Sci 11. https://doi.org/10.3389/fenvs.2023.1269416
65. World Bank (2023) Leveraging resource wealth during the low carbon transition, no 27, Washington
66. Gunaratna KL (2018) Urban primacy. In: Towards equitable progress. Springer, Singapore, pp 91–107. https://doi.org/10.1007/978-981-10-8923-7_8
67. Şakaci BK, Soumanou A, Özkaya S (2021) Benin on the axis of the law of the primate city: the case of Cotonou. Uluslararası Yönetim Akad Derg 4(3):493–508. https://doi.org/10.33712/mana.984399
68. Zhang J, Shuang Chen S, Gao Q, Shen Q, Kimirei IA, Mapunda DW (2020) Morphological characteristics of informal settlements and strategic suggestions for urban sustainable development in Tanzania: Dar es Salaam, Mwanza, and Kigoma. Sustainability 12(9):3807. https://doi.org/10.3390/su12093807
69. Ouma S, Cocco Beltrame D, Mitlin D, Beth Chitekwe-Biti B (2024) Informal settlements: domain Report. African Cities Research Consortium, The University of Manchester, Manchester

70. Anzilotti E, Appavoo J (2025) How African cities can develop to become green, resilient, and inclusive. UrbanShift. Available at: https://www.shiftcities.org/post/how-african-cities-can-develop-become-green-resilient-and-inclusive. Accessed 23 Jun 2025
71. Aref A, Fallentine A, Zahran S (2024) The state of urbanization, demographic changes, and family dynamics in Africa. J Comp Fam Stud 54(3):283–297. https://doi.org/10.3138/jcfs-083-2023
72. Gutu Sakketa T (2023) Urbanisation and rural development in sub-Saharan Africa: a review of pathways and impacts. Res Globaliz 6:100133. https://doi.org/10.1016/j.resglo.2023.100133
73. Echendu AJ (2025) A portrait of the urban demographic profile of an African City—Port Harcourt, Nigeria. Urban Science 9(5):178. https://doi.org/10.3390/urbansci9050178
74. OECD, United Nations Economic Commission for Africa, and African Development Bank (2022) Africa's urbanisation dynamics 2022: the economic power of Africa's cities. OECD. https://doi.org/10.1787/3834ed5b-en
75. IPCC (2022) Climate change 2022: impacts, adaptation and vulnerability. Contribution of working Group III to the sixth assessment report of the intergovernmental panel on climate change. https://doi.org/10.1017/9781009157926.
76. Awuah KGB (2018) Urban development and governance in Nigeria: challenges, opportunities and policy direction. Int Dev Plan Rev 40(1):27–49. https://doi.org/10.3828/idpr.2018.1
77. Burdett R, Hall S (2017) The SAGE Handbook of the 21st Century City. SAGE Publications Ltd, pp 1–730
78. Güneralp B, Lwasa S, Masundire H, Parnell S, Seto KC (2017) Urbanization in Africa: challenges and opportunities for conservation. Environ Res Lett 13(1):015002. https://doi.org/10.1088/1748-9326/aa94fe
79. IEA (2023) Africa Energy Outlook 2022. International Energy Agency, pp 1–250
80. Kassouri Y, Okunlola OA (2022) Analysis of spatio-temporal drivers and convergence characteristics of urban development in Africa. Land Use Policy 112:105868. https://doi.org/10.1016/j.landusepol.2021.105868
81. Emily MM, Muyengwa G (2021) Maintenance of municipality infrastructure: a case study on service delivery in Limpopo Province at South Africa. Am J Oper Res 11(6):309–323
82. Mazele O, Amoah C (2022) The causes of poor infrastructure management and maintenance in South African municipalities. Prop Manage 40(2):192–206
83. Lindfield M, Inkoom KBD (2022) Financing low carbon and resilient cities in Africa. African Development Bank
84. Koranteng C, Simons B, Essel C (2019) Climate responsive buildings: a comfort assessment of buildings on KNUST campus, Kumasi. J Eng, Des Technol 17(5):862–877. https://doi.org/10.1108/JEDT-03-2019-0054
85. Hazeleger W, van den Hurk BJ, Min E, Van Oldenborgh GJ, Petersen AC, Stainforth DA, Vasileiadou E et al (2015) Tales of future weather. Nat Clim Chang 5(2):107–113
86. Ameer B, Krarti M (2024) Impact of building design and operating strategies on urban heat island effects—part I: model development and validation. J Eng Sustain Build Cities 5(3). https://doi.org/10.1115/1.4066053
87. Andoni H, Wonorahardjo S (2018) A review on mitigation technologies for controlling urban heat island effect in housing and settlement areas. IOP Conf Ser: Earth Environ Sci 152:012027. https://doi.org/10.1088/1755-1315/152/1/012027
88. Dioha MO, Abraham-Dukuma MC, Dato P (2025) Unearthing the reality of 'Zombie energy systems' in Africa's energy transition. Environ Res: Energy 2(1):013001
89. Ramluckun R, Malumbazo N, Ngubevana L (2024) A review of the energy policies of the BRICS countries: the possibility of adopting a just energy transition for South Africa. Sustainability 16(2):703
90. Tareke KM, Baraki GA (2024) Affordability and accessibility of condominium housing in urban Ethiopia using a combined transportation and housing cost (CHT) index model: implication for sustainable urban infrastructure, policy and development. Int J Urban Sustain Dev 16(1):459–482. https://doi.org/10.1080/19463138.2024.2429393

91. Arceo A, O'Brien W, Touchie M (2024) Ten questions concerning the environmental impacts of housing built form. Build Environ 256:111490. https://doi.org/10.1016/j.buildenv.2024.111490
92. Ekpo A (2019) Housing deficit in Nigeria: issues, challenges and prospects. Econ Financ Rev 57(4):13
93. Dovey K, Shafique T, Van Oostrum M, Chatterjee I (2021) Informal settlement is not a euphemism for 'slum': what's at stake beyond the language? Int Dev Plan Rev 43(2):139–150. https://doi.org/10.3828/idpr.2020.14
94. Ritchie H, Samborska V, Roser M (2024) Urbanization. Our World in Data
95. Sastrodiningrat MR, Bisri M, Ahmad Imron Rozuli S (2024) Building planning policy innovation: integrated solutions for transforming slum areas into sustainable settlements. Nongye Jixie Xuebao/Trans Chin Soc Agric Mach 55(4)
96. Ganiyu BO, Fapohunda JA, Haldenwang R (2017) Sustainable housing financing model to reduce South Africa housing deficit. Int J Hous Mark Anal 10(3):410–430. https://doi.org/10.1108/IJHMA-07-2016-0051
97. Intesar A, Parvez MS (2024) Living with vulnerability: triple burden through the eyes of urban slum women in Bangladesh. Soc Sci Human Open 10:101014. https://doi.org/10.1016/j.ssaho.2024.101014
98. Mumah J, Kabiru C, Izugbara C, Mukiira C (2014) Coping with unintended pregnancies: narratives from adolescents in Nairobi's Slums
99. Nasreen A, Singh AK (2020) Field work practice in urban slums: Issues and challenges in a megapolis of India. Soc Work Soc 18(2)
100. Albuquerque N, Correia Guedes M (2021) Cities without slums and the right to the city: slums in Subsaharan Africa. In: Sayigh A (ed) Renewable energy and environmental sustainability, vol 6, p 24. https://doi.org/10.1051/rees/2021022
101. UNEP (2023) Building materials and the climate: constructing a new future. United Nations Environment Programme, GlobalABC, Yale Center for Ecosystems, and +Architecture
102. Ahsan MM, Zulqernain M, Ahmad H, Wajid BA, Shahzad S, Hussain M (2019) "Reducing the operational energy consumption in buildings by passive cooling techniques using building information modelling tools. Int J Renew Energy Res 9(1):343–353
103. Habib E (2009) The role of government and NGOs in slum development: the case of Dhaka City. Dev Pract 19(2):259–265. https://doi.org/10.1080/09614520802689576
104. Ali M (2021) Urbanisation and energy consumption in Sub-Saharan Africa. Electr J 34(10):107045. https://doi.org/10.1016/j.tej.2021.107045
105. Azimi Fereidani N, Rodrigues E, Gaspar AR (2021) A review of the energy implications of passive building design and active measures under climate change in the Middle East. J Clean Prod 305:127152. https://doi.org/10.1016/j.jclepro.2021.127152
106. Nagasue M, Kitagawa H, Asawa T, Kubota T (2024) A systematic review of passive cooling methods in hot and humid climates using a text mining-based bibliometric approach. Sustainability 16(4):1420. https://doi.org/10.3390/su16041420
107. Lokko M, Manu FW, Mboup N, Etman MA, Raugei M, Niang I, Ametepe K et al (2024) Comparing the whole life cycle carbon impact of conventional and biogenic building materials across major residential typologies in Ghana and Senegal. Sustain Cities Soc 106:105332. https://doi.org/10.1016/j.scs.2024.105332
108. Ampratwum G, Agyekum K, Adinyira E, Duah D (2021) A framework for the implementation of green certification of buildings in Ghana. Int J Constr Manag 21(12):1263–1277. https://doi.org/10.1080/15623599.2019.1613207
109. Yen Y-C, Nixon K, De La Rue Du Can S, Covary T, Lutz J, Dekenah M (2024) Residential water heating demand side management (DSM)—South Africa. Department of Mineral Resouces and Energy, South Africa; and US Department of Energy. https://doi.org/10.2172/2368544
110. Agyekum K, Kissi E, Danku JC (2020) Professionals' views of vernacular building materials and techniques for green building delivery in Ghana. Scientific African 8:e00424. https://doi.org/10.1016/j.sciaf.2020.e00424

111. Babayomi OO, Olubayo B, Denwigwe IH, Somefun TE, Adedoja OS, Somefun CT, Olukayode K et al (2023) A review of renewable off-grid mini-grids in Sub-Saharan Africa. Front Energy Res 10. https://doi.org/10.3389/fenrg.2022.1089025
112. Stritzke S, Jain P (2021) The sustainability of decentralised renewable energy projects in developing countries: learning lessons from Zambia. Energies 14(13):3757. https://doi.org/10.3390/en14133757
113. Besner R, Mehta K, Zörner W (2023) How to enhance energy services in informal settlements? Qualitative comparison of renewable energy solutions. Energies 16(12):4687. https://doi.org/10.3390/en16124687
114. Mensah JK (2022) Electricity and informal settlements: Towards achieving SDG 7 in developing countries. Energy Res Soc Sci 93:102844. https://doi.org/10.1016/j.erss.2022.102844
115. Conway D, Robinson B, Mudimu P, Chitekwe T, Koranteng K, Swilling M (2019) Exploring hybrid models for universal access to basic solar energy services in informal settlements: case studies from South Africa and Zimbabwe. Energy Res Soc Sci 56:101202. https://doi.org/10.1016/j.erss.2019.05.012
116. UCL (2025) Urban infrastructures and the on/offgrid city—Cape Town. Bartlett Faculty of the Built Environment. Available at: https://www.ucl.ac.uk/bartlett/construction/research/virtual-research-centres/engineering-international-development-centre/urban-infrastructures-and-onoffgrid-city-cape-town. Accessed 26 Jun 2025
117. Arku G (2009) Rapidly growing African cities need to adopt smart growth policies to solve urban development concerns. Urban Forum 20:253–270. https://doi.org/10.1007/s12132-009-9047-z
118. Pietersen JM (2023) National development policies of south africa since democracy: are we there yet? J Public Adm 58(2):462–477. https://doi.org/10.53973/jopa.2023.58.2.a16
119. Alrashed F, Asif M, Burek S (2017) The role of vernacular construction techniques and materials for developing zero-energy homes in various desert climates. Buildings 7(1):17. https://doi.org/10.3390/buildings7010017
120. Ciancio D, Beckett C (2013) Rammed earth: an overview of a sustainable construction material. In: Proceedings of the 3rd sustainable construction materials and technologies conference, pp 1–9
121. Alabi M (2024) Sustainable infrastructure and climate change resilience
122. Hanafi WHH (2021) Compressed stabilized earth block: environmentally sustainable alternative for villages housing. J Eng Appl Sci 68(1):20. https://doi.org/10.1186/s44147-021-00017-9
123. Green Africa Foundation (2018) The Greenmark standard for green buildings
124. Olujobi OJ (2024) Nigeria's climate change act 2021: a pathway to net-zero carbon emission, energy security and sustainability. Environ Sci Pollut Res 31(25):36834–36848. https://doi.org/10.1007/s11356-024-33347-1
125. Sweerts B, Longa FD, van der Zwaan B (2019) Financial de-risking to unlock Africa's renewable energy potential. Renew Sustain Energy Rev 102:75–82. https://doi.org/10.1016/j.rser.2018.11.039
126. Jung H (2020) Development finance, blended finance and insurance. Int Trade, Polit Dev 4(1):47–60. https://doi.org/10.1108/ITPD-12-2019-0011
127. Clark R, Reed J, Sunderland T (2018) Bridging funding gaps for climate and sustainable development: Pitfalls, progress and potential of private finance. Land Use Policy 71:335–346. https://doi.org/10.1016/j.landusepol.2017.12.013
128. Adegun OB, Adedeji YMD (2017) Review of economic and environmental benefits of earthen materials for housing in Africa. Front Arch Res 6(4):519–528. https://doi.org/10.1016/j.foar.2017.08.003
129. Ebekozien A, Aigbavboa CO, Samsurijan MS (2023) Appraising alternative building technologies adoption in low-cost housing provision to achieving Sustainable Development Goal 11. Eng, Constr Arch Manag 31(13):41–58

FSC
www.fsc.org

MIX
Papier aus verantwortungsvollen Quellen
Paper from responsible sources
FSC® C105338

If you have any concerns about our products,
you can contact us on
ProductSafety@springernature.com

In case Publisher is established outside the EU,
the EU authorized representative is:
Springer Nature Customer Service Center GmbH
Europaplatz 3, 69115 Heidelberg, Germany

Printed by Libri Plureos GmbH
in Hamburg, Germany